Second International Symposium on Polymer Electrolytes

Proceedings of the Second International Symposium on Polymer Electrolytes held at the Centro Didattico degli Istituti Biologici, University of Siena, Italy, 14–16 June 1989

ORGANISING COMMITTEE

Professor G. Adembri
University of Siena, Italy

Dr F. Bonino
University of Rome, Italy

Dr F. Croce
University of Rome, Italy

Dr S. Passerini
University of Rome, Italy

Professor M. Scotton
University of Siena, Italy

Dr A. Selvaggi
University of Rome, Italy

SCIENTIFIC COMMITTEE

Professor B. Scrosati (Chairman)
University of Rome, Italy

Professor G. C. Farrington (Chairman)
University of Pennsylvania, USA

Professor L. Alcacer
1st. Tecnico, Lisbon, Portugal

Dr M. Armand
EBSEEG Grenoble, France

Dr A. Roggero
ENIRICERCHE, Milan, Italy

Professor C. A. Vincent
University of St. Andrews, Scotland

Under the auspices of:
—The International Society of Electrochemistry
—The Electrochemistry Division of the Italian Chemical Society

Sponsored by:
—The University of Siena
—The University of Rome
—Consiglio Nazionale delle Ricerche, CNR
—US Army European Office

Second International Symposium on Polymer Electrolytes

Edited by

B. Scrosati

Department of Chemistry, University of Rome, Italy

ELSEVIER APPLIED SCIENCE
LONDON and NEW YORK

ELSEVIER SCIENCE PUBLISHERS LTD
Crown House, Linton Road, Barking, Essex IG11 8JU, England

Sole Distributor in the USA and Canada
ELSEVIER SCIENCE PUBLISHING CO., INC.
655 Avenue of the Americas, New York, NY 10010, USA

WITH 35 TABLES AND 244 ILLUSTRATIONS

© 1990 ELSEVIER SCIENCE PUBLISHERS LTD
Except pp. 151–64, 421–31 © 1990 UKAEA

British Library Cataloguing in Publication Data

International Symposium on Polymer Electrolytes (2nd: 1989:
  University of Siena
  Second International Symposium on Polymer Electrolytes.
  1. Polyelectrolytes
  I. Title II. Scrosati, Bruno
  541.372
  ISBN 1-85166-473-4

Library of Congress Cataloging-in-Publication Data

International Symposium on Polymer Electrolytes (2nd: 1989:
  University of Siena)
    Second International Symposium on Polymer Electrolytes/edited by
  B. Scrosati.
    p. cm.
    Proceedings of a conference sponsored by the University of Siena
  ... et al., and held at the Centro didattico degli istituti
  biologici, University of Siena, Italy, June 14–16, 1989.
    Includes bibliographical references.
    ISBN 1-85166-473-4
    1. Polyelectrolytes—Congresses. I. Scrosati, Bruno.
  II. Università di Siena. III. Title.
  QD382.P64I58 1989
  541.3'72—dc20   90-2853
                                                             CIP

Photoset and printed by The Universities Press (Belfast) Ltd.

Foreword

Polymer electrolytes, i.e. complexes of high molecular weight polymers with metal salts, were first introduced by P. V. Wright in 1975 (*Br. Polym. J.,* **7** (1975) 319) and proposed as material of interest for the development of electrochemical devices by M. Armand in 1978 (Second International Meeting on Solid Electrolytes, St. Andrews, Scotland, Sept. 1978).

These new compounds which are capable of offering good ionic conductivities still retaining all the properties and characteristics of the common plastic materials, attracted the attention of many scientists throughout the world. The driving force was initially the prospect of using them for the realization of thin-layer, flexible, high-energy batteries. Particularly relevant in this respect was the effort carried out within a research project sponsored by the Commission of the European Communities, the so-called Anglo-Danish project, which in a few years produced impressive results, showing that the concept of a rechargeable, high-energy battery based on polymer electrolytes was indeed feasible.

Following the Anglo-Danish project, many laboratories became interested in polymer electrolyte batteries and related aspects. The list includes large research and development groups (e.g. Harwell Laboratory in the UK, Mead Imaging in the USA., Hydro-Québec Laboratory in Canada, Energy Research Laboratory in Denmark), as well as many laboratories in universities.

As a result of this consistent and diversified interest, relevant progress has been achieved in the field, to the point that prototypes of polymer batteries having values of energy density almost one order of magnitude greater than the conventional lead acid battery and a cyclic life comparable with that of the nickel cadmium battery, are presently under test and demonstration.

The discovery of polymer electrolytes not only stimulated interest in applications but also promoted attention on the fundamental aspects of this new class of ionic conductors. In fact, it was soon clear that the mechanism of ion diffusion and transport in the polymer electrolytes, the reciprocal role of the metal salt and the polymer host, and the characteristics of the electrode interfaces, were all intriguing aspects which deserved quite a systematic and thorough investigation to be properly described and fully understood. Consequently, the interest in the field grew consistently in the last decade, with participation of scientists of various cultural backgrounds, such as polymer chemists, electrochemists, physicists, electrical engineers. Many papers have been published on the subject and a few books (*Polymer Electrolyte Reviews 1 and 2* ed. MacCallum & Vincent, Elsevier Applied Science, London 1987; *Electrochemical Science and Technology of Polymers,* ed. Linford, Elsevier Applied Science, London 1987; *Electro-Responsive Molecular and Polymer Systems, Vol. 1,* ed. Skotheim Marcel Dekker, New York, 1988) are now available for comprehensive overviews of the knowledge achieved in the field.

Considering this situation, it was felt necessary to organize a Symposium entirely devoted to the subject of polymer electrolytes and the first of the series (First International Symposium on Polymer Electrolytes, ISPE 1) was held at St Andrews, Scotland in June 1987. The need for such an event which offered the unique opportunity of cross-fertilization of ideas between scientists of various extractions, was clearly demonstrated by the large response to the St Andrews Symposium which was enthusiastically attended by over 100 delegates from various laboratories of the world. It was obvious from the contributions and the related discussions that the field of polymer electrolytes was greatly expanding with many aspects of their fundamentals and applications still requiring further comprehension and development. Therefore, it was proposed to continue the series with the organization of a second Symposium (ISPE 2) which was held in Siena in June 1989 under the auspices of the International Society of Electrochemistry and of the Electrochemistry Division of the Italian Chemical Society, with financial sponsorships by the University of Siena, the University of Rome, the Italian National Council of

Research (CNR), Eniricerche, the European Research Office of the US Army and Mead Imaging.

Like ISPE 1, ISPE 2 also attracted participants and contributions from many laboratories in the world, with an increase, however, of about 50% in attendance, this confirming the growing interest in the field.

The programme of the Siena Symposium was divided into seven main sessions. The first, named 'Theory and Physical Properties' concerned the description of spectroscopic techniques directed to the study of the physical properties of polymer electrolytes. The eight contributions presented in the session covered investigations based on Brillouin and Raman Scattering, Near-edge X-ray Absorption Fine Structure (NEXAFS), Dielectric Relaxation and Electron Spin Resonance (ESR) measurements.

The second session, named 'New Syntheses', was devoted to the description of synthetic procedures to obtain materials for the development of novel polymer electrolytes. The session included ten presentations which described various routes for the realization of new polymer hosts having high chain mobility and/or low level of crystallinity at low temperature, as well as of new salts with improved solute characteristics. Comb-shaped, crosslinked and blend structures were proposed as polymer hosts alternative to linear poly(ethylene oxide), PEO, and plasticizing perfluorosulfonamide salts were proposed as alternative solutes to classical lithium salts.

The third session, named 'New Electrolytes' dealt with the description of new complexes having enhanced transport properties. The five presentations reported original results on systems having room temperature conductivities several orders of magnitude higher than those of standard PEO-based electrolytes.

The fourth session, named 'Polyvalent, Protonic and Composite Electrolytes' was the last devoted to the bulk properties of the systems. The fourteen presentations described the physical and the electrochemical properties of electrolytes using polyvalent cation salts. Results were reported on systems based on divalent (including zinc, lead, copper, cadmium, mercury and calcium) and trivalent (europium, neodymium and lanthanum) salts. Attention was also devoted to proton conductors and to crystallization processes.

The final three sessions of the symposium were devoted to electrode processes and applications. The fifth session, named 'Interfaces and

Electrodes', included presentations on the behaviour of the negative (lithium) electrode and of the positive (intercalation) electrodes in polymer electrolyte cells, as well as on the properties and characteristics of fully polymeric interfaces based on the contact between polymer electrolytes and polymer electrodes.

In the sixth session, named 'Batteries', the most recent achievements in the development of high energy lithium polymer electrolyte batteries were discussed and demonstrated.

Finally, the seventh session named 'Electrochromic Displays', included four presentations which showed that polymer electrolytes may be of consistent importance also for the progress of advanced technology in the field of optical information and of energy control in buildings and automobiles.

Thanks to the co-operation of the authors, the extended versions of the majority of the cited presentations are collected in this volume which then reports in detail the proceedings of the Second International Symposium on Polymer Electrolytes, ISPE 2.

We hope that this volume may supply to those who were unable to be in Siena most of the information that was provided and discussed at the Symposium. Furthermore, we think that the volume, collecting papers which cover wide and diversified aspects of the present research and development in polymer electrolytes, may also be of interest to all those who are or plan to be involved in this stimulating and exciting field of research.

BRUNO SCROSATI

Contents

Theory and Physical Properties

ix

Electrochromic Displays

Optical Spectroscopy of Polymer Electrolytes

M. Kakihana, J. Sandahl, S. Schantz & L. M. Torell

Department of Physics, Chalmers University of Technology, S-412 96 Gothenburg, Sweden

ABSTRACT

Ion association and local chain mobility have been studied in polymer-salt complexes using light—Raman and Brillouin—scattering techniques. It is found that the amount of ion pairs increases with salt concentration, temperature and polymer molecular weight. The T-behaviour is attributed to free volume dissimilarities as is also the decreasing solubility with increasing molecular weight. The local chain dynamics is found to dramatically slow down with increasing salt content due to transient crosslinks. We conclude that it is the internal polymer flexibility rather than ion–ion interactions which governs the overall conductivity behaviour, the rapid σ-increase with increasing T and the drastic σ-drop at large concentrations. Results are given for PPO– and PEO–NaCF$_3$SO$_3$ complexes.

INTRODUCTION

The properties which are at focus in optimizing the performance of a polymeric electrolyte are, on the one hand, the number and mobility of the dissociated ions of the dopant salt and on the other, the internal flexibility of the host polymer.[1] Unfortunately there are few methods by which these properties and their impact on the overall conductivity can be determined. We demonstrate how two light scattering techniques can be applied for quantitative determinations; Raman scattering for determining the number of free charge carriers; Brillouin scattering for studies of the elastic properties and the local chain flexibility. So far the investigations are restricted to studies of

1

polyethers (PPO and PEO). Results will be summarized for NaCF$_3$SO$_3$-polymer complexes based on previously reported data[2-4] and with new data on the effect of temperature, molecular weight of host polymer, and exchange of host polymer.

One way to increase the number of charge carriers and thereby the ionic conductivity in polymer electrolytes is to choose the dopant salt and the host polymer such that the ion–ion interactions in the complex are minimized. However, so far little is known about ion–ion association effects in salt–polymer complexes. Vibrational spectroscopy of other salt solutions shows that various ion associations can be determined from observations of the molecular vibration frequencies. For salt–polymer complexes there are a few vibrational studies reported in which the effect of ion association was observed, namely Raman investigations of complexes based on poly(ethylene oxide)[5-8] and poly(propylene oxide),[9] even though the spectra did not allow for quantitative determinations. We have recently demonstrated that, using high resolution Raman spectroscopy in combination with spectral deconvolution techniques, various ion associations present in polymer electrolytes can be determined and also the amount of ions participating in each kind of constellation.[2,3] In a high resolution Raman spectrum the various ion–ion associations are observed as splittings of the internal ionic vibrational frequencies.[2,3] For identical ions present for instance in the form of free ions, ion pairs and ion multiplets, the splitting results in different Raman modes each corresponding to the respective surroundings. The bands of the symmetric modes are then ideal to be chosen for a detailed analysis since then the splitting cannot be due to the lifting of any degeneracy but is only due to the different surroundings of the ions. In the present study we report results from such investigations of NaCF$_3$SO$_3$-doped PPO and PEO systems, in which the effect of temperature, salt concentration, and host polymer molecular weight on the various ion interactions have especially been investigated.

Apart from reducing the number of ions available for conduction by ion–ion interactions, their mobility may be severely reduced through crosslinking phenomena. It has been observed that transient crosslinking occurs in salt–polymer complexes via the dissociated cations which act as crosslinking centres between oxygens of adjacent polymer chains. The effect increases the viscosity and stiffens the network. The reduced internal flexibility of the host polymer affects in turn the mobility of the diffusing charge carriers. The stiffening can be followed using the Brillouin scattering technique as recently shown in this laboratory.[3,4] In Brillouin scattering, which is the inelastic light scattered from thermally excited sound waves in a material, the frequency shift of the Brillouin component is related to the sound velocity and the width of the component to sound attenuation. Therefore, from the position and width of the Brillouin lines the elastic properties are obtained, which in turn determine the characteristic structural relaxation times of the system. In the present study the technique has been used to investigate the effect of transient crosslinks on the local dynamics in complexes of $PPO\text{-}NaCF_3SO_3$ of various concentrations and in a wide temperature range.

The aim of the study is twofold. On the one hand we demonstrate how light scattering methods—Raman and Brillouin—can be used for characterizing salt–polymer complexes. On the other hand we aim at resolving the much debated issue whether it is the number or the mobility of the charge carriers which is of major importance for a high performance polymer electrolyte.

LIGHT SCATTERING APPARATUS

Raman scattering. The experimental set-up includes an argon ion laser operating at 488·0 nm, a Spex double monochromator (model 1403) with 1800 lines/mm holographic gratings, a cooled photomultiplier and photon counting facilities. A continuous flow cryostat (Oxford CF1204) was used for all temperatures investigated. The temperature stability was better than $\pm 0\cdot 1$ K. Every recorded spectrum, stored in a Spex datamate DM1B, is the result of the repeated scanning and accumulation of, on average, 20 spectra. The overall resolution of the spectrometer, mainly restricted by the slit widths, was set to give a value better than $2 \, cm^{-1}$.

Brillouin scattering. Single mode lasers (Ar^+ and Kr^+ lasers) were used as light sources. The light, vertically polarized to the scattering plane, was incident on the sample situated in a thermostat. Right angle scattering geometry was employed. The scattered light was frequency resolved by a triple-passed piezo-electrically scanned Fabry–Perot interferometer and photon counting techniques were used for detection of the signal. The spectral free ranges of the interferometer, 18–21·7 GHz, were chosen to optimize the resolution of every spectrum. Each recorded spectrum was the result of about 20 000 repeated scans of ~1·5 spectral free ranges. During scanning the finesse of the interferometer was optimized by a Burleigh DAS 10 stabilization system. The overall finesse of the system was kept at about 35 for all recorded spectra. The temperature was controlled within ±1 K and the temperature stability was better than ±0·1 K during a measurement.

RAMAN SCATTERING AND THE CONCENTRATION OF FREE CHARGE CARRIERS

The Spectra

Raman spectra of PEO and PPO based $NaCF_3SO_3$ complexes were recorded in the concentration range $5:1 \leq O:M \leq 30:1$. Here O refers to the oxygen of the host polymer and M to the dopant salt, i.e. O:M is the number of monomer units per dopant cation. Spectra were obtained over a wide temperature range of about 180–360 K, i.e. from temperatures below the glass transition up to temperatures typical of applications. The appearance of distinct, closely spaced components within the band contour of the symmetrical mode of the SO_3-stretch of the various complexes demonstrates clear evidence of ion interactions in all the investigated samples (see Fig. 1). The spectral bands were analysed in detail using Marquardt's non-linear least-square algorithm. The vibrational peaks were assumed to be of Lorentzian shape. The instrumental function was measured to be Gaussian. The observed band was therefore fitted to Lorentzians convoluted with a Gaussian instrumental function. The various frequency modes obtained from the analyses of the split band were attributed to the presence of dissociated 'free' ions, ion pairs and ion–ion multiplet formations, respectively, as discussed in previous reports.[2,3]

Temperature Dependence of Ion Associations

Figure 1 shows that the ion–ion association increases rapidly with both concentration and temperature. Similar temperature dependencies were found for salt complexes of different concentrations as demonstrated in Fig. 2, where the integrated intensity of the Raman modes corresponding to the 'free' ions is shown for complexes of concentra-

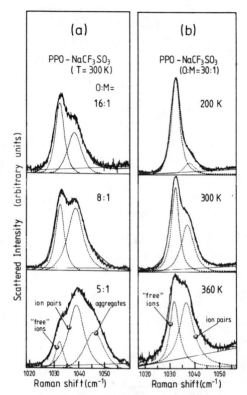

Figure 1. The SO$_3$ symmetric stretching mode for PPO(4000)–NaCF$_3$SO$_3$ at different concentrations (a) and temperatures (b). Dashed lines correspond to the components of the two-(or three-) Lorentzian envelope fitted to the spectrum (smoothed solid line). The components attributed to 'free' ions, ion pairs and ions in aggregates are indicated. Linear solid lines represent luminescence backgrounds.

tion 16:1 and 30:1 respectively. It can be seen that the intensity of the 'free' ion mode is more or less constant at temperatures below the glass transition temperature T_g. Then above T_g the amount of dissociated ions n decreases dramatically with temperature and is found to follow an Arrhenius behaviour, i.e. it can formally be represented by the equation

$$n = N' \exp(A/k_B T)$$

where A is found to be ~40 meV. The present observations thus

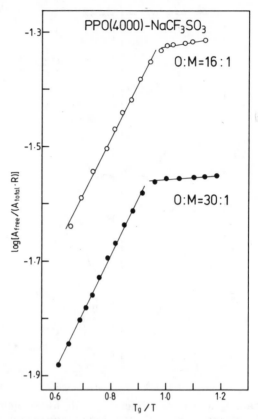

Figure 2. Arrhenius plot of the amount of 'free' ions versus inverse temperature, calculated from the normalized integrated intensity of the Raman mode of free ions for PPO(4000)–NaCF$_3$SO$_3$ complexes. Results from two concentrations, O:M (= R) = 16(○) and 30 (●) respectively, are shown. As a reference the total integrated intensity A_{total} of the SO$_3$ symmetric stretch was used.

demonstrate the failure of applying the weak electrolyte model to polymer electrolytes. For a weak electrolyte solution the number of dissociated ions is expected to increase with temperature whereas the reverse phenomenon is observed in the present case. In fact, one can hardly expect weak electrolyte behaviour to hold for electrolytes based on polymers. For polymer solubility in general there is convincing evidence in the literature that the temperature dependence can be explained by differences in free volume between polymer and solvent.[10] This 'free volume dissimilarity' is mainly due to the inescapable difference of size of chain length between polymer and solvent molecules, in this case the dissociated salt. As the temperature is raised the demands for free volume are much more limited for the structurally restricted long chain polymer than for the dissolved cations and anions. The effect will subsequently result in phase separation which is observed to take place in all polymer solutions at elevated temperatures.[10] Free volume dissimilarities are therefore likely to explain the increased ion–ion interaction observed in the present study. This may in turn explain the salt precipitation phenomena reported for many salt–polymer complexes at elevated temperatures.[11]

From the above it is clear that weak electrolyte behaviour, which has been suggested to hold for polymer electrolytes, can hardly be present in these systems based on the physics of polymers in general. However, since it is an often present phenomenon in 'small-molecule' electrolytes and also in the field of solid state ionics, weak electrolye theory may have been introduced by convenience for ion conducting polymers as well to account for the conductivity behaviour. It is, thus, observed that in salt–polymer complexes the conductivity σ is rapidly increasing with temperature and generally the σ versus T relation is expressed in a Vogel–Tamman–Fulcher (VTF) behaviour. Some precautions are then needed, since it has recently been reported that at elevated temperatures and for samples equilibrated long enough for salt precipitation to be fully developed, the conductivity in fact drops from the VTF curve.[11] The presently observed increasing ion–ion interaction with temperature may then initiate the reported salt precipitation. The resulting reduction of the conductivity is however a minor change of the overall conductivity increase which occurs when the temperature is raised from T_g to the upper limit of the present investigated range, i.e. $T/T_g \approx 1.3$. We therefore conclude that the number of charge carriers, which we observe to be significantly reduced with increasing temperature, has little influence on the conductivity behaviour. The dramatic conductivity increase with increasing temperature has therefore to be sought in an increased ion mobility as discussed in a following section.

Host Polymer Molecular Weight and Ion Association

In Fig. 3 we show how the amount of dissociated ions is affected by the host polymer molecular weight. Below the glass transition temperature T_g it can be seen that the low molecular weight polymer PPO 400 is

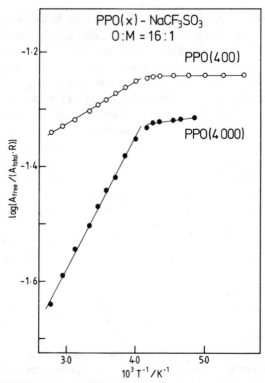

Figure 3. Arrhenius plot of the amount of free ions calculated from the normalized integrated intensity of the Raman mode of free ions for PPO–NaCF$_3$SO$_3$ complexes. Results from two polymer molecular weights, 400 (○) and 4000 (●) respectively, are shown.

slightly more efficient than the higher molecular weight PPO 4000 at dissolving the salt. Also, increasing the temperature above T_g the amount of free ions is more rapidly decreasing in PPO 4000. This indicates reduced solubility with increasing molecular weight and the effect can be ascribed to the free volume dissimilarity of the polymer and the dissolved salt as was also the case for the temperature dependence discussed above. A short chain polymer has on the average larger free volume than that of a long chain since there are less bonding restrictions in the short chain system.[10] Hence, decreasing the molecular weight decreases the free volume difference between polymer and salt and increases polymer solubility as is demonstrated in Fig. 3.

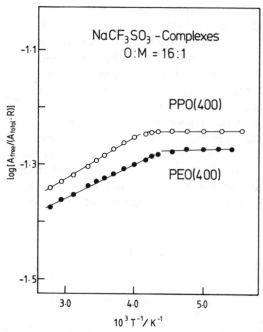

Figure 4. Arrhenius plot of the amount of free ions calculated from the normalized integrated intensity of the Raman mode of free ions for NaCF$_3$SO$_3$–polymer complexes; the host polymer being respectively PPO (O) and PEO (●). The molecular weight is 400.

Host Polymer Exchange and Ion Associations

An exchange of the host polymer from PPO to PEO of similar molecular weight has little influence on the solubility, as can be seen in Fig. 4. Note however that the salt is slightly more dissolved in PPO than in PEO in contrast to predictions based on discussions of the dielectric constant.[12] The effect is at present subject to a more detailed investigation.

Concentration Dependence of the Ion Associations

Next we focus on the concentration dependence on the ion–ion association, as demonstrated in Fig. 1. From the figure it can be seen that the ion pairing effects are significant over the whole concentration range studied. Also, at the higher concentrations $O:M \leq 8$ the dopant salt is largely present in ion–ion formations, either in pairs or in multiple aggregates, and only a limited amount is available for conduction. It is then to be noted that in the same concentration range a drop in the conductivity has in fact been reported.[13]

To compare the amount of 'free' ions with the conductivity for increasing salt concentration we calculate the integrated intensities of the corresponding Raman modes. Results are shown in Fig. 5 together with reported conductivity data.[13] Figure 5 shows that a maximum in the number of 'free' ions occurs at a concentration of $O:M \approx 10$, i.e. in the vicinity of the reported conductivity maximum.[13] At large concentrations a decrease of the amount of 'free' ions is observed as is

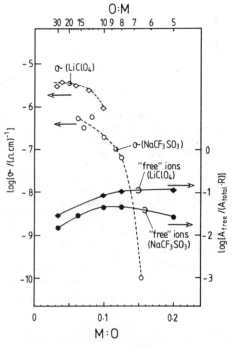

Figure 5. Concentration dependence of the normalized integrated intensity of the Raman mode of free ions in complexes of PPO(4000)–NaCF$_3$SO (●) and PPO(4000)–LiClO$_4$ (◆). Dashed lines represent ionic conductivity data for PPO–NaCF$_3$SO$_3$(○, Ref. 13) and PPO–LiClO$_4$ (◇, Ref. 17) respectively.

also the case of the conductivity. Note however that while the conductivity is reduced by about four orders in magnitude, the amount of free ions is only dropping by a factor of two. The carrier number is thus of less importance for the conducting properties in this concentration range. Again we expect that it is the mobility rather than number which is of major influence, see discussion below. Differences between the conductivity of different complexes can however be attributed to different solvating abilities. This is demonstrated in Fig. 5 in which also results for a $LiClO_4$–PPO complex[14] have been included for comparison. The considerably weaker ion–ion interaction in the $LiClO_4$–PPO complex and the resulting larger amount of free ions available in this complex may well explain the higher conductivity as compared to the PPO–$NaCF_3SO_3$ system.[15]

BRILLOUIN SCATTERING AND THE LOCAL CHAIN FLEXIBILITY

Sound Absorption and Transient Ionic Crosslinking

Brillouin spectra of PPO–$NaCF_3SO_3$ complexes were obtained in the O:M concentration range 16:1–5:1 and in a temperature range 300–480 K. From a study of the concentration dependence of the sound attenuation it was found that the position in temperature of the sound absorption maximum, $T_{\alpha_{max}}$, changes with salt content in a manner similar to the glass transition temperature T_g,[4,16] see Fig. 6. In the inset of Fig. 6 the absorption data of the complexes of various concentrations are plotted and shown to follow the same curve when using $T_{\alpha_{max}}$ as a scaling parameter. The observation of similar $T_{\alpha_{max}}$ and T_g behaviour indicates that both phenomena reflect basically the same process, i.e. the slowing down of the segmental polymer motions as the salt concentration increases. The number of free ions as measured by Raman scattering (see above) is also included in Fig. 6 and is found to exhibit a concentration dependence similar to T_g and $T_{\alpha_{max}}$. This implies that it is mainly the dissociated ions which stiffen the system by transiently crosslinking the polymer chains. Note, however, the deviations at larger salt concentrations in Fig. 6. Therefore, in the high concentration range single-ion crosslinks cannot alone explain the observed segmental slowing down, which also implies that ion–ion formations may act as crosslinking centres.

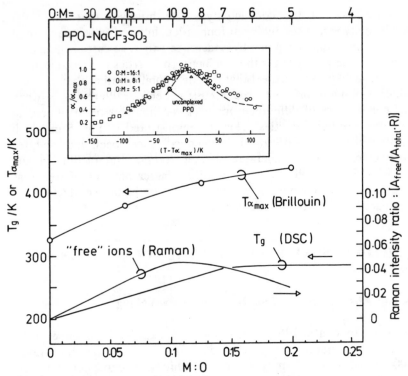

Figure 6. Data of $T_{\alpha_{max}}$ (Ref. 4), the glass transition temperature T_g (Ref. 16), and the normalized integrated intensity of the Raman mode of free ions versus salt concentration in PPO(4000)–NaCF$_3$SO$_3$ complexes. $T_{\alpha_{max}}$ is defined as the temperature at which maximum sound absorption α is observed. Inset (from Ref. 4) shows normalized sound absorption data versus $T - T_{\alpha_{max}}$ for PPO(4000)–NaCF$_3$SO$_3$ of concentrations O:M = 16(O), 8($\triangle$) and 5($\square$) together with results for uncomplexed PPO(4000) (dashed curve).

Structural Relaxation Times and Transient Ionic Crosslinking

The characteristic time of the segmental mobility, i.e. the structural relaxation time, can be obtained from the frequency shift of the Brillouin lines subject to maximum damping by using the peak condition $\omega\tau \approx 1$ and $\omega = 2\pi\nu$.[4] Values so obtained are 2.92×10^{-11} s for O:M = 16:1, 3.09×10^{-11} s for O:M = 8:1 and 3.18×10^{-11} s for O:M = 5:1 and the corresponding peak temperatures are respectively

368, 403 and 429 K. Using these data and assuming a VTF behaviour
of the temperature dependence of the relaxation time, i.e.

$$\langle \tau \rangle = \tau_0 \exp B/(T - T_0)$$

the B parameter was determined to be 1098, 1114 and 1208 for
$0:M = 16:1$, $8:1$ and $5:1$ respectively. In the calculations the relaxation time in the high temperature limit (τ_0) was taken to be identical to
that of uncomplexed PPO.[4] The various T_0 values were chosen to be
30 K below the experimentally observed glass transition temperatures
presented in Fig. 6, as is reported to be the case for uncomplexed
PPO.[3] The resulting $\langle \tau \rangle$ versus T relations are shown in Fig. 7. It can

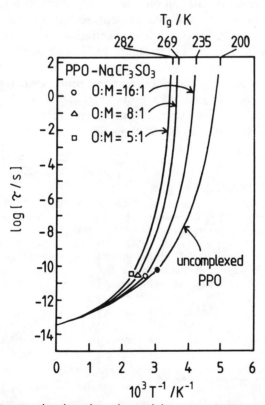

Figure 7. Structural relaxation times (τ) versus inverse temperatures
(T^{-1}) for PPO(4000)–NaCF$_3$SO$_3$ complexes of concentrations O:M =
16(O), 8(△) and 5(□), together with results for uncomplexed PPO(4000)
(●). Data for T_g (Ref. 16) are shown at the top of the figure.

be seen in Fig. 7 that the introduction of salt dramatically decreases the time scale for local segmental motion; the average structural relaxation times are shifted towards higher temperatures and longer times as the salt concentration—and thereby the density of crosslinks—increases.

DISCUSSION

Number and Mobility versus Conductivity

To examine the suggested coupling between the local dynamics of the polymer backbone and the ion diffusion, we present in Fig. 8(a) data for ionic conductivity[13] and the present results for the structural relaxation time plotted versus salt concentration. Also, to clarify the relation between number and mobility of 'free' charge carriers and conductivity we include in Fig. 8(a) present Raman results for the number of 'free' ions. Raman data represent room temperature values whereas the σ- and $\langle \tau \rangle$-curves are obtained at $T = 333$ K. However, the number of 'free' ions changes only a few tens in percent from 293 K to 333 K (see Fig. 2) which allow for an adequate comparison.

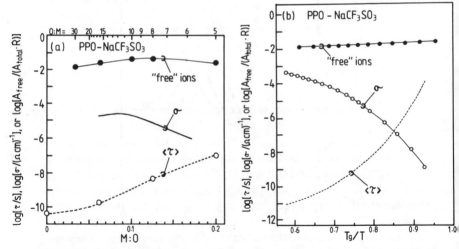

Figure 8. Arrhenius plots of (a) concentration and (b) temperature dependence of the average structural relaxation time $\langle \tau \rangle$ (present Brillouin results), the ionic conductivity σ (Refs 18, 13), and the number of free ions (present Raman result for the integrated intensity of the free ion mode $[A_{free}/(A_{total} \cdot R)]$) in PPO(4000)–NaCF$_3SO_3$ complexes.

Figure 8(a) shows that as salt is complexed to the polymer there is a huge slowing down of the structural relaxation time (about three orders in magnitude for the largest salt concentration 5:1 studied). The decreased local mobility negatively affects the ionic conductivity, which is reported to drop more than one order in magnitude when increasing the salt content from 10:1 to 6:1. In the same concentration range the amount of 'free' ions is also decreasing, due to ion pairing, however only by about 25%, whereas the relaxation time increases between one and two orders. We therefore conclude that reduced segmental motion is the dominant factor governing the conductivity drop in salt–polymer complexes at high salt content.

Finally, we compare the temperature dependence of the number of free charge carriers, the conductivity, and the local polymer mobility (see Fig. 8(b)). It is obvious from the figure that again it is the flexibility of the host polymer chain which is of major impact for the conducting properties. The structural relaxation becomes about seven orders in magnitude more rapid as the temperature increases from $1.1T_g$ to $1.7T_g$. In the same range the conductivity is increasing dramatically, however slightly less (about 5–6 orders in magnitude) than the change of the relaxation time. The difference may be attributed to the number of charge carriers which is reduced by a factor of 3 in the same temperature range due to the formation of ion pairs and multiplets. Thus, ion pairing effects, although significant, seem to play a minor role for the temperature behaviour of the ionic conductivity.

CONCLUSION

In conclusion we find that the conducting properties of a given polymer electrolyte are mainly governed by the dynamics of the host polymer, the temperature and salt concentration dependence of the polymer internal flexibility explain the rapid conductivity increase at elevated temperatures and the dramatic conductivity drop at large salt content. Ion–ion interactions, even though considerably temperature and concentration dependent, are of less importance for the overall conductivity behaviour. Ion association effects may however explain differences in the conductivity between different complexes. They may also explain deviations from Vogel–Tamman–Fulcher behaviour at elevated temperatures. In contrast to predictions ion–ion associations' are observed to increase with temperature (rapidly and in an Arrhenius manner) and increase with molecular weight of the host polymer. The phenomena can be attributed to differences in free volume demands of the host polymer and the dopant salt.

ACKNOWLEDGEMENTS

Financial support by the Swedish National Board for Technical Development and partial support from Axel and Margaret Ax:son Johnsson's Foundation are gratefully acknowledged.

REFERENCES

1. For reviews see for instance MacCallum, J. R. & Vincent, C. A. (p. 23) and Ratner, M. A. (p. 235) In *Polymer Electrolyte Reviews*, ed. J. R. MacCallum & C. A. Vincent. Elsevier, Amsterdam 1987.
2. Schantz, S., Sandahl, J., Börjesson, L., Torell, L. M. & Stevens, J., *Solid State Ionics*, **28–30** (1988) 1047.
3. Torell, L. M. & Schantz, S. In: *Polymer Electrolyte Reviews-2*, ed. J. R. MacCallum & C. A. Vincent. Elsevier, Amsterdam, 1989.
4. Torell, L. M. & Angell, C. A., *Brit. Polym. J.*,**20** (1988) 173; Sandahl, J., Schantz, S., Börjesson, L., Stevens, J. & Torell, L. M., *J. Chem. Phys.*, **91** (1989) 655.
5. Papke, B. L., Ratner, M. A. & Shriver, D. F., *J. Phys. Chem. Solids*, **42** (1981) 493.
6. Papke, B. L., Dupon, R., Ratner, M. A. & Shriver, D. F., *Solid State Ionics*, **5** (1981) 685.
7. Papke, B. L., Ratner, M. A. & Shriver, D. F., *J. Electrochem. Soc.*, **129** (1982) 1434.
8. Dupon, R., Papke, B. L., Ratner, M. A., Whitmore, D. H. & Shriver, D. F., *J. Am. Chem. Soc.*, **104** (1982) 6247.
9. Teeters, D. & Frech, R., *Solid State Ionics*, **18–19** (1986) 271.
10. For a review see Patterson, D., *Macromolecules*, **2** (1969) 672.
11. Wintersgill, M. C., Fontanella, J. J., Greenbaum, S. G. & Adamic, K. J., *Brit. Polym. J.*, **20** (1988) 195.
12. Cheradame, H. & Le Nest, J. F. In: *Polymer Electrolyte Reviews*, ed. J. R. MacCallum & C. A. Vincent. Elsevier, Amsterdam, 1987, p. 118.
13. Armand, M. B., *Ann. Rev. Mater. Sci.*, **16** (1986) 245.
14. Kakihana, M. & Schantz, S. In: *Second International Symposium on Polymer Electrolytes*, ed. B. Scrosati. Elsevier Applied Science, London, UK, p. 23.
15. Fontanella, J. J., Wintersgill, M. C., Smith, M. K., Semancik, J. & Andeen, C. G., *J. Appl. Phys.* **60** (1986) 2665.
16. Stevens, J. & Schantz, S., *Polym. Comm.*, **29** (1988) 330.
17. Watanabe, M., Ikeda, J. & Shinokara, I., *Polym. J.*, **15** (1983) 65.
18. MacLin, M. & Angell, C. A., *J. Phys. Chem.*, **92** (1988) 2083.

Poly(Ethylene Oxide) Complexed with KI. An X-ray Absorption Study

X. Q. Yang, J. Chen, T. A. Skotheim

Brookhaven National Laboratory, Upton, New York 11973, USA

Y. Okamoto

Polytechnic University, 333 Jay Street, Brooklyn, New York 11201, USA

&

M. L. denBoer

Hunter College of CUNY, 695 Park Ave., New York, New York 10021, USA

ABSTRACT

Polymer–salt complexes of poly(ethylene oxide) (PEO) and KI have been studied with the technique of Near-Edge X-ray Absorption Fine Structure of the K-edge of potassium. The changes observed in the spectra as a function of temperature are compared with model systems. The results suggest that the oxygen complexation of the potassium ion is reduced at elevated temperatures.

INTRODUCTION

The relationship between ionic conductivity and the microscopic structure of polymer–salt complexes is not well understood. In the poly(ethylene oxide) (PEO) salt systems the alkali and the oxygen atoms in the ether group form a complex in which the oxygen atoms form a 'cage' around the cation, dissociating the anion–cation pair,

17

resulting in ionic conductivity.[1] On the other hand, this complexation also bonds the cation to the polymer chain and restricts its mobility. Therefore, studying the complexation is crucial for understanding the ionic conductivity.

However, the absence of long-range order in the amorphous phase responsible for conductivity precludes investigation of the microscopic structure using conventional X-ray diffraction. X-ray absorption spectroscopy, on the other hand, especially near-edge X-ray absorption fine structure (NEXAFS) is sensitive to local structural properties such as interatomic spacings and coordination number, even in the absence of long-range order, and it is therefore well suited to a study of the amorphous conducting phase. In particular, by using the alkali atom as an absorber, the structure of the cation–polymer complex can be studied.

We report here the first systematic study of the temperature dependence of the NEXAFS above the K-edge of potassium in the PEO-KI complex in the temperature range between 25°C and 100°C. The observed NEXAFS spectra exhibit important changes in intensity with temperature and KI salt concentration, which we interpret by comparison with reference systems.

RESULTS AND DISCUSSION

The measurements were performed at beam line X23B of the National Synchrotron Light Source at Brookhaven National Laboratory, using a Si(200) double crystal monochromator. PEO with an average molecular weight of 600 000 was complexed with KI and cast as a film about 100 μm thick on Kapton tape and dried thoroughly under vacuum. The absorption was computed as $\ln(I/I_0)$, where I_0 is the X-ray flux incident on the sample and I is the transmitted flux, both measured by ionization chambers using a mixture of helium and nitrogen gas. Spectra were obtained with the samples held at 25, 50, 75 and 100°C.

Figure 1 shows the NEXAFS spectra of PEO/KI with an oxygen/potassium ratio of 4:1 at four different temperatures. As the temperature is increased, a relative change takes place between the peaks labelled B and C. The lower energy peak, B, increases with increasing temperature, becoming the dominant peak at 100°C. The feature labeled A remains unchanged. Similar effects were observed at intermediate concentrations of 8:1 oxygen/potassium ratio.

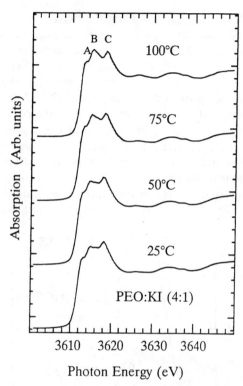

Figure 1. Near-edge X-ray absorption spectrum of the K-edge of potassium in PEO₄/KI at several temperatures as indicated.

The interpretation of the spectral changes is complicated by the fact that it is not appropriate to assign individual spectral features to particular transitions for potassium NEXAFS spectra, as is commonly done for low Z atoms such as carbon and oxygen.[2] In order to relate these changes to the microscopic structure of the PEO/KI complex, we examine the spectra of selected reference systems.[3] The reference compounds include potassium iodide (KI), potassium t-butoxide $((CH_3)_3COK)$, potassium acetate $(KC_2H_3O_2)$, and dibenzo-18-crown-6/KI, in which the potassium is coordinated with oxygen atoms ranging from 3 to 6.

Figure 2 shows the spectra of the three reference systems. The potassium t-butoxide forms a tetrameric structural unit with potassium and oxygen atoms at alternate corners of slightly distorted cubes surrounded by the *t*-butyl groups along the diagonals.[4] The potassium ion is coordinated with three oxygen atoms at the nearest corners of the cube with a 2·56 Å bond length. No detailed study of the potassium acetate is available. However, an X-ray diffraction study of the similar compound sodium acetate shows that the Na ion is surrounded by six oxygens at distances of 2·35–2·67 Å.[5] Two of these six oxygens belong to the same acetate ion while the other four belong to four different acetate ions. Assuming that the crystal structure is similar, in potassium acetate the K ions are coordinates with two nearest oxygens and four next nearest oxygens. In the dibenzo-18-crown-6/KI complex, the K ion, located at the centre of a nearly co-planar ring of six oxygens,[6] is coordinated with six oxygens. It is apparent from Fig. 2, that as the oxygen coordination number

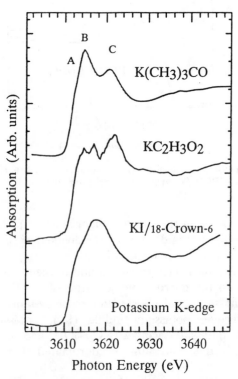

Figure 2. Near-edge X-ray absorption spectrum of the *K*-edge of potassium in several reference compounds: $(CH_3)_3COK$, $C_2H_3O_2K$ and dibenzo-18)-crown-6/KI at 25°C.

decreases from 6 to 3, the intensity of the higher energy peak, C, decreases relative to that of the lower energy peak B. Similar changes in the near-edge white line features above the potassium K-edge as a function of oxygen coordination number have been observed by Spiro et al.[3]

By comparing the near-edge spectra of PEO.KI at different temperatures with those of the model compounds, we conclude that the observed relative intensity changes of the near-edge peaks are due to breaking of the symmetry of the local electric field in the vicinity of the K-ion, caused by a reduction in the oxygen coordination number with increasing temperature. The reduction in coordination number does not, however, lead to salting out of KI, the absence of which can be determined both by X-ray absorption and X-ray diffraction.[1]

In conclusion, we have found that the primary effect of increasing the temperature of the PEO/KI complex, particularly at high salt concentrations, is to reduce the oxygen coordination number of the K ion. This reduces the number of K—O bonds and chain interlinking and is responsible for the lower activation energy for ionic conductivity of this system at elevated temperatures. These results also demonstrate the power of the NEXAFS technique in the investigation of amorphous systems. A full discussion of these issues can be found in Ref. 7.

ACKNOWLEDGEMENT

This work was supported by the US Department of Energy, Division of Materials Science under Contract No. DE-AC02-76CH00016.

REFERENCES

1. Yang, X. Q., Chen, J., Harris, C. S., Skotheim, T. A., denBoer, M. L., Mei, H., Okamoto, Y. & Kirkland, J., *Mol. Cryst. Liq. Cryst.,* **160** (1988) 89.
2. Stohr, J., Outka, D. A., Baberschke, K., Arvanitis, D. & Horsley, J. A., *Phys. Rev.,* **B36** (1987) 2976.
3. Spiro, C. L., Wong, J., Lytle, F. W., Greegor, R. B., Maylotte, D. H. & Lamson, S. H., *Fuel,* **65** (1986) 327.
4. Weiss, E., Alsdorf, H., Kuhr, H. & Grutzmacher, H.-F., *Chem. Ber.,* **101** (1986) 3777.

5. Hsu, L.-Y. & Nordman, C. E., *Acta Cryst.*, **C39** (1983) 690.
6. Weber, E. & Vögtle, F. In: *Topics in Current Chemistry-98*, ed. F. L. Boschke *et al.* Springer-Verlag, Berlin, Heidelberg, New York, 1981, p. 1.
7. Yang, X. Q., Chen, J. M., Sotheim, T. A., Okamoto, Y., Kirkland, J. & denBoer, M. L., *Phys. Rev.* **B40** (1989) 7948.

Temperature Dependence of the Charge Carrier Generation in Polymer Electrolytes; Raman and Conductivity Studies of Poly(Propylene Oxide)–LiClO$_4$

M. Kakihana, S. Schantz, B.-E. Mellander & L. M. Torell

Department of Physics, Chalmers University of Technology, S-412 96 Gothenburg, Sweden

ABSTRACT

Raman scattering and ionic conductivity measurements have been carried out over a wide temperature range on low molecular weight poly(propylene oxide) (PPO) complexed with LiClO$_4$ of salt concentration O:M = 30 (where O:M is the PO:Li ratio). The Raman spectra reveal a splitting of the ClO$_4^-$ symmetric stretching vibrational mode which is attributed to different environments for the anion. A two-component band analysis of this mode leads to the identification of dissociated anions and contact ion pairs. Above the glass transition temperature the number of 'free' ions decreases with increasing temperature in an Arrhenius-type behaviour. A comparison with the carrier concentration for a PPO–NaCF$_3$SO$_3$ electrolyte has been performed and the observed difference in 'free' ion concentration between the two complexes is related to the difference in the measured ionic conductivity.

INTRODUCTION

Much interest has recently been focused on the ion–ion interactions in solvent-free polymer electrolytes. The importance of ion-pairing effects on the conductivity behaviour in these systems has been

23

stressed by several authors,[1-5] however, so far little is known about the phenomenon. There are only a limited number of reports on direct observations of ion association. Vibrational spectroscopy has then been used to study some poly(propylene oxide) (PPO)[6-9] and poly(ethylene oxide) (PEO)[10] based complexes.

In the various theoretical approaches to polymer electrolytes, a first attempt to incorporate ion-pairing effects was introduced by superimposing an Arrhenius behaviour on the commonly used Vogel–Tamman–Fulcher conductivity expression[11]

$$\sigma = \sigma_o \exp\left[-B/(T - T_o) - E_{ion}/k_B T\right] \qquad (1)$$

The above relation includes both ion pairing and activated ion displacements through the activation energy E_{ion} by

$$E_{ion} = E' + W/2\varepsilon \qquad (2)$$

where E' is the activation energy involved in the transfer of an ion and W is the dissociation energy of the salt molecule in a medium of dielectric constant ε. Later Watanabe et al. suggested a similar expression for the conductivity.[12] It is then to be noted that these two approaches predict an exponential increase of the number of free ions with temperature, i.e. the ion association is decreasing with increasing temperature.

Contrary to the above it has also been reported that ion–ion interactions are observed to be very temperature sensitive and, are in fact, increasing rapidly with increasing temperature.[13,14] It is then suggested that the ion pairing may be a precursor to salt precipitation, an effect which has been observed in many polymer–salt systems at elevated temperatures.[9,13,14] It is also to be noted that above some critical temperature the conductivity is decreasing with increasing temperature for some complexes (PPO–NaI and PPO–KSCN), if the samples are equilibrated long enough at each temperature.[14] The conductivity drop is then attributed to the salt precipitation which is observed to be initiated at approximately the same critical temperature.

The aim of the present study is, on one hand, to resolve the conflicting reports on the sign of the temperature dependence of ion pair formation and, on the other hand, to relate the ion association effects to conducting properties. We have chosen the electrolyte PPO–LiClO₄ of salt concentration O:M = 30 (where O:M is the PO:Li ratio) for a detailed study of ion association phenomena.

EXPERIMENTAL

The procedure for preparation of PPO–LiClO$_4$ complexes has been described elsewhere.[7] In the present study the composition O:M = 30 was used. A low molecular weight PPO of 4000 was chosen as the host polymer. The PPO–salt complex was kept in a dry He atmosphere in a light scattering pyrex cell of rectangular shape.

The Raman spectral band of the symmetric ClO$_4^-$ stretch with components at 930 and 938 cm^{-1} was chosen for a thorough investigation. The band was fitted to two Lorentzians convoluted with a Gaussian instrumental function.

The experimental set-up for the Raman scattering measurements is described in Ref. 7 and for the conductivity measurements in Ref. 15.

RESULTS AND DISCUSSION

The Raman Spectra

The unperturbed ClO$_4^-$ anion has tetrahedral (T$_d$) symmetry. The representation of the normal modes is A$_1$ + E + 2F$_2$, where all the four vibrational modes are Raman active.[16] Three of the modes were observed in the Raman spectra of the present study. The assignments of the modes follow those of a previous study of PPO–LiClO$_4$ complexes in this laboratory,[7] in which complexes of different concentrations were investigated. In the present study we focus on the temperature dependence, and Raman spectra of PPO–LiClO$_4$ complexes of salt concentration O:M = 30 recorded over the temperature range 200–340 K. This corresponds to a range from below the glass transition (T_g is ~240 K as indicated by the present Raman results, see the discussion) up to the highest temperatures possible while still keeping a non-degraded sample.

The ClO$_4^-$ Symmetric Stretch

The most dramatic change of the anion vibrational modes, when changing the temperature, was observed for the perchlorate symmetric stretch. This rather narrow mode, recorded at a frequency of 930 cm^{-1}

at low temperatures, was clearly broadened at the high frequency side for increasing temperatures, which indicates the growth of a second component at 938 cm^{-1}. The whole contour was well fitted to two Lorentzian curves, each convoluted with a Gaussian function measured to be the instrumental bandshape (see Fig. 1). For temperatures below the glass transition the narrow component at 930 cm^{-1}

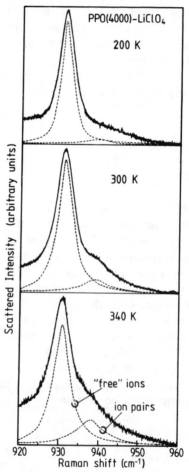

Figure 1. The ClO_4^- symmetric stretching mode at different temperatures for PPO(4000)–LiClO$_4$ of O : M ratio of 30. Dashed lines correspond to the two components of the two-Lorentzian envelope fitted to the spectrum (smoothed solid line).

stays approximately constant in intensity at a level of ~96% of the total salt peak area. As the temperature is raised above T_g, the broad component at 938 cm^{-1} increases rapidly in intensity at the expense of the 930 cm^{-1} peak. At the highest temperature measured (340 K) the dominant peak has decreased to a value of ~78% of the whole band.

Free Ions and Contact Ion Pairs

Since the symmetric stretching vibration at ~930 cm^{-1} is nondegenerate, the appearance of distinct components in the experimental profile indicates different environments for the anion rather than splitting due to lifting of the degeneracy.[6,7] Following earlier Raman results of this system,[7] the predominant and narrow component is attributed to 'free' ClO_4^- anions and the broader peak to anions paired with cations. In the former study the broad component was assigned to solvent separated ion pairs. However, in retrospect, the definition of solvent separated ion pairs, originating from the work by Fuoss,[17] seems inappropriate in the case of very large solvent molecules. Following Fuoss, two ions of opposite charge are called solvent separated if they are separated by a shell of solvent molecules (generally made up of *small* solvent molecules).[17] Then, the larger the solvent molecule the weaker is the Coulomb attraction between the ions and finally at some critical length the 'pair' cannot be distinguished from solvated ions. In the present case of a PPO salt solution it is generally believed that the solvation of the salt in the polyethers involves coordination of the cation to the oxygens of the polymer backbone.[1] The anions, on the other hand, are weakly solvated in polymer electrolytes[5] and are therefore likely to be distributed at random along the polymer chain. The possible 'solvent separated configuration' in the case of polymer–salt complexes is then an ether-oxygen separating the cation from the anion. However, the anion should be repulsed by the electronegative ether-oxygen and a more probable site for the anion, apart from being in direct contact with the cation, is close to the weakly positively charged CH_3— group or the OH— end group of the polymer chain. In the present system it is then likely that the number of ions in solvent separated configurations is very low and that only ions which are in direct contact with each other can be distinguished spectroscopically from 'free' ions. This is also supported by Szwarc, who claims that solvent separated ion pairs may only exist in those solvents in which at least one of the ions possesses a tight solvation shell.[18]

Temperature Dependence of the Carrier Generation

As the temperature increases the Raman spectra show that the component attributed to ion pairs increases in intensity at the expense of the narrow 'free' ion component (see Fig. 1). This means that the ion association in fact *increases* with temperature, in contrast to the theoretical prediction of eqn 1.

To estimate the amount of dissociated charge carriers, we calculate the ratio of the integrated Raman intensity of the 'free' ion component (at 930 cm^{-1}) to the intensity of the total band (centred at 930 and 938 cm^{-1}). In order to check for any non-linear dependence of the scattering cross-section the ratio of the total area of the symmetric stretching anion mode to the area of the 2933 cm^{-1} band of PPO was also calculated. This ratio, corrected for the Bose–Einstein thermal factor, was found to be constant and independent of temperature within the experimental error. Figure 2 shows results of calculations of the logarithm of the normalized intensity of the 'free' ion component versus inverse temperature. The figure demonstrates that above the glass transition temperature the number of dissociated ions decreases in an Arrhenius fashion, i.e. the temperature dependence can formally be represented by the equation

$$n = N' \exp [A/k_B T] \qquad (3)$$

where N' is a constant. The formal activation energy, A, was found to be ~15 meV. We now compare this expression for the carrier generation with that given in eqn (1)

$$n = N_o \exp [-W/2\varepsilon k_B T] \qquad (4)$$

which is based on a dissociation hypothesis valid for weak liquid electrolytes and proposed by Barker & Thomas[19] to be valid also for polymer electrolytes. Our results show that the exponent which expresses the ion association must be of *opposite* sign. It is then interesting to note that the behaviour of eqn (4) has been reported for a PPO–LiClO$_4$ electrolyte (cross-linked PPO) of approximately the same salt concentration.[12] In the latter study the number of charge carriers was, rather indirectly, estimated from conductivity and time-

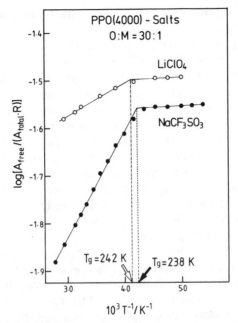

Figure 2. Arrhenius plot of the amount of 'free' ions measured from the intensity of the Raman mode of 'free' ions for PPO(4000)–LiClO$_4$ (O) and –NaCF$_3$SO$_3$ (●) complexes with O:M ratio (R) of 30. The arrows indicate the respective glass transition temperature (T_g) for the two complexes.

of-flight mobility measurements. However, it has recently been stressed that some precaution is necessary when interpreting conductivity data for systems not equilibriated long enough.[14] In fact, careful measurements of PPO–NaI and PPO–KSCN complexes show that after the samples were equilibriated for at least 30 min the conductivity for temperatures above $T \approx T_g + 80$ K begins to decrease[14] (T_g being the glass transition temperature). The observation was interpreted in terms of increasing ion–ion association and the phenomenon is suggested to initiate the salt precipitation observed at higher temperatures. In the present Raman study the sample was equilibrated for 30 min and the scanning time was approximately 5 h for each temperature measured. A further extension of the equilibrium time and/or the scanning time did not significantly change the spectrum. The recorded intensity profiles can therefore be taken as the equilibrium values.

Comparison with Other Complexes and with Conductivity Data

Next we compare the temperature dependence of the number of dissociated anions in PPO–LiClO$_4$ with that in a PPO–NaCF$_3$SO$_3$ electrolyte (PPO4000 and O:M = 30) previously studied in this laboratory[8] (see Fig. 2). It can be seen in Fig. 2 that at low temperatures the number of 'free' charge carriers is approximately constant in both systems. Then, rather abruptly, at temperatures of 238 K and 242 K in the NaCF$_3$SO$_3$ and LiClO$_4$ complexes respectively, the number of dissociated anions starts to decrease with increasing temperature. It seems reasonable to attribute the cross-over temperatures to T_g values of the respective system, since it is expected that the ratio of dissociated ions and ion pairs is more or less constant due to the high energy barriers for ionic transport in the glassy state. Then, as the temperature is raised above T_g the ions can move more freely and the probability to form ion pairs should change, resulting in a change of the 'free' ion concentration, which is also observed.

It is obvious from Fig. 2 that the ion–ion association in PPO is stronger for Na$^+$CF$_3$SO$_3^-$ than for Li$^+$ClO$_4^-$. This is in accordance with recent Raman results from higher salt concentration complexes of PPO–NaCF$_3$SO$_3$ and PPO–LiClO$_4$ in the range O:M = 30–5 investigated at a constant temperature (295 K).[6,7] Also, we find that the temperature dependence of the ion association in the two complexes is quite different (see Fig. 2); a considerably larger decrease of the amount of dissociated charge carriers with increasing temperature is observed in the PPO–NaCF$_3$SO$_3$ complex.

To relate the observed difference in 'free' charge carrier concentration to the ionic conductivity of the two systems, we plot in Fig. 3 conductivity data obtained in this laboratory together with the Raman results for the 'free' ion concentration. The conductivity data can be taken as equilibrium values since the samples were kept for at least 30 min at each temperature and no change, within the experimental accuracy, could be observed for any further extension of the waiting time. To overcome the mobility contribution to the conductivity, we use in Fig. 3 a reduced representation with T_g as a scaling parameter, assuming a close relation between the ionic mobility and T_g. (It is generally believed that the ionic migration is closely linked to the local segmental motions of the host polymer,[2,5] which in turn is related to the value of T_g.) A note of precaution is however necessary, since in this case the difference in conductivity is rather small and therefore

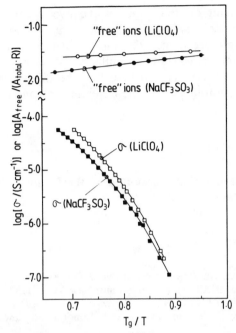

Figure 3. T_g-reduced Arrhenius plot of the ionic conductivity σ and the amount of 'free' ions (from the intensity of the Raman mode of 'free' ions) for PPO(4000)–LiClO$_4$ (open symbols, $\square$ and $\bigcirc$, respectively) and –NaCF$_3$SO$_3$ (closed symbols, $\blacksquare$ and $\bullet$, respectively) complexes with O : M ratio (R) of 30.

sensitive to the T_g values chosen. Furthermore, the assumption of a mobility reduction by T_g scaling might be over-simplified since the two salts have different cations and anions and all the four different ions may have different mobilities.

With this in mind it is still interesting to note in Fig. 3 the similar temperature behaviour of the difference in the conductivity and 'free' ion concentration of the two systems. Also, notice that the conductivity values for PPO–LiClO$_4$ are higher than those for the PPO–NaCF$_3$SO$_3$ complex. At room temperature the PPO–LiClO$_4$ complex has conductivity about 1·6 times larger than that of the PPO–NaCF$_3$SO$_3$ complex and this agrees very well with the difference in the number of 'free' charge carriers at the same temperature. Then decreasing the temperature reduces the difference both in conductivity and in the amount of 'free' ions of the two systems. In total the difference in conductivity is reduced with about ~0·2 log units over the measured temperature range. The difference in the amount of 'free' ions over the same range shows a similar but slightly smaller reduction (see Fig. 3). We therefore conclude that the difference in 'free' ion concentration may well explain the difference in conductivity between the two systems.

ACKNOWLEDGEMENT

This program is financially supported by the National Swedish Board for Technical Development.

REFERENCES

1. Ratner, M. A., In: *Polymer Electrolyte Reviews-1*, ed. J. R. MacCallum & C. A. Vincent. Elsevier, London and New York, 1987, p. 235.
2. Watanabe, M. & Ogata, N., In: *Polymer Electrolyte Reviews-1*, ed. J. R. MacCallum & C. A. Vincent. Elsevier, London and New York, 1987, p. 39.
3. Gorecki, W., Andreani, R., Berthier, L. & Armand, M. B., *Sol. Stat. Ion.*, **18–19** (1986) 295.
4. Killis, A., Le Nest, J. F., Gandini, A. & Cheradame, H., *Macromolecules*, **17** (1984) 63.
5. MacCallum, J. R. & Vincent, C. A. (eds), *Polymer Electrolyte Reviews-1*. Elsevier, London and New York, 1987, p. 23.
6. Schantz, S., Sandahl, J., Börjesson, L., Torell, L. M. & Stevens, J. R., *Sol. Stat. Ion.*, **28–30** (1988) 1047.
7. Schantz, S., Torell, L. M. & Stevens, J. R., *J. Appl. Phys.*, **64** (1988) 2038.
8. Kakihana, M., Schantz, S. & Torell, L. M., *J. Chem. Phys.* (in press).
9. Teeters, D. & Frech, R., *Sol. Stat. Ion.*, **18–19** (1986) 271.
10. Dupon, R., Papke, B. L., Ratner, M. A., Whitmore, D. H. & Shriver, D. F., *J. Am. Chem. Soc.*, **104** (1982) 6247.
11. Cheradame, H. In: *IUPAC Macromolecules*, ed. H. Benoit & P. Rempp. Pergamon Press, New York, 1982, p. 351.
12. Watanabe, M., Sanui, K., Ogata, N., Kobayashi, T. & Ohtaki, Z., *J. Appl. Phys.*, **57** (1985) 123.
13. Greenbaum, S. G., Adamic, K. J., Pak, Y. S., Wintersgill, M. C., Fontanella, J. J., Beam, D. A. & Andeen, C. G., *Proc. Electrochem. Soc. Symp. on Electro-Ceramics and Solid State Ionics*, Honolulu, 1987.
14. Wintersgill, M. C., Fontanella, J. J., Greenbaum, S. G. & Adamic, K. J., *Brit. Polym. J.*, **20** (1988) 195.

15. Svantesson, P. A., Albinsson, I. & Mellander, B.-E., to be published.
16. Herzberg, G., *Molecular Spectra and Molecular Structure*. D. Van Nostrand Co. New York, 1956.
17. Sadek, H. & Fuoss, R. M., *J. Am. Chem. Soc.*, **76** (1954) 5905.
18. Szwarc, M., In: *Ions and Ion Pairs in Organic Reactions-1*, ed. M. Szwarc. Wiley-Interscience, New York, London, Sydney and Toronto, 1972, p. 1.
19. Barker, R. E. Jr & Thomas, C. R., *J. Appl. Phys.*, **35** (1964) 3203.

Dielectric Relaxation Studies on Polymer Electrolytes

S. G. Greenbaum

Department of Physics, Hunter College of CUNY, 695 Park Ave., New York, New York 10021, USA

&

J. J. Wilson, M. C. Wintersgill, J. J. Fontanella

US Naval Academy, Annapolis, Maryland 21402-5026, USA

ABSTRACT

Studies of dielectric relaxation (DR) in poly(propylene oxide) (PPO) and ionic conductivity in PPO containing lithium or sodium salts reveal that the activation energy and activation volume associated with the glass transition, or α-relaxation, in PPO are quite similar to the conductivity activation parameters in PPO–salt complexes. It is also found that both the dielectric loss in PPO and the electric modulus in PPO–salt complexes can be fitted to the Stretched Exponential (SE) decay function in which the SE parameter, β, is approximately temperature-independent and, again, quite similar in value for the two processes.

Finally new results for amorphous oxymethylene linked poly(ethylene oxide) and two of its Na–salt complexes are presented.

INTRODUCTION

Dielectric relaxation (DR) is a delayed polarization resulting from a changing electric field in a linear insulating system. There have been a great many studies of this phenomenon in a wide variety of materials

35

and there are several general books on the topic such as that by Daniel[1] and, more specifically, books on DR in polymers such as that by Hedvig.[2] The term dielectric relaxation is not strictly applicable to polymer electrolytes above the glass transition temperature because of the motion of charges associated with ionic conductivity. However, what is meant by polarization in insulators is the induced dipole moment per unit volume and this can still be identified in an ionic conductor and thus classified as dielectric relaxation. The difficulty of studying dielectric phenomena in ionic conductors is that both dipole reorientation and ionic conductivity are detected via energy loss and thus it is often difficult to separate the two.

One way of observing DR directly in polymer electrolytes is to cool the material below the glass transition temperature, where the ionic conductivity is very low. A review of low temperature dielectric relaxation studies in polymer electrolytes has recently been given[3] and, except for new results on amorphous PEO complexes, those studies will not be discussed here. Dielectric phenomena in polymer electrolytes above T_g can be observed directly, however, if the applied frequency is sufficiently high that the contribution of the mobile ions to the imaginary part of the dielectric constant is greatly reduced. There have been some interesting results recently in the radio and microwave frequency region which have identified ionic contributions to the overall DR in addition to relaxations attributable to polymer segmental motion.[4,5]

Dielectric relaxation behavior of the host polymer (without ions) has been shown to yield important insight into ionic transport phenomena occurring in the corresponding polymer–salt complex.[3,6] Several examples of these studies will be reviewed. Finally electrical conductivity data for polymer electrolytes, when recast in the formalism of the electric modulus,[6] provide a physically revealing description of the corresponding electrical relaxation which is quite similar to the dielectric relaxation observed in the host polymers. Some recently obtained results on poly(propylene oxide) (PPO) reported elsewhere[6] will be reviewed and new studies of amorphous, oxymethylene-bridged poly(ethylene oxide) (PEO) will be presented. These results will also highlight the observation that polymer electrolytes are found to exhibit stretched exponential (SE) behavior as formulated by Kolrausch,[7] and Williams & Watts,[8] which has been observed in a wide variety of disordered systems.

RELATION BETWEEN HOST DR AND IONIC CONDUCTIVITY OF COMPLEX

Amorphous polymers generally undergo a DR process, referred to as the α-relaxation, which is characterized by large-scale segmental motions (e.g. ~5–10 or more repeat units) that are associated with the glass transition. Although it is well known that significant ionic conductivity in amorphous polymer electrolytes (and in the amorphous regions of mixed phase materials) occurs only above T_g, some of the most compelling evidence for the specific role that the α-relaxation plays in the ion transport process comes from DR and conductivity measurements on the host polymer and its complex, respectively. Dielectric loss peaks associated with the glass transition in PPO have been measured and fitted by the Havriliak–Negami function for the complex capacitance:

$$C^* = C - \frac{iG}{\omega} = \frac{D}{[1 + (i\omega\tau_o)^{(1-\alpha)}]^\beta} \tag{1}$$

where C and G are the real capacitance and conductance, respectively, determined by techniques described elsewhere.[9] Data for three temperatures, taken from Ref. 9, are shown in Fig. 1. The peak frequency

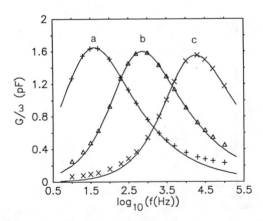

Figure 1. Dielectric loss peaks for PPO at (a) 215·8 K (+); (b) 221·8 K ($\triangle$); (c) 230·8 K ($\times$). The solid lines are the best-fit Havriliak–Negami curves (from Ref. 3).

38 S. G. Greenbaum *et al.*

positions gleaned from the best fit Havriliak–Negami curves were then fitted by the VTF equation which is commonly employed in describing the dynamics of amorphous polymers, including the ionic conductivity of polymer electrolytes:[10]

$$\omega_p = AT^{1/2} \exp\left[-E_a/k(T - T_0)\right] \tag{2}$$

The ionic conductivity of the complex $PPO_8LiCF_3SO_3$ also follows a VTF temperature dependence given by eqn (2), with the conductivity σ in place of ω_p. It was found[9] that the activation parameters E_a and T_0 were very close for both sets of measurements. Specifically $E_a = 0.089$ eV, $T_0 = 173$ K; $E_a = 0.086$ eV, $T_0 = 214$ K for the DR and conductivity data, respectively. The higher T_0 value of the complex is attributable to its elevated glass transition temperature, relative to uncomplexed PPO. T_0 is typically 35–45 K below T_g. To emphasize the correlation between both sets of data the VTF conductivity curve is plotted in Fig. 2, and the peak frequencies associated with the α-relaxation in PPO are included in the same plot, after appropriate scaling.

Although slight discrepancies between segmental motion as probed by Brillouin scattering and conductivity (which are probably due to ionic aggregation, and local motions that don't result in ion transport)

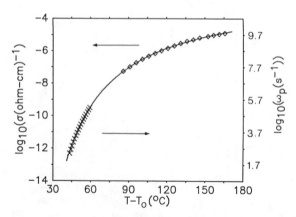

Figure 2. Electrical conductivity data for $PPO_8LiCF_3SO_3$ and peak position for the α-relaxation in PPO ($\times$) plotted versus temperature relative to T_0. The solid line is the VTF equation best fit to the conductivity data only ($\diamond$) (from Ref. 3).

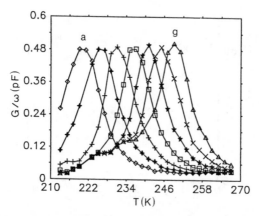

Figure 3. Dielectric loss peaks for PPO at seven pressures at 1000 Hz in the region of the glass transition. The curves are: (a) 0·0001 (1 atm.) ($\Diamond$); (b) 0·03 (+); (c) 0·06 (+); (d) 0·09 ($\Box$); (e) 0·12 ($\star$); (f) 0·15 ($\times$); (g) 0·21 GPA ($\triangle$). Straight line segments connect the experimental points (from Ref. 3).

have been reported by Kakihana et al.,[11] Fig. 2 illustrates the strong similarities between the two processes.

Additional correlations between the α-relaxation of the host and the conductivity of the complex have been observed by investigating their pressure dependencies.[12] The effect of hydrostatic pressure on the α-relaxation is evident from the data in Fig. 3 (from Ref. 3) which shows the DR loss peaks shifting to higher temperature with increasing pressure. A shift in T_g of 17 K/kbar can be deduced from the data in Fig. 3.[3] A comparable T_g shift has been observed in the PPO–salt complexes indirectly by conductivity measurements.[9] In particular, the ionic conductivity decreases and the ^{23}Na NMR linewidth increases in a manner consistent with a pressure induced elevation in the T_g of the complex.

A useful quantity that can be extracted from the isothermal pressure data is the activation volume. Although the α-relaxation and conductivity follow the previously mentioned VTF behavior, for simplicity it is assumed that an 'Arrhenius activation volume' ΔV_A can be defined over a limited temperature region as[3]

$$\Delta V_A = -kT \left(\frac{\partial \ln \alpha}{\partial P} \right)_T \tag{3}$$

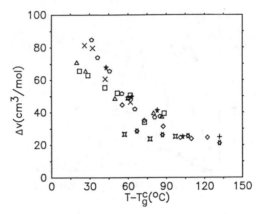

Figure 4. Activation volume versus temperature relative to the central T_g for the α-relaxation in PPO ($\times$); and for the conductivity in PPO-salt (8:1) complexes (denoted by salt): $LiCF_3SO_3$ ($\diamond$); $LiClO_4$ ($\square$); LiI ($\triangle$); $LiSCN$ ($\square$); $NaClO_4$ ($\varhexagon$); NaI ($\bigstar$); $NaSCN$ ($\star$); ^{14}C and [22]Na tracer ($+$) diffusion measurements (from Ref. 13), on NaSCN (from Ref. 3).

where $\alpha = \omega_p$ for DR measurements and $\alpha = \sigma$ for conductivity measurements. Activation volumes obtained from conductivity data for a variety of PPO–salt complexes and from dielectric loss peak frequency measurements in PPO are plotted as a function of $T - T_g^c$, where T_g^c is the central glass transition temperature determined by differential scanning calorimetry, in Fig. 4. Included in the plot are [22]Na tracer diffusion data from Bridges & Chadwick.[13] That all of the data follow the same 'universal' trend is again suggestive of the central role played by the α-relaxation in the ion transport process.

The decrease in ΔV_A with increasing T is primarily a consequence of the Arrhenius description of processes which exhibit VTF behavior. The conductivity and DR activation volumes vary from roughly 25 to 80 cm^3/mol. These values serve to highlight the large volume changes associated with cooperative segmental motion that must accompany the α-relaxation and, of course, ion transport. Recent work by Watanabe & Ogata[14] on rigid polymer backbones containing PEO side chains demonstrates that a minimum number of EO segments ($\sim$3–7) are necessary for ion transport. This result is consistent with the large activation volumes displayed in Fig. 4.

STRETCHED EXPONENTIAL BEHAVIOR OF DIELECTRIC AND ELECTRICAL RELAXATIONS

Relaxation phenomena in disordered systems often do not reflect simple exponential behavior with a single relaxation time (i.e. Debye process). Attempts to model the relaxation include the assumption of a distribution of relaxation times or, alternatively, an altogether different decay function. A popular example of the latter is the SE decay expressed as

$$\phi(t) = \exp\left[-(t/\tau)^\beta\right] \tag{4}$$

The popularity of the SE formalism is partly attributable to its wide applicability to systems characterized by spatial or energetic disorder, and to the often nonphysical and arbitrary nature of relaxation time distributions that otherwise must be invoked to fit the data.[15]

The complex dielectric constant $\varepsilon^* = \varepsilon' - i\varepsilon''$, where ε' and ε'' are the real and imaginary parts, respectively, is related to the decay function via:

$$\varepsilon^* = \varepsilon_\infty + (\varepsilon_s - \varepsilon_\infty)\mathscr{L}(-d\phi/dt) \tag{5}$$

where ε_∞ and ε_s are, respectively, the high and low frequency limits of the dielectric constant and $\mathscr{L}$ is the Laplace transform. A useful consequence of this expression is that it is directly adaptable to the experimental data, which is taken in the frequency domain. Figure 5 shows dielectric loss peaks in PPO at three temperatures. The data are similar to those shown in Fig. 1, except that $\varepsilon'' = G/\omega C_0$, where C_0 is the geometric capacitance, is displayed directly. Also, in contrast to the Havriliak–Negami fit to the data in Fig. 1, the data in Fig. 5 are fitted to the SE decay function utilizing the procedure originated by Moynihan et al.,[16] and adapted to polymer electrolytes in Ref. 6.

It is clear that although the fit is quite good in the vicinity of the loss peaks and, in fact, comparable in overall quality to the four parameter Havriliak–Negami fits in Fig. 1, there is considerable deviation in the wings of the loss curves. One can, in fact, improve the wing fit, but only at the expense of the quality of the peak fit. For reasons which will be explained later, the best fit SE curves were generated by utilizing only the nine experimental data points closest to the peak frequency for each temperature, rather than all seventeen data points. The best fit SE parameters for the dielectric loss associated with the α-relaxation in PPO are listed in Table 1.

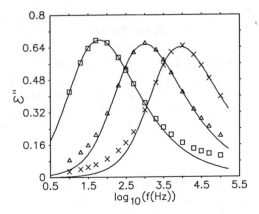

Figure 5. Imaginary part of the dielectric constant versus $\log_{10}(f(\text{Hz}))$ for PPO at: 216·8 K ($\square$); 222·8 K ($\triangle$); and 228·8 K ($\times$). The solid lines are the best-fit SE curves to only the central nine data points.

Relaxations associated with electrical (in particular ionic) conductivity can be probed by utilizing the complex electric modulus $M^* = M' + iM''$, where M' and M'' are the real and imaginary parts, respectively, and M^* is defined as $1/\varepsilon^*$, the reciprocal of the complex dielectric constant.[16] Peaks observed in plots of M'' versus frequency are generally associated with the true electrical relaxation of the material under study whereas ε'' in a conducting system has a frequency dependence that is sensitive to the particular nature of the electrodes (i.e. blocking or nonblocking) used in the experiment.

The electric modulus is related to the SE decay function via

$$M^* = M_s[1 - \mathcal{L}(-\mathrm{d}\phi/\mathrm{d}t)] \tag{6}$$

Table 1. Best-Fit Stretched Exponential Parameters for Dielectric Relaxation Associated with the α-Relaxation in Pure PPO, Utilizing only the Nine Experimental Points Closest to the Loss Peak

T(K)	β_{SE}	τ_{SE} $(10^{-4}\,\text{s})$	$\varepsilon_s - \varepsilon_\infty$	Std dev. (10^{-4})
216·8	0·489	18·73	2·53	23
222·8	0·484	1·15	2·47	50
228·8	0·483	0·130	2·59	45

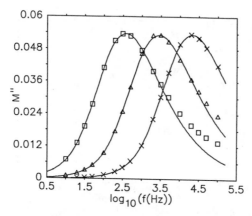

Figure 6. Imaginary part of the electric modulus versus $\log_{10}(f(\text{Hz}))$ for PPO_8NaClO_4 at: 294·4 K ($\square$); 300·4 K ($\triangle$); and 308·4 K ($\times$). The solid lines are the best-fit SE curves to only the cental nine data points.

where M_s is the high frequency limit of M'.[16] Following the procedure described in Refs 6 and 16 (the same as for ε'' in PPO), best fit SE curves were generated for M″. Figure 6 displays the frequency dependence of M'' for PPO_8NaClO_4. Once again, only the nine experimental data points closest to the peak frequency at each temperature were utilized to generate the SE curves. The values of the SE parameter β for several PPO-salt complexes, all with an 8:1 repeat-unit/salt ratio, and at several reduced temperatues $(T - T_g)$ are plotted in Fig. 7. Included in the plot are the β values gleaned from the ε'' data for PPO (Table 1). With the exception of two of the complexes exhibiting low values, all of the β parameters fit into a relatively narrow range ($\sim$0·47–0·52) and do not exhibit a noticeable temperature dependence. It is important to note that SE curves generated from the full set (seventeen points) of data yield β values that are somewhat different and show a tendency to increase slightly with increasing temperature. It has been pointed out in Ref. 6 that the overall quality of the nine-point fit is as good as the seventeen-point fit (and superior in the vicinity of the peak frequency). The former curves also appear to yield a more physically meaningful result, namely that SE relaxation processes ought to exhibit a temperature independent β. This last point has been discussed by Börjesson et al.[17] As for the anomalously low β values obtained for the NaI and $LiCF_3SO_3$ complexes, there is no satisfactory explanation for this result at the

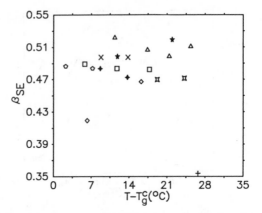

Figure 7. Best-fit SE parameter for only the central nine data points (four on each side of the peak) versus reduced temperature (relative to the central T_g) for: electrical relaxation in PPO ($\square$); electric modulus in PPO containing the following salts in an 8:1 ratio: NaClO$_4$ ($\triangle$); LiClO$_4$ ($\times$); (LiI ($\star$);   NaCF$_3$SO$_3$ ($\hookleftarrow$);   NaI ($\diamond$);   LiSCN ($\blacklozenge$);   LiCF$_3$SO$_3$ (+);   and NaSCN ($\bowtie$) (modulus data from Ref. 6).

present time, although Xue & Angell[18] have also reported low β values in some dilute PPO–salt complexes.

Finally, the observation that the α-relaxation in PPO can be successfully fitted by SE curves yielding roughly the same β parameter as the ionic conductivity relaxation in PPO complexes provides yet another indication of the intimate relationship between the two processes.

LOW TEMPERATURE DR IN AMORPHOUS PEO COMPLEXES

The final section of this paper concerns new DR data on an amorphous form of PEO and two of its Na-salt complexes. It has been shown by Nicholas *et al.*[19] that the crystalline nature of PEO can be suppressed by incorporating evenly but irregularly spaced units into the PEO chain. One such preparation, oxymethylene-linked PEO and its complexes with NaCF$_3$SO$_3$ and NaI (both with a repeat-unit/Na ratio of 9:1) were the subject of a recent investigation in our laboratory.[20] In that work it was shown that partial crystallinity is

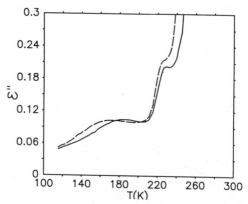

Figure 8. Imaginary part of the dielectric constant of amorphous PEO versus temperature at 100 Hz (dashed line) and 1000 Hz (solid line).

retained in the modified PEO (hereafter referred to as a-PEO) up to about 30°C while both Na-salt complexes are completely amorphous.

Figure 8 displays ε'' in a-PEO as a function of temperature at 100 and 1000 Hz. On the basis of previous DR measurements on high molecular weight PEO,[3] the broad 100 Hz peak centered at about 170 K is attributed to the γ-relaxation, which is observed in a wide variety of polymers and reflects the motion of only small segments of the polymer. The γ-relaxation peaks are observed more clearly in the a-PEO complexes; Figs 9 and 10 show the results for a PEO$_9$NaI and

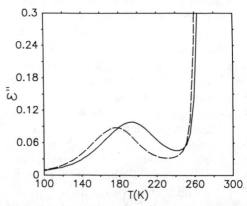

Figure 9. Imaginary part of the dielectric constant of a-PEO$_9$NaI versus temperature at 100 Hz (dashed line) and 1000 Hz (solid line).

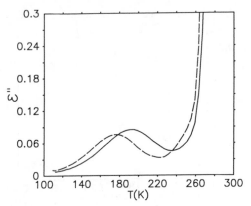

Figure 10. Imaginary part of the dielectric constant of a-PEO$_9$NaSCN versus temperature at 100 Hz (dashed line) and 1000 Hz (solid line).

a-PEO$_9$NaSCN, respectively. Although both complexes exhibit an upward shift of about 10 K from the peak in uncomplexed a-PEO, this is much smaller than the shift in T_g associated with formation of the complex (38 K and 46 K for the triflate and iodide salts, respectively[20]). Results obtained earlier for PEO complexes generally show little effect of the ions on the γ-relaxation of the host polymer, unless there is some structural distortion of the chain segments induced by the presence of large ions (e.g. K).[3] As pointed out in the introduction, the dielectric loss associated with the glass transition (α-relaxation) is masked by the conductivity for materials containing mobile ions. Therefore the rapid increase in ε'' shown in both Figs 9 and 10 reflects the ionic conductivity. It is of interest then to examine the data in Fig. 8 above the γ-relaxation peak. The 100 Hz 'shoulder' located at 225 K is clearly associated with the glass transition, whereas the steep rise in ε'' above this temperature reflects the presence of impurity ions in the material. It is noteworthy that the higher values of ε'' in the complexes are comparable to the higher ε'' values in a-PEO, but occur at a significantly higher temperature than in the latter. While this result merely reflects the upward shift in T_g upon complex formation, it also demonstrates that only a small number of impurity ions (down at least an order of magnitude from the ion concentration in the complexes) are sufficient to yield relatively high ionic conductivity. This last point highlights the predominant role of polymer segmental motion in the overall ionic conductivity.

CONCLUSIONS

The relationship between ionic conductivity in polymer–salt complexes and the α-relaxation associated with the glass transition in the host polymer (PPO) has been explored. Close similarities are found between the temperature and pressure dependencies of the α-relaxation loss peak in PPO and the ionic conductivity in the PPO–salt complex.

The DR loss peaks in PPO and the electric modulus in PPO–salt complexes have been fitted by the stretched exponential (SE) decay function. The SE parameter, β, is generally confined to values between 0·47 and 0·52, with no discernible temperature dependence.

Finally, new DR data on a modified amorphous form of PEO and two of its sodium salt complexes show that the γ-relaxation is affected somewhat by the presence of the ions.

ACKNOWLEDGEMENTS

This work was supported in part by the US Office of Naval Research, the US Naval Academy Research Council, and the PSC-CUNY Faculty Research Award Program.

REFERENCES

1. Daniel, V. V., *Dielectric Relaxation*. Academic Press, London, 1967.
2. Hedvig, P., *Dielectric Spectroscopy of Polymers*. Adam Hilger Ltd, Bristol, 1977.
3. Wintersgill, M. C. & Fontanella, J. J., Low frequency dielectric properties of polyether electrolytes. In: *Polymer Solid Electrolytes 2*, ed. J. R. MacCallum & C. A. Vincent. Elsevier Applied Science, London, 1989, p. 43–59.
4. Ansari, S. M., Brodwin, M., Stainer, M., Druger, S. G., Ratner, M. A. & Shriver, D. F., Conductivity and dielectric constant of the polymeric solid electrolyte, $PEO_8NH_4SO_3CF_3$ in the 100 Hz to 10^{10} Hz range. *Solid State Ionics*, **17** (1985) 101–6.
5. Gray, F. M., Vincent, C. A. & Kent, M., Dielectric studies of PEO-based polymer electrolytes using time domain spectroscopy. *Solid State Ionics*, **28–30** (1988) 936–40.
6. Wilson, J. J., Smith, M. K., Fontanella, J. J., Wintersgill, M. C., Mazaud, P., Greenbaum, S. G. & Siddon, R. L., Application of the stretched

exponential to electrical studies of poly (propylene oxide)–salt complexes. *Solid State Ionics*, submitted.

7. Kohlrausch, R., *Prog. Ann. Phys.*, **12** (1847) 393.
8. Williams, G. & Watts, D. C., *Trans. Farad. Soc.*, **66** (1970) 80–5.
9. Fontanella, J. J., Wintersgill, M. C., Smith, M. K., Semancik, J. & Andeen, C. G., The effect of high pressure on the electrical relaxation in PPO and electrical conductivity in PPO complexed with lithium salts. *J. Appl. Phys.*, **60** (1986) 2665–71.
10. MacCallum, J. R. & Vincent, C. A. (eds), *Polymer Solid Electrolytes 1*, Elsevier Applied Science, London, 1987.
11. Kakihana, K., Sandahl, J., Schantz, S. & Torell, L. M., Optical spectroscopy of polymer electrolytes. In: *Second International Symposium on Polymer Electrolytes*, ed. B. Scrosati. Elsevier Applied Science, London, UK, pp. 23–33.
12. Greenbaum, S. G., Adamic, K. J., Pak, Y. S., Wintersgill, M. C., Fontanella, J. J., Bean, D. A. & Andeen, C. G. In: *Proceedings of the Electrochemical Society Symposium on Electro-Ceramics and Solid State Ionics*, ed. H. Tuller. 1988, Vol. 88–3, pp. 211–8.
13. Bridges, C. & Chadwick, A. V., Activation volumes for diffusion in polyether electrolytes. *Solid State Ionics*, **28–30** (1988) 965–8.
14. Watanabe, M. & Ogata, N., Ionic conductivity of polymer electrolytes and future applications. *Brit. Polym. J.*, **20** (1988) 181–92.
15. Schlesinger, M. F. & Bendler, J. T., Stretched exponential relaxation. *Physics Today* (January 1989) 531–2.
16. Moynihan, C. T., Boesch, L. P. & Laberge, N. L., Decay function for the electric field relaxation in vitreous ionic conductors. *Phys. Chem. Glasses*, **14** (1973) 122–5.
17. Börjesson, L., Torell, L. M. & Stevens, J. R., *Polymer*, **30** (1989) 370–2.
18. Xue, R. & Angell, C. A., High ionic conductivity in PEO.PPO block polymer + salt solutions. *Solid State Ionics*, **25** (1987) 223–30.
19. Nicholas, C. V., Wilson, D. J., Booth, C. & Giles, J. R. M., Oxymethylene-linked poly(ethylene oxide) electrolytes. *Brit. Polym. J.*, **20** (1988) 289–93.
20. Wintersgill, M. C., Fontanella, J. J., Pak, Y. S., Greenbaum, S. G., Al-Mudaris, A. & Chadwick, A. V., Electrical conductivity, differential scanning calorimetry and nuclear magnetic resonance studies of amorphous poly(ethylene oxide) complexed with sodium salts. *Polymer*, **30** (1989) 1123–6.

Thermal and Thermo-oxidative Degradation of Polymer Electrolytes Based on Poly(Ethylene Oxide)

Luigi Costa, Ali M. Gad,* Giovanni Camino‡

Dipartimento di Chimica Inorganica, Chimica Fisica e Chimica dei Materiali dell'Università di Torino, Via P. Giuria 7, 10125 Torino, Italy

&

G. Gordon Cameron, Malcolm D. Ingram, M. Younus Qureshi

Department of Chemistry, University of Aberdeen, Aberdeen AB9 2UE, UK

ABSTRACT

The thermal and thermo-oxidative behaviour of polymer electrolyte PEO complexes with NaSCN and LiClO$_4$ was assessed by thermogravimetry, differential scanning calorimetry and identification of degradation products.

Both NaSCN and LiClO$_4$ reduce the thermal stability of PEO in an inert atmosphere and modify its mechanism of degradation to volatile products, whereas the salts protect the polymer against thermal oxidation. Mechanisms are proposed to interpret these effects of the salts on PEO.

INTRODUCTION

Although much is known about the morphology and mechanism of ion migration in polymer electrolytes derived from polyethers,[1,2] little is

* Present address: Department of Materials Science, Institute of Graduate Studies and Research, Alexandria University, Alexandria, Egypt.
‡ To whom correspondence should be addressed.

49

known about their thermal and thermo-oxidative stability, or their
long-term stability under normal usage or storage conditions. We
report here results of a study of the comparative stabilities of
poly(ethylene oxide) (PEO, ex-BDH, nominal MW 6×10^5) and PEO
complexes with NaSCN and $LiClO_4$. These salts were chosen because:
(i) their complexes with PEO are well documented, and (ii) their
anions have contrasting chemical characteristics (the perchlorate ion is
powerfully oxidizing and the thiocyanate mildly reducing).

EXPERIMENTAL

Complexes of the desired composition were prepared by the standard
cosolution method with dry acetonitrile. Differential scanning calorimetry (DSC) and dynamic thermogravimetry (TG) were conducted
with a Du Pont DSC cell and 951 thermogravimetric analyser
respectively, both driven by a 1090 thermal analyser unit at a heating
rate of $10°C \, min^{-1}$ in a flowing atmosphere of N_2 or air.

The major volatile products of degradation were collected by
carrying out degradations in sealed tubes under N_2. Programmed
heating at $10°C \, min^{-1}$ was carried out to the predetermined temperature required to produce an approx. 50% weight loss at 410°C and
310°C for NaSCN and $LiClO_4$ complexes (4:1) respectively. The
volatile products were then analysed by the GC-FTIR technique.

RESULTS AND DISCUSSION

Thermal Decomposition under N_2

Figure 1 shows TG traces for pure PEO and its complexes
([EO units]/[M^+] = 8) with NaSCN and $LiClO_4$. In each case the PEO
decomposition occurs in a single step. The pure polymer volatilizes
almost completely, while the complexes leave a quantity of residual
salt. Both salts exert an overall destabilizing effect on the polymer.

In the case of the NaSCN complex the influence of the salt seems
relatively mild. However, more detailed studies of PEO.NaSCN
complexes,[3] including isothermal TG and product analysis, reveal that
the salt destabilizes the polymer more than is suggested by Fig. 1 and
also alters the degradation mechanism. The TG trace for a LiSCN

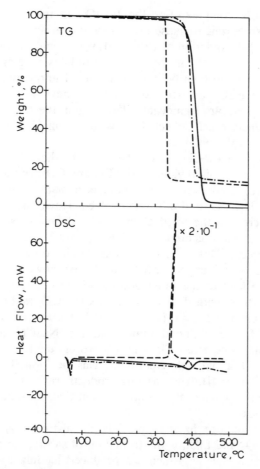

Figure 1. TG and DSC traces of PEO (——), PEO.NaSCN (—•—) and PEO.LiClO$_4$ (– – –) under N$_2$. Heating rate 10°C min^{-1}; [EO units]/[M$^+$] = 8.

complex is superimposable on that of the NaSCN complex of the same composition.

The destabilizing effect of LiClO$_4$ on the PEO is profound. This complex undergoes a very rapid, almost explosive, decomposition at c. 320°C, that is about 80°C below the onset of degradation of pure PEO or the thiocyanate complex.

The DSC traces for these degradations (Fig. 1) reveal that decomposition is mildly endothermic for pure PEO and slightly exothermic for the NaSCN complex. For the LiClO$_4$ complex the reaction is sufficiently exothermic to cause instrumental overheating.

The thermal degradation of PEO and of its complexes with NaSCN and LiClO$_4$ gives a mixture of volatile products ranging from small gaseous molecules to high boiling fragments. The mixtures of gaseous products from PEO and from its complexes have several features in common but there are significant differences in the relative quantities of these products which reflect the diverse chemical influences of the dissolved salt on the polymer.

Pure PEO evolves CH$_2$O, CH$_3$CHO and ether chain fragments, together with traces of ethylene, CH$_4$, CO and CO$_2$. In the presence of NaSCN, as reported previously,[3] there is a marked increase in the amounts of CH$_4$, CO$_2$ and ethylene evolved. However, as the NaSCN content is increased the yield of aldehydes decreases so that acetaldehyde for example was not detected as a product of degradation from the 4:1 complex. This is probably due to the direct interaction of NaSCN with acetaldehyde which we have shown by a separate experiment to lead to the rapid decomposition of the latter, even at room temperature, with the formation of CH$_4$, CO$_2$ and other gases.

In the presence of LiClO$_4$ (4:1 complex) greater yields of ethylene, CO$_2$ and CH$_4$ are obtained than from the NaSCN complex and CH$_3$CHO is again detected. Since in this process the LiClO$_4$ is reduced rapidly to LiCl a parallel experiment was performed with a LiCl complex ([EO/Li = 4]). The gaseous products were identical to those evolved from the perchlorate complex except that the yield of CO$_2$ was substantially less.

A common feature in the pyrolysis chemistry of all three salts NaSCN, LiClO$_4$ and LiCl, is the formation of substantial quantities of CH$_4$, CO$_2$ and ethylene, which are produced in only trace amounts from pure PEO. The very large yield of CO$_2$ from the LiClO$_4$ complex can be accounted for by the direct attack on the polymer of the powerfully oxidizing perchlorate ion. However, there must be an alternative route for the production of CO$_2$ since both the NaSCN and LiCl complexes also evolve this gas in appreciable amounts. Similarly, the formation of ethylene and CH$_4$ is not adequately explained in terms of the mechanisms, discussed previously, for the decomposition of pure PEO.[3]

It seems likely that the degradation mechanism is affected primarily by the coordination of the cation to the oxygen atoms, resulting in a weakening of the adjacent C—O bonds of the polymer skeleton:

$$\text{\textasciitilde\textasciitilde CH}_2\text{—CH}_2 \underset{3}{\Big|} \underset{2}{\Big|} \text{CH}_2\text{—}\underset{1}{|}\text{CH}_2 \diagdown_O \diagup \text{CH}_2\text{—CH}_2 \diagdown_O\text{\textasciitilde}$$

(1)

with M^+, X^- complexed below the central oxygen.

In pure PEO, scission of the C—C bond (1) and the C—O bonds (2, 3) should be equally probable on account of the similar bond energies. Whereas in the complex, because of the interaction with M^+, scission could occur more readily at bonds 2 or 3 leading to an increase of the ratio of C_2 over C_1 terminated macroradicals as compared to pure PEO. This could explain the increase in the amount of ethylene in the case of complexes since ethylene is a typical product of beta scission of such C_2 terminated macroradicals.[3]

$$\text{\textasciitilde\textasciitilde CH}_2\text{—CH}_2\diagdown_O\diagup\text{CH}_2\text{—CH}_2^\cdot \longrightarrow \text{\textasciitilde}_O\diagup\text{CH}_2\text{—CH}_2\diagdown_{O^\cdot} + \text{CH}_2\text{=CH}_2$$

(2)

Furthermore, the increased amounts of CH_4 and CO_4 observed in the complexes might tentatively be explained by fragmentation of oxygen terminated radicals originating at the complexed structural units. For example, by scission of two bonds of type (3), a $C_2H_4O_2$ diradical is split from the polymer chain:

$$\text{\textasciitilde}_O\diagup\text{CH}_2\text{—CH}_2\diagdown_O\diagup\text{CH}_2\text{—CH}_2\diagdown_O\diagup\text{CH}_2\text{—CH}_2\diagdown_O\text{\textasciitilde}$$

$$M^+ \quad X^-$$

(3)

$$\downarrow$$

$$\text{\textasciitilde}_O\diagup\text{CH}_2\text{—CH}_2^\cdot \quad ^\cdot{O}\diagup\text{CH}_2\text{—CH}_2\diagdown_{O^\cdot} \quad ^\cdot\text{CH}_2\text{—CH}_2\diagdown_O\text{\textasciitilde}$$

$$M^+ \quad X^-$$

Methane and CO_2 can be stoichiometrically derived from the diradical, assuming that further fragmentation is favoured by the temperature and catalytic action of the salt:

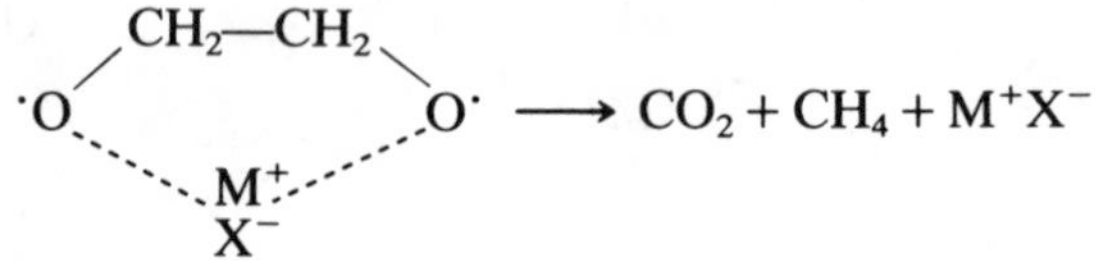

$$\cdot O \underset{M^{+}\,X^{-}}{\overset{CH_2{-}CH_2}{\diamond}} O\cdot \longrightarrow CO_2 + CH_4 + M^+X^- \qquad (4)$$

An alternative ionic mechanism could be proposed to explain the increase of formation of ethylene from the complexes:

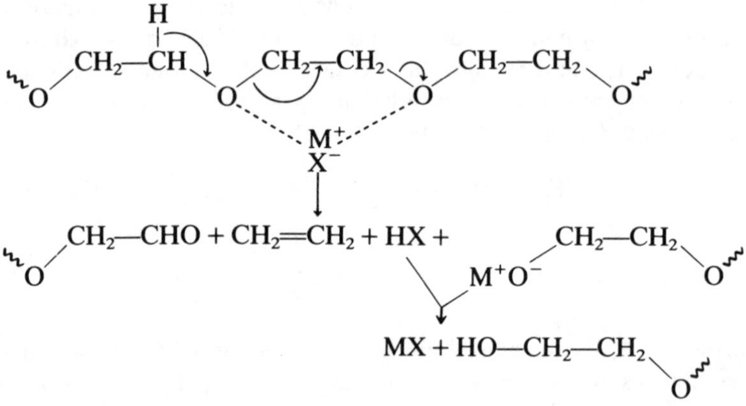

$$(5)$$

CH_2—CHO + CH_2=CH_2 + HX +

MX + HO—CH_2—CH_2

However, it is not easy to explain the formation of CO_2 and CH_4 on this basis.

Furthermore, in an experiment to check the occurrence of this mechanism, a complex of PEO and $NaOCH_3$ ([EO/Na = 4]) was prepared and pyrolyzed. Methanol, which would be produced if the ionic mechanism took place ($X = CH_3O$), was not detected in the degradation products and consequently we feel that the radical mechanism above is the more likely.

Thermal Oxidation

In the presence of oxygen there is a dramatic change in the relative stabilities of the samples (Fig. 2).

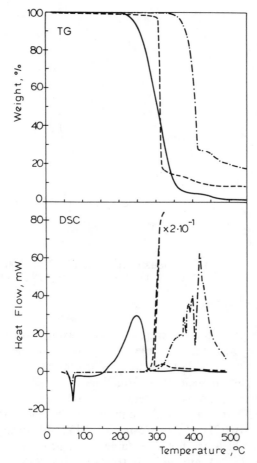

Figure 2. TG and DSC traces as in Fig. 1 but under air.

In terms of the onset of weight loss, the pure PEO is now the least stable and begins a fairly slow volatilization from c. 200°C. The NaSCN complex is the most stable showing a TG trace which, apart from the small second stage of decomposition, is almost superimposable on the corresponding trace in Fig. 1. Similar behaviour was found for the LiCl complex. The LiClO$_4$ complex is also more stable in air than is pure PEO, although once initiated (at a somewhat lower temperature than in N$_2$) the weight loss from the sample is very rapid.

The DSC traces show for: (i) pure PEO, a broad, steady exothermic

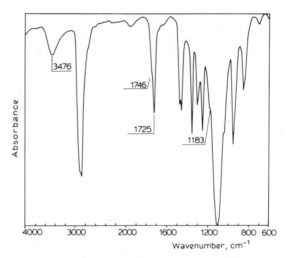

Figure 3. IR spectrum of PEO heated in air at 150°C for 200 min. Peaks due to oxidation: 3476 cm^{-1} (OH); 1746 cm^{-1} (ester); 1725 and 1183 cm^{-1} (formate).

process, (ii) the perchlorate complex, a sharp exotherm again causing instrumental overheating, (iii) the thiocyanate complex, a strong irregular exotherm. The last of these is irreproducible, the jagged nature reflecting intumescent char formation.

The onset of oxidation is apparent in the IR spectrum (Fig. 3) on heating pure PEO at 150°C for 200 min in air. The spectrum contains well-defined bands (OH at 3476, ester at 1746, and formate at 1725 and 1183 cm^{-1}) which are not detectable when PEO complexes with NaSCN, LiCl or LiClO$_4$ are similarly treated.

These data show that thermal oxidation of PEO is somewhat inhibited by the presence of the salts. As far as the mechanism of oxidation is concerned, it seems probable that hydroperoxides play an important role in the thermal oxidation, as they do in the photo-oxidation of pure PEO,[5] thus: (see page 57). It is unlikely that the salt functions directly as an oxygen scavenger since no metal oxides M$_2$O have been detected in the pyrolysis residue. In the case of thiocyanates a measure of antioxidant activity could arise through elimination of hydroperoxides by direct reaction with the SCN$^-$ ion. Indeed, a separate experiment showed that t-butyl hydroperoxide is rapidly decomposed by NaSCN at room temperature.

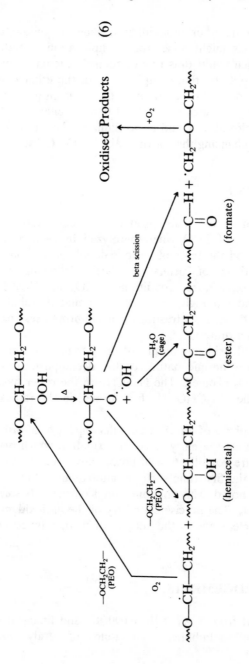

(6)

However, some other antioxidant mechanism could operate with the other salts. This might arise from complexation of radicals with the metal ion which could thus reduce radical reactivity towards oxygen. or it might simply be that complexation of the ether oxygen with the metal cation reduces the reactivity of the adjacent methylene groups to peroxidation by enhancing the C—H bond strength. Thus, the overall effect of complexation would be to diminish the strength of the C—O bonds while enhancing that of the adjacent C—C and C—H bonds.

CONCLUSIONS

The formation of the main products of degradation and the rapid decline in MW of PEO when pyrolyzed in vacuum or N_2 can be accounted for on the basis of various depolymerization and hydrogen transfer reactions of primary radicals following random chain scission.[3,4] In oxygen-free conditions $LiClO_4$ and NaSCN modify this mechanism and although both exert a similar destabilizing effect on the C—O bond, their contrasting chemical properties ensure that their modes of action differ.

The powerful oxidizing nature of the perchlorate ion is manifest in the strongly exothermic nature of the decomposition and the large amounts of CO_2 evolved. The NaSCN, on the other hand, effectively decomposes the CH_3CHO which is formed in the radical decomposition of PEO.

In an atmosphere of air all these salts appear to afford a measure of protection against the early stages of oxidation which bring about the low-temperature volatilization of uncomplexed PEO. As far as we are aware, this is the first time such an 'antioxidant' effect of alkali metal salts has been noted. This phenomenon is particularly surprising with a salt like $LiClO_4$. The precise chemistry of the antioxidant effect of the salt is still unclear and is the subject of further investigations in our laboratories.

ACKNOWLEDGEMENTS

A travel grant from NATO [RG 090/87] and financial support from Ministero della Pubblica Istruzione of Italy are gratefully acknowledged.

Ali M. Gad haṣ carried out this work with the support of the 'TWAS Italian Awards Scheme for Research and Training in Italian Laboratories'.

REFERENCES

1. See for example: Robitaille, C., Marques, S., Boils, D. & Prud'homme, J., Thermal properties of poly(ethylene oxide) complexed with NaSCN and KSCN. *Macromolecules,* **20** (1987) 3023–34.
2. Cameron, G. G., Harvie, J. L., Ingram, M. D. & Sorrie, G. A., Ion migration in liquid polymer electrolytes. *Brit. Polym. J.,* **20** (1988) 199–202.
3. Cameron, G. G., Ingram, M. D., Qureshi, M. Y., Gearing, H. M., Costa, L. & Camino, G., The thermal degradation of poly(ethylene oxide) and its complex with NaSCN. *Eur. Polym. J.,* **25** (1989) 779–84.
4. Grassie, N. & Perdomo Mendoza, A., Thermal degradation of polyether-urethanes: Part 1. Thermal degradation of poly(ethylene glycols) used in the preparation of polyurethanes. *Polym. Deg. Stab.,* **9** (1984) 155–65.
5. Gauvin, P., Lemaire, J. & Sallet, D., Photo-oxidation de Polyéther-*bloc*-polyamides, 3: Propertiétés des hydroperoxides dans les homopolymerés des polyéthers correspondants. *Makromol. Chem.,* **188** (1987) 1815–24.

ESR Studies of Divalent Copper in Polymer Electrolytes

K. J. Adamic, F. J. Owens, S. G. Greenbaum

Department of Physics, Hunter College of CUNY, 695 Park Ave., New York, New York 10021, USA

&

M. C. Wintersgill, J. J. Fontanella

Physics Department, US Naval Academy, Annapolis, Maryland 21402-5026, USA

ABSTRACT

Electron spin resonance (ESR) spectra have been obtained in poly(propylene oxide) (PPO) based solid electrolytes, in particular the compositions $(PPO)_8NaClO_4$ containing a small amount of Cu^{2+} spin probe, and $(PPO)_nCu(CF_3SO_3)_2$, where $n = 9, 12, 27$. Temperature-dependent shifts in the parallel (to the magnetic field) hyperfine component reflect motions associated with the glass transition and yield approximately Arrhenius activation energies for the motional correlation time, which increases with increasing copper salt concentration. The ionic conductivity of the 9:1 complex exhibits a free volume-type temperature depenence, although with much less curvature than is usually observed in amorphous systems.

INTRODUCTION

The use of deliberate paramagnetic impurities as motional probes of polymers is well documented.[1] The impurities give rise to electron spin resonance (ESR) spectra whose parameters change with temperature. These spectral changes can often be exploited to yield valuable information concerning polymer chain dynamics. Recently it has been

61

shown that copper complexes with large hyperfine and g-tensor anisotropies, which produce larger spectral shifts than do nitroxide probes over a given temperature range, can be employed productively in these studies.[2] Two polymer electrolyte systems based on poly(propylene oxide) (PPO) are the subject of this work. The first $(PPO)_8NaClO_4$ containing less than 1 wt% copper benzoyl acetonate (CuBA) as an impurity probe, and the second complex is $(PPO)_nCu(CF_3SO_3)_2$, where $n = 9$, 12 and 27, in which the paramagnetic ion itself constitutes a significant part of the material.

PPO and its complexes are completely amorphous and thus free of the complications associated with the multiphase nature of PEO-based materials. The $(PPO)_8NaClO_4$ complex has been studied extensively by other techniques including ^{23}Na NMR, DSC and complex impedance,[3] and is of interest in this investigation in order to demonstrate the efficacy of CuBA as a motional probe of polymer electrolytes and to provide reference spectra for the results obtained for the $PPO:Cu(CF_3SO_3)_2$ complexes. The latter class of materials, on the other hand, is the subject of recent interest in copper-containing polymer electrolytes in particular[4,5] and divalent metal salt–PEO complexes in general.[6]

EXPERIMENTAL

The starting material was PAREL-58 (Hercules, Inc.) which is a high molecular weight elastomer containing approximately 95% PPO and 5% allyl glycidyl ether, and will hereafter be referred to simply as PPO. The PPO and salt were dissolved in appropriate solvents, anhydrous methanol for $NaClO_4$ and acetonitrile for $Cu(CF_3SO_3)_2$, cast onto teflon dishes and dried in a roughing vacuum at 60°C for at least 72 h. Approximately 0·5% by weight CuBA was added to the solution containing the $NaClO_4$ complex prior to the vacuum treatment. No special measures (other than the vacuum treatment and subsequent handling of samples in a dry N_2 environment) were employed to eliminate water from the samples. The resulting materials were sticky, transparent films that were generally amber in color with a light green tint.

Electron spin resonance first-derivative spectra were obtained with an IBM/Bruker ER-220D EPR spectrometer equipped with a second set of magnetic field modulation coils which allowed the acquisition of second-derivative spectra. Temperatures were maintained to $\pm 2°C$ with an N_2 flow-through dewar. The DSC and complex impedance details have been described elsewhere.[3]

RESULTS AND DISCUSSION

DSC thermograms for $(PPO)_9Cu(CF_3SO_3)_2$ are shown in Fig. 1(a) as prepared, and (b) quenched to $-98°C$ from $102°C$). A strong feature characteristic of a glass transition centered at about $12°C$ is present in (a). This implies that the sample is completely amorphous and that the copper is strongly complexed as evidenced by the increase in T_g over that of pure PPO ($-60°C$). The DSC thermogram for the quenched sample (b) shows a small shift in T_g to lower temperature which suggests that the material is somewhat unstable at higher temperatures. This result is borne out by sample decomposition occurring at $\approx 117°C$. Prior to decomposition, the sample changes color from light blue-green to increasingly darker shades of brown. DSC results for $(PPO)_{27}Cu(CF_3SO_3)_2$ reveal a central T_g very close (within $5°C$) to that of pure PPO. The central T_g of $(PPO)_8NaClO_4$ has previously been reported to be approximately $4°C$.[3]

Figure 2 displays room temperature X-band ESR spectra for $(PPO)_8NaClO_4$ doped with CuBA (I) and $(PPO)_9Cu(CF_3SO_3)_2$ (II). The Cu^{2+} powder spectrum (I) is characterized by large g anisotropy

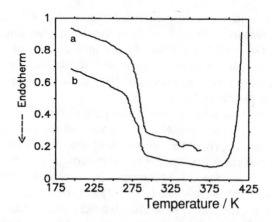

Figure 1. DSC thermograms for $(PPO)_9Cu(CF_3SO_3)_2$: (a) as prepared; (b) quenched from $102°C$.

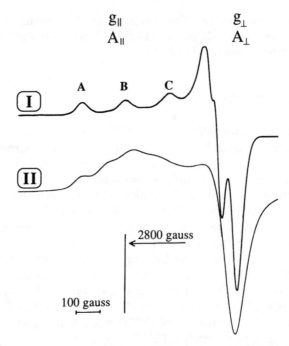

Figure 2. Room temperature X-band ESR spectra of $(PPO)_8NaClO_4$:CuBA (I) and $(PPO)_9Cu(CF_3SO_3)_2$ (II).

leading to the intense feature centered at roughly 3300 gauss $(g_\perp)$ and the weaker feature with well-resolved hyperfine components at lower fields $(g_{\parallel})$. Given the relatively high concentration of Cu^{2+} in the triflate complex, it is somewhat surprising to observe the presence of resolved hyperfine lines in spectrum (II) although substantial dipolar broadening is clearly indicated.

The onset of tumbling motion of the paramagnetic species results in an inward shift of the parallel edge extrema of the powder spectrum. The separation between the lowest field line and and the next to lowest field line measures the hyperfine parameter $A_{\parallel}$. Figure 3 shows the low field portion of spectrum (I), containing the hyperfine components A and B (referred to Fig. 2) at various temperatures. The inward shift of the extrema as the temperature is raised is clearly indicated, and the plots of the inward shift of line A versus temperature for all samples are displayed in Fig. 4. The data in Fig. 4

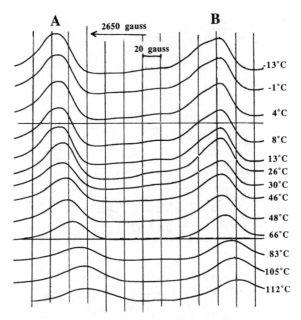

Figure   3. Low-field   hyperfine   lines   (first   derivative)   in
(PPO)$_8$NaClO$_4$:CuBA at various temperatures.

suggest an abrupt increase in motion in the vicinity of the corresponding glass transition for each sample.

A second feature of the data for the $Cu(CF_3SO_3)_2$ complexes shown
in Fig. 4 is the apparent increase in slope with increasing salt
concentration in the enhanced motion region (above T_g) of each
sample. This effect is demonstrated more clearly in Fig. 5 in which the
square of the shift in line position from its value at T_g is plotted in an
Arrhenius fashion. According to calculations by Lee & Brown,[7] the
inward shift of the hyperfine lines resulting from dynamical averaging
of the hyperfine tensor is inversely proportional to the square root of
the motional correlation time τ. It is therefore not surprising to
observe the Arrhenius behavior depicted in Fig. 5. The data tabulated
in Fig. 5 are for the line shifts measured above T_g and corrected for
small sub-T_g temperature dependencies. It is interesting to note that the
activation energies extracted from the plots vary by nearly a factor of
three in going from the dilute (27:1) to concentrated (9:1) samples. In
view of the fact that T_g versus salt concentration effects have already

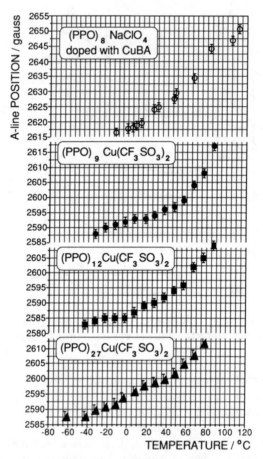

Figure 4. Lowest hyperfine line position versus temperature for all samples.

been taken into account there appears to be an additional contribution to the motional energy barrier arising from higher salt concentrations, perhaps due to strong cation–anion interactions which are known to exist in polymers with high concentrations of monovalent salts.

It is important to mention two qualifications to the above analysis. First, the dynamical results of Lee & Brown[7] do not necessarily predict that the lowest hyperfine line (A) and the next lowest line (B) would have different temperature dependencies and hence different activation energies which has, in fact, been observed experimentally (although

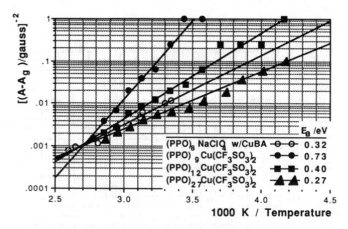

Figure 5. Arrhenius plots of $(A - A_g)^2$ where A is the magnetic field position of the lowest field hyperfine line and A_g is the value at $T = T_g$.

the trend in E_a versus concentration is the same for line B). Second, the motional averaging of the Cu^{2+} ESR hyperfine interaction does not necessarily imply copper ion transport, but could as well result from polymer segmental and/or anion motion with associated directional averaging of the Cu^{2+} ligands.

The half-height linewidth of the hyperfine line A for $(PPO)_8NaClO_4 : Cu^{2+}$ is plotted as a function of temperature in Fig. 6. Line broadening, which is also a consequence of slow tumbling motion, is clearly indicated although the temperature at which the broadening occurs is somewhat higher than T_g. This result demonstrates that the broadening mechanism involves higher frequency combinations of spectral parameters.

A typical complex impedance plot for the $(PPO)_9Cu(CF_3SO_3)_2$ is shown in Fig. 7. Over the frequency range of the measurements (10 Hz–100 kHz), only a single arc was observed. The arc was well represented by a section of a depressed circle. This behavior is similar to that observed for monovalent ions in PPO).[3] However, the temperature variation of the bulk resistance, as determined from the complex impedance analysis and shown in Fig. 8, is significantly different from the monovalent salt/PPO complexes. Specifically, much less curvature associated with free volume behavior is observed than expected for a material with such a high glass transition temperature.

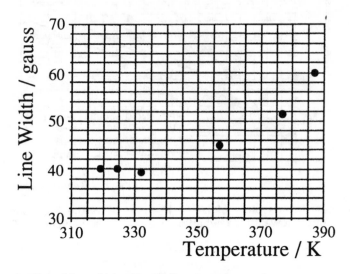

Figure 6. Plot of linewidth of hyperfine component A versus temperature in Cu^{2+}-doped $(PPO)_8NaClO_4$.

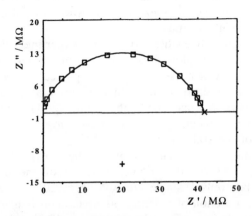

Figure 7. Complex impedance plot for $(PPO)_9Cu(CF_3SO_3)_2$ at 49°C for a frequency range of 10 Hz–100 kHz. The horizontal intercept (x) yields the bulk resistance. The solid line is the best-fit depressed arc (Cole–Cole equation) and the center of the circle is shown by the plus (+) sign.

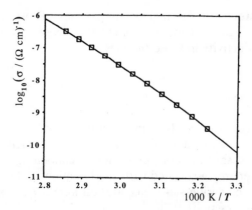

Figure 8. Arrhenius plot for the electrical conductivity. The solid line is the best-fit VTF expression.

For the comparison purpose, the VTF expression:[3]

$$\sigma = \frac{A}{\sqrt{T}} \exp\left(-\frac{E_a}{k(T - T_0)}\right)$$

was best-fit to the data and a very large E_a, 0·36 eV, and a very low T_0, 172·6 K, were obtained. (The preexponential is given by $\log_{10}(A) = 5·01 \, \text{S cm}^{-1} \text{K}^{1/2}$.) The value of T_0 is particularly interesting. It is about 120 K below the central glass transition temperature, $T_g = 295$ K, in contrast to most other materials for which T_0 is about 50 K below T_g. At least part of this behavior can be attributed to the relatively small (280–350 K) temperature range of the present experiment (the temperature range was limited because the sample decomposes at higher temperatures), and it has been shown previously that fitting only the lower temperature portion of a VTF curve yields high values of E_a and low values of T_0.[8] However, it is doubtful that that effect could account entirely for the exceedingly large value of $T_g - T_0$ observed for the present material. It is interesting to note that some PEO:Cu^{2+} complexes have been shown to exhibit free volume behavior (i.e. more curvature) typical of most amorphous polymer electrolytes.[4]

While the frequency variation of the complex impedance was well behaved for the present sample (which was clear and amber with a light green tint), other samples showed different behavior. In particular, for some samples the conductance was higher than expected for blocking electrodes at the highest frequencies. Those samples were black or dark brown. It was concluded that there was at least some electronic conductivity in those samples.

CONCLUSIONS

PPO-based polymer electrolytes incorporating Cu^{2+} either as a dopant or as a major constituent yield anisotropic and strongly temperature dependent ESR spectra. For compositions containing $Cu(CF_3SO_3)_2$, the activation energy associated with reorientation of the Cu^{2+} ligands increases with increasing salt concentration above T_g, possibly due to cation–anion interactions.

The temperature dependence of the ionic conductivity shows substantially less curvature (on an Arrhenius plot) than is usually observed in amorphous polymer electrolytes.

ACKNOWLEDGEMENTS

This work was supported in part by the US Office of Naval Research and the PSC-CUNY Faculty Research Award Program.

REFERENCES

1. Berliner, L. J. (ed.), *Spin Labeling, Theory and Application.* Academic Press, New York, 1976.
2. Owens, F. J., Paramagnetic resonance of transition complexes as probes of motion during glass-to-rubber transitions in polymers. *J. Macromol. Sci.,* **B28** (1989) 407–15.
3. Greenbaum, S. G., Pak, Y. S., Wintersgill, M. C., Fontanella, J. J., Schultz, J. W. & Andeen, C. G., NMR, DSC, DMA, and high pressure electrical conductivity studies in PPO complexed with sodium perchlorate. *J. Electrochem. Soc.,* **135** (1988) 235–8.
4. Abrantes, T. M. A., Alcacer, L. J. & Sequeira, C. A. C., Thin film solid state polymer electrolytes containing silver, copper and zinc ions as charge carriers. *Solid State Ionics,* **18, 19** (1986) 315–20.
5. Bonino, F., Pantaloni, S., Passerini, S. & Scrosati, B., New poly(ethylene oxide)-$Cu(CF_3SO_3)_2$ polymer electrolytes. *J. Electrochem. Soc.,* **135** (1988) 1961–5.
6. Hug, R. & Farrington, G. C., Solid polymeric electrolytes formed by

poly(ethylene oxide) and transition metal salt. *J. Electrochem. Soc.*, **135** (1988) 524–8.

7. Lee, S. & Brown, I. M., Experimental application of the new orientational dynamic theory of ESR fine-structure centers in an amorphous matrix. *Phys. Rev.*, **B34** (1986) 1442–8.

8. Greenbaum, S. G., Adamic, K. J., Pak, Y. S., Wintersgill, M. C. & Fontanella, J. J., NMR, DSC and electrical conductivity studies of MEEP complexed with $NaCF_3SO_3$. *Solid State Ionics*, **28–30** (1988) 1042–6.

Synthesis and Processing of Polymer Electrolyte Hosts

G. E. Wnek, K. Gault, J. Serpico, C.-Y. Yang, G. Venugopal & S. Krause

Department of Chemistry, Polymer Science and Engineering Program, Rensselaer Polytechnic Institute, Troy, New York 12180-2590, USA

ABSTRACT

Three specific issues concerning ionically conductive polymers represent a focus of recent activity in our laboratory. These are (1) the development of a more detailed picture of the coupling of segmental polymer motions to ion motion, (2) the synthesis of new polymer hosts having high chain flexibility and requisite polarity to dissolve large amounts of electrolytes, and (3) the use of electric and magnetic fields as processing aids to alter the morphology of polymer electrolyte 'alloys'.

INTRODUCTION

The ability of certain synthetic organic polymers, most notably polyethers, to chelate alkali metal ions has generated considerable interest in the use of such materials as solid-state ionic conductors.[1,2] The prototypical polymer is poly(ethylene oxide), PEO. Early proposals[3] suggested that alkali metal ions are solvated in PEO helices (crystalline complexes) and that transport was primarily via hopping between solvation sites within the helix. While the model is in accord with crystal structures of certain complexes (those containing small cations such as Li^+ and Na^+), it is known that larger ions (e.g. Rb) form amorphous complexes with PEO.[4] The picture which emerges from these data is that conduction occurs principally in the mobile, amorphous ($T_g = -60°C$) regions in PEO. However, the formation of

73

crystalline complexes with melting points higher than that of PEO (65°C) is important for solid electrolyte applications since useful conductivities (10^{-4} S/cm) are obtained at 100°C, i.e. well above the T_m of PEO itself. Thus, the regions of crystalline complex act as physical crosslinks, maintaining the integrity of the material in the appropriate temperature range. The data also suggest that the highest room temperature conductivities will be realized with polymer hosts having low T_g, necessitating backbones which are highly flexible and devoid of strong intermolecular forces. This expectation has been borne out by the discoveries that salt complexes of appropriately modified polyphosphazenes[5] and silicones,[6,7] polymers having among the lowest T_g known, afford conductivities of about 10^{-4} S/cm at room temperature.

We have been interested in correlations between ion transport and chain mobility in polymer/salt complexes in an effort to rationally design new polymer electrolyte hosts. Preliminary results from an ESR spin probe study are reported here. We next mention briefly a new approach to the functionalization of silicones, developed in our laboratory, which may be useful in the synthesis of new hosts. Finally, we describe a novel approach to modulate the morphology of blends of PEO and polystyrene using electric fields.

ESR SPIN PROBE STUDIES OF PEG/NaSCN COMPLEXES

The ESR spectra of nitroxide radicals provide information on the rotational correlation time of the nitroxide, and thus these molecules can be used as probes of microscopic viscosity. We have used the ESR spin probe approach to examine the relationship between probe mobility and molecular weight in a series of molten polyethylene gylcols (M_n 200–10 000, Aldrich)) containing NaSCN (O/M = 40). As it is believed that ion transport is governed by segmental motions, the ultimate objective of this work is to seek a relationship between ionic conductivity and microscopic viscosity.

Samples were prepared in a dry box from pre-dried PEG (80°C, 24 h) and NaSCN (120°C, 24 h). No solvent was used. As PEGs of molecular weights greater than about 1000 Daltons are solids, these polymers were heated above their melting points prior to addition of NaSCN (O/M = 40). The ESR spin probe was 4-hydroxy-2,2,6,6-tetramethylpiperidine-1-oxyl (TMPO) and was added to all samples

(O/TMPO = 2000). Conductivities were measured from 37°C to 120°C using a GenRad RLC Capacitance Bridge at 200 Hz and a FISHERbrand conductivity cell. All data reported below are for melts of PEG/NaSCN so as not to complicate the picture with possibilities for ion transport in amorphous and crystalline regions. ESR spectra of the melts were obtained using an IBM ER 200 spectrometer equipped with a Bruker VT-1000 temperature controller. Correlation times were determined as outlined by Törmälä et al.[8]

We find that the conductivities (σ) of the PEG/NaSCN melts are essentially independent of molecular weight beyond about 4000 Daltons (Fig. 1(A)). A likely interpretation is that the contribution to free volume by chain ends becomes negligible at this molecular weight, and the probe begins to sample local, segmental motions. The TMPO correlation times (τ) of these samples also become invariant with molecular weight above about 4000 Daltons (Fig. 1(B)), suggesting that the microscopic viscosity sampled by the probe also controls ion transport. The independence on both σ and τ on molecular weight is apparent in the log–log plot in Fig. 1(C). Note that the data points for samples above about 2000 Daltons fall on a reasonably straight line. These data suggest a rather direct dependence of conductivity on segmental motion, but a number of salts at several concentrations must be tested to see if this relationship is general.

SYNTHESIS OF NEW CLASSES OF SILICONES BEARING POLAR GROUPS

Polysiloxanes have received much attention as specialty polymers since their commercial application in the 1940s and are by far the most important of the inorganic backbone polymers. Special interest in these systems has developed as a result of their unique properties, which include low glass transition temperatures, good thermal and oxidative stability, low surface energies, excellent biocompatibility, and high gas permeabilities.

The low T_g of polysiloxanes has recently prompted efforts to investigate their use as polymer electrolytes which are capable of transporting dissolved ions.[6,7] Numerous studies have shown that ion transport is coupled to backbone motions, and hence polymers with low T_g are expected to have higher ion mobilities. However, the non-polar nature of typical polysiloxanes precludes dissolution of salts, and thus they

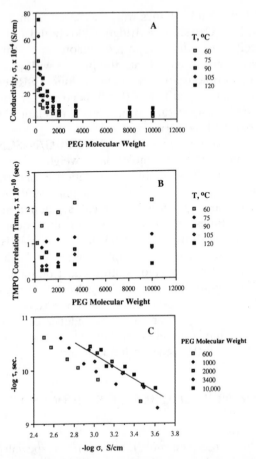

Figure 1. (A) Conductivity versus number-average molecular weight for PEG/NaSCN complexes (O/Na = 40:1) at various temperatures; (B) TMPO rotational correlation time versus number-average molecular weight for samples described in A; (C) plot of log conductivity (σ) versus log correlation time (τ) using data from A and B.

must be suitably modified in order to be useful electrolytes. A popular approach is to carry out a hydrosilation of a poly(methyl-hydrosiloxane) with a terminal olefin of an oligoether which can complex alkali metal cations).[6,7]

We wish to report a new, apparently versatile method of functionalizing polysiloxanes with polar groups. Our approach involves hydrosilation of polymethylhydrosiloxane with olefins bearing silyl ketene acetal moieties which can be subsequently reacted with numerous highly polar electrophiles.[9] The resulting materials may have large dielectric constants over a broad frequency range, and may serve as new electrolyte hosts which are able to increase ion pair dissociation compared with oligoether pendants. The synthesis and selected properties of one new polysiloxane is briefly discussed.

As a model system, we have investigated reactions of dimethylmethoxy(trimethylsilyl)ketene acetal with various electrophiles (Scheme 1). The structures of the products were confirmed using ^{1}H NMR and mass spectrometry. Each reaction proceeds to at least 95% conversion.

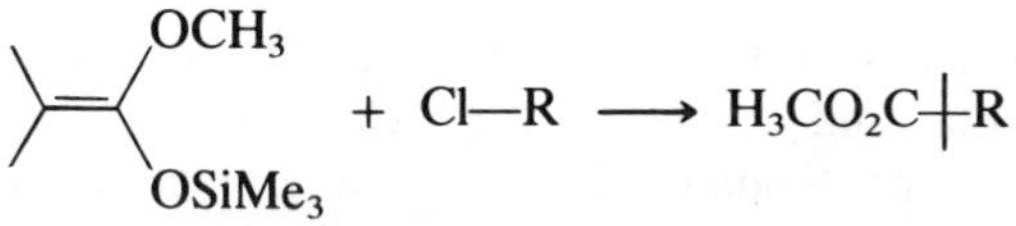

$$\underset{OSiMe_3}{\overset{OCH_3}{\diagdown}} + \text{Cl—R} \longrightarrow H_3CO_2C\text{—}R$$

R = SPhNO$_2$, SPh(NO$_2$)$_2$, OCPhNO$_2$

Scheme 1

Scheme 2 illustrates our approach for the synthesis of polysiloxanes bearing polar pendants. The commercially available (Aldrich) terminal olefin ethyl 2-methyl-4-pentenoate was first transformed into the corresponding silyl ketene acetal by treatment with LDA in THF followed by reaction with trimethylsilyl chloride. Polymethylhydrosiloxane (PMHS, Petrarch) was hydrosilylated with this olefin using chloroplatinic acid as catalyst. The extent of reaction was followed by IR spectroscopy. After about 22 h in refluxing toluene, the extent of reaction was at least 95%. Toluene and unreacted silyl ketene acetal were then removed under reduced presssure for 24 h. The resulting polymer (in dry ether) was added to a solution of 4-nitrobenzene sulfenyl chloride in dry ether. The solution was refluxed for 30 min, during which time a yellow dispersion (the desired polymer) appeared. The ether was removed under reduced pressure. Unreacted 4-nitrobenzene sulfenyl chloride was removed by washing the product with ether several times.

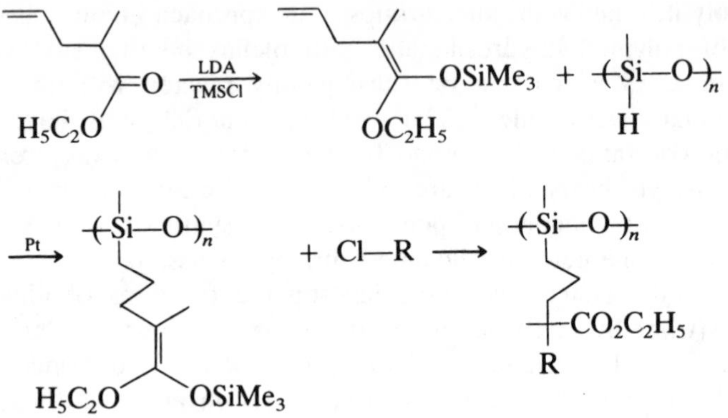

Scheme 2

Copolymers are readily prepared by hydrosilation with a mixture of the ketene acetal and the ester from which it was derived, namely ethyl 2-methyl-4-pentenoate. Reaction with electrophiles such as 4-nitrobenzene sulfenyl chloride occurs only at the ketene acetal sites.

Dielectric constants were measured at room temperature from 0·1 to 100 kHz using a Balsbaugh cell and GenRad capacitance bridge. Glass transition temperatures were measured using a Perkin–Elmer DSC-2.

Table 1 summarizes our recent DSC and dielectric constant data for this series of copolymers.

The dielectric constant at 0·1 kHz increases steadily as the concentration of the highly polar *p*-nitrothiophenoxy groups increases. However, ε for sample 1 is diminished by almost a factor of 2 at 100 kHz, and is actually the lowest of any sample containing *p*-nitrothiophenoxy groups. This material has the highest T_g, and we believe the dipoles can no longer respond effectively at the highest a.c.

Table 1

Sample	% *p*-nitrothiophenoxy pendants	ε(0·1 kHz)	ε(100 kHz)	T_g (°C)
1	90	12·74	6·52	−14
2	70	9·91	8·96	−32
3	50	8·69	8·11	−45
4	30	7·90	7·20	−65
5	0 (R = H)	4·98	4·83	−84

frequencies. We believe that the steady increase in T_g from samples 5 to 1 is principally the result of increased dipole-dipole interactions between siloxane chains. It is evident that the attainment of high dielectric constants over a broad frequency range will require a subtle balance between a high dipole concentration and high chain flexibility. We believe that siloxanes are particularly well suited to meet this balance, and are working toward optimizing their dielectric properties. We note that sample 1 dissolves about 20 mole% tetraethylammonium chloride, and the complex has a conductivity of about 10^{-4} S/cm at 100°C. We anticipate that the conductivity will be sensitive to a trade-off between the dielectric constant and T_g, both of which increase with increasing concentration of polar groups.

MODULATION OF THE MORPHOLOGY OF PEO-CONTAINING POLYMER BLENDS USING ELECTRIC FIELDS

Alloys or blends of different polymers typically exhibit phase separation because of low entropies of mixing. A characteristic morphology when one of the components is in excess is shown in Fig. 2(A). This sample of polystyrene/PEO (10 wt% PEO) was prepared by casting from cyclohexanone. Note that the minor phase (PEO) is distributed as spheres or droplets within the polystyrene matrix.

It occurred to us that it might be possible to obtain unusual morphologies if blends are cast in the presence of electric fields, using the field as a processing aid.[10] It is well known that particles having a dielectric constant different from that of their suspending medium will aggregate in electric fields by a process termed mutual dielectrophoresis.[11] Very different morphologies are indeed observed when polystyrene/PEO films are formed in the presence of a d.c. electric field (10 kV/cm, Fig. 2(B). The 'columns' or 'fibers' appear to form along the field lines as expected. We are presently examining the influence of molecular weight, composition, solvent and electric field strength on morphology.

A long-range goal is to prepare polymer blends having anisotropic electrical, optical and/or mechanical properties. We note that the approach should be applicable to a large number of incompatible polymer pairs.

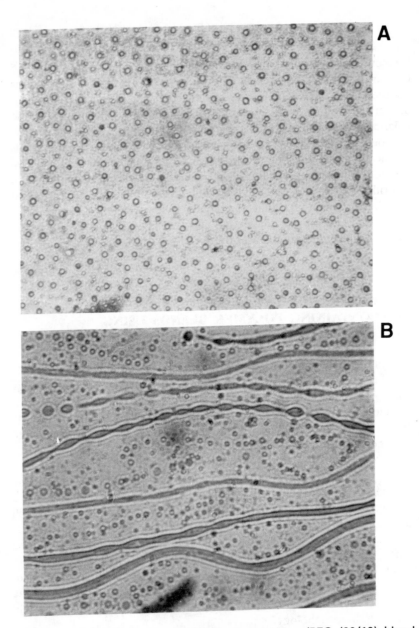

Figure 2. (A) Optical micrograph of a polystyrene/PEO (90/10) blend cast from cyclohexanone on a glass microscope slide; (B) a blend of the same composition, cast on a slide coated with two parallel Cu strips with an electric field of 10 000 V/cm applied during the solvent evaporation process (about 30 min). The distance between the Cu strips is 2 mm. In the photographs, approximately 1 cm = 25 μm. Note that the field direction is parallel to the 'fibers' in B.

ACKNOWLEDGEMENTS

We are grateful to the donors of the Petroleum Research Fund, administered by the American Chemical Society, and to the Defense Advanced Research Projects Agency and the Rensselaer Science Initiatives Program for support of this work. The efforts of Barbara Masi in the ESR and conductivity studies are deeply appreciated.

REFERENCES

1. Vicent, C. A., *Prog. Solid St. Chem.*, **17** (1987) 145.
2. Ratner, M. A. & Shriver, D. F., *Chem. Rev.*, **88** (1988) 109.
3. Armand, M. B., Chabagno, J. M. & Duclot, J. M. In: *Fast Ion Transport in Solids*, ed. Vashishta, Mundy & Shenoy. North-Holland, Amsterdam, 1979, 131.
4. Papke, B. L., Ratner, M. A., Austin, P. & Shriver, D. F., *J. Electrochem. Soc.*, **129** (1982) 1694.
5. Blonsky, P. M., Shriver, D. F. & Allcock, H. R., *J. Am. Chem. Soc.*, **106** (1984) 6854.
6. Fish, D., Khan, I. M. & Smid, J., *Makromol. Chem. Rapid Commun.*, **7** (1986) 115.
7. Spindler, R. & Shriver, D. F. In: *Conducting Polymers: Special Applications*, ed. L. Alcacer, Reidel, Dordrecht, 1987, p. 151.
8. Törmäla, P., Lattila, H. & Lindberg, J. J., *Polymer*, **14** (1973) 481.
9. Colvin, E. W., *Silicon in Organic Synthesis*, Keriger, Malabar, FL, Chap. 17, 1985.
10. Venugopal, G., Krause, S. & Wnek, G. E., *J. Polym. Sci. Polym. Lett. Ed.*, accepted.
11. Pohl, H. A., *Dielectrophoresis.* Cambridge University Press, Cambridge, 1978, pp. 38–43.

Polyelectrolytes with Sterically Hindered Anionic Charges

T. F. Yeh, H. Liu, Y. Okamoto

Polytechnic University, 333 Jay Street, Brooklyn, New York 11201, USA

&

H. S. Lee, T. A. Skotheim

Brookhaven National Laboratory, Upton, New York 11973, USA

ABSTRACT

The central problem in the development of single phase polymer electrolytes with exclusive cation conduction has been inadequate ion mobility due to extensive ion pairing between the mobile cation and the covalently attached anion. We have developed a new class of single-ion conducting polymers, or polyelectrolytes, based on highly flexible polysiloxane backbones and attached sterically hindered phenolate anions. The combination of a highly delocalized and sterically enclosed anionic charge facilitates charge separation and consequently enhanced cation mobility.

INTRODUCTION

For many applications of polymer solid electrolytes, such as to lithium batteries, undesirable local concentration gradients occur due to anion mobility, with consequently deleterious effects on the power output of the device. This problem has opened up an area of research in developing single-ion conductors, or polyelectrolytes, where the ionic charges of one polarity are chemically bonded to the polymer backbone. In principle, therefore, the polyelectrolytes will have unity transference number for the un-bonded ion species. The same requirements for chain flexibility to assist the ion motion hold for polyelectrolytes as for dual-ion conductors. The choice of ion pair is also critical to achieve adequate dissociation of the salt.

Several attempts to synthesize single-ion conductors have appeared, with most accounts describing specific cation conductors.[1-5] In general, only low conductivities have been achieved, usually in the 10^{-9}– 10^{-7} S/cm range, at room temperature. Higher conductivities have been achieved by using poly(ethylene glycol) as plasticizer.[1] For long term stability, however, single-phase materials are required.

In the absence of solvents, tight ion pairing is always likely to occur in such systems as the screening of the counterionic charge is reduced by the bonding constraints. This significantly reduces the mobility of the charge carriers, leading to systems with room temperature conductivities which are too low for practical applications. Nevertheless, these systems represent important advances in a direction which will lead to new generations of highly conducting specialized polymer electrolytes.

We have taken a new approach to the design of single-ion conducting polymers. The design approach has two essential aspects: (1) A highly flexible polymer backbone is necessary for local segmental motion to aid in the transport of the ions. This requirement is identical to that for dual-ion conductors. (2) The anionic charge, which is covalently attached to the polymer backbone, incorporates both steric hindrance and delocalization to prevent ion pairing.

With polyelectrolytes based on polysiloxane backbones and sterically hindered phenol compounds, the room temperature conductivities are generally in the 10^{-5} S/cm range, depending on the cation. The highest conductivities were achieved with K^+ ions and the lowest with Li^+ ions, clearly demonstrating that the dissociation is a function of the size of the cation.

POLYMER SYNTHESIS

Previous attempts to develop exclusive cation conducting polymers have been based on covalent attachment of carboxylate and sulfonate groups to the polymer backbone or side groups.[1-5] With a relatively localized bonded anionic charge, the electrostatic screening required for efficient charge separation becomes inadequate. The conductivity, consequently, is lower than that of dual-ion conductors with equivalent ion pairs.

Sterically hindered phenols, such as 2,6-di-*t*-butyl phenol, have been used in a variety of applications in organic synthesis.[6,7] Among the

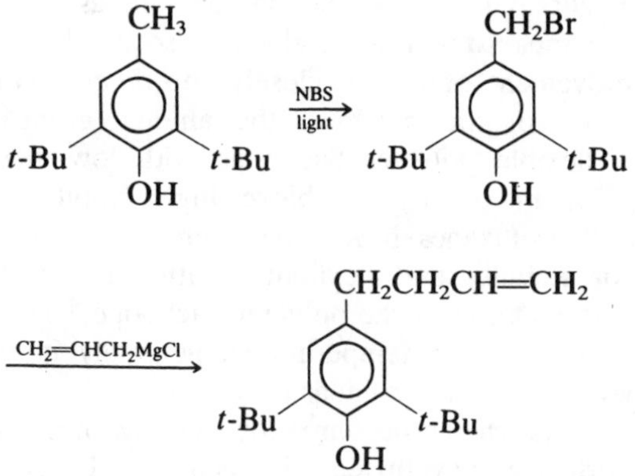

(I)

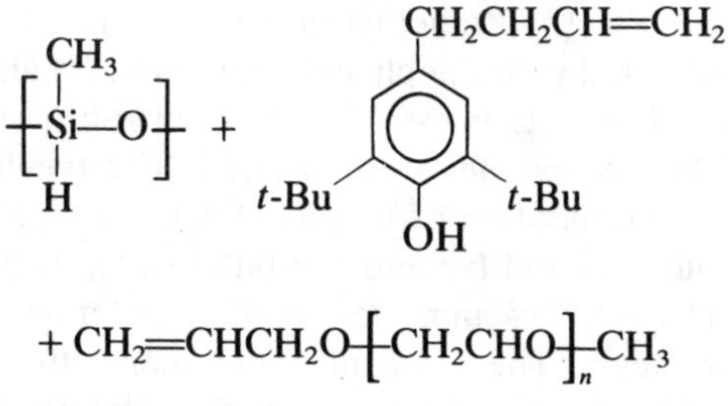

(II)

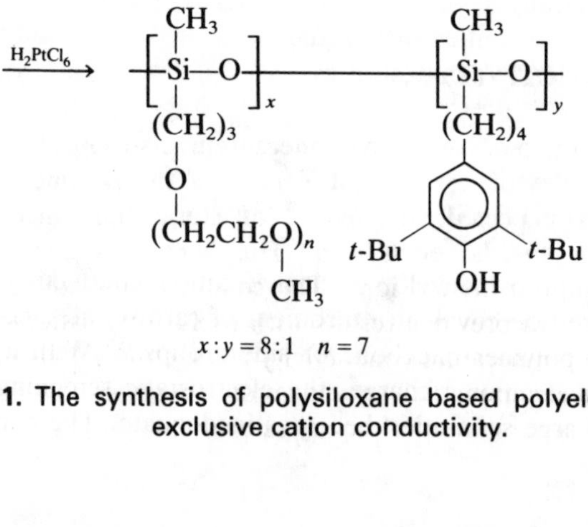

$x:y = 8:1 \quad n = 7$

Scheme 1. The synthesis of polysiloxane based polyelectrolytes with exclusive cation conductivity.

noteworthy properties of the compound are the stability of the phenol radical and the absence of normal phenol properties.[8] Geometrical calculations indicate that bulky *t*-butyl substituents in 2,6-di-*t*-butyl phenoxide effectively separate alkali metal cations such as K^+ and Na^+ from the phenoxide oxygen. The resulting charge separation would be expected to result in more mobile ions as well as a higher charge carrier density compared with more closely associated ion pairs.

Since the movement of ions is closely coordinated with the local segmental motion of the polymer, the anionic groups should be attached to a flexible polymer backbone with low glass transition temperature (T_g) in order to achieve high mobility at ambient temperatures. Polysiloxanes have long been known for the unique combination of stability and flexibility, with an exceptionally low energy barrier to rotation of the polymer backbone.[9] This is reflected in the low glass transition temperatures, generally in the -120 to $-130°C$ temperature range for linear polymers.

Scheme 1 shows the procedure for the synthesis of a comb polysiloxane with oligo-oxyethylene side chains and attached 2,6-di-*t*-butyl phenol moiety. Poly(hydromethylsiloxane) (PHMS) (Petrarch) was partially hydrosilylated with the phenol compound (I) and the allyl ether (II) using H_2PtCl_6 as catalyst at $90°C$ under nitrogen atmosphere until the SiH and allyl compounds were reacted. The typical ratio of $x:y$ was $8:1$. For the preparation of the alkali salts, the polymer was neutralized with *n*-butyl Li and Na and K *t*-butyloxide. The cast films were dried at $60°C$ under vacuum for at least $24\,h$ prior to the conductivity measurements. The T_g of the copolymer without salt was about $-100°C$, while the Li, Na and K salts had T_g about $-80°C$.

IONIC CONDUCTIVITY

Figure 1 shows the a.c. conductivity measurements using stainless steel blocking electrodes of Li, Na and K salts of the polymer shown in Scheme 1. As expected, the effect of steric hindrance is most pronounced for the larger K^+ ion. The Na^+ falls somewhere in between the K^+ and the Li^+ ions. The conductivities are remarkably high compared with previous results with exclusively cationic conductors. The room temperature conductivity is a factor of $100-1000$ times that previously reported.[1-5] The activation energy is low, indicating efficient charge separation at all temperatures.

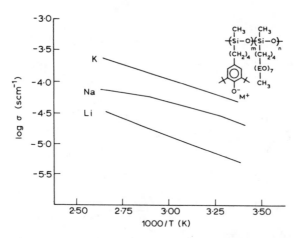

Figure 1. Temperature dependence of the a.c. conductivity of exclusive cation conducting polymers with 2,6-di-*t*-butyl phenol groups covalently bound to a polysiloxane backbone as in Scheme 1.

Previous studies using ^{23}Na NMR of sterically hindered phenolate anions attached to nylon-1 backbones have shown that the charge separation is efficient.[10] The ratio of mobile to bound Na ions reached 75% at 40°C and remained constant above that temperature. This implies that the availability of charge carriers at or near room temperature is adequate for high conductivity.

Figure 2 shows the effect of increasing the steric hindrance at the 2,6 positions of the phenolate anion. The conductivity of the Li complex is practically unchanged and the conductivity of the K complex is marginally increased. The largest effect of the additional steric hindrance is in the substantially increased conductivity of the Na complex. This is expected from geometrical calculations which shows that the enhanced screening effect of the added ethyl groups primarily affects Na$^+$ ions. The di-*t*-butyl groups are adequate for screening K$^+$ ions.

Adding steric hindrance beyond the additional ethyl groups does not appear to afford appreciable increase in the conductivity of the Li complex. This is believed to be due to a decrease in the effective dielectric constant in the increasingly more hydrocarbon-like environment. There thus appears to be an optimum size of the steric groups for maximum conductivity. In terms of approaching specific Li$^+$ ion conductivities of the order of 10^{-4} S/cm at room temperature, the

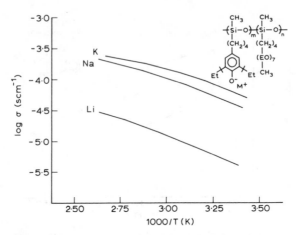

Figure 2. Effect on the temperature dependence of the a.c. conductivity of additional steric hindrance at the 2,6 position of the phenolate anion.

present approach may be inadequate. However, for devices based on Na^+ ion mobility, e.g. insertion batteries, the present approach provides adequate bulk conductivities. Studies are presently underway to optimize the conductivity and determine the electrochemical stability of this class of electrolytes.

ACKNOWLEDGEMENT

This work was supported by the US Department of Energy, Division of Materials Science under contract No. DE-AC02-76CH00016.

REFERENCES

1. Hardy, L. C. & Shriver, D. F., *J. Am. Chem. Soc.*, **107** (1985) 3823.
2. Bannister, D. J., Davies, G. R., Ward, I. M. & McIntyre, J. E., *Polymer*, **25** (1984) 1291.
3. Tsuchida, E., Kobayashi, N. & Ohno, H., *Macromolecules*, **21** (1988) 96.
4. Ganapathiappan, S., Chen, K. & Shriver, D. F., *Macromolecules*, **21** (1988) 2299.
5. Zhou, G., Khan, J. M. & Smid, J., *Polymer Preprints*, **30** (April 1989).
6. Coppinger, G. M. & Campbell, T. W., *J. Am. Chem. Soc.*, **75** (1953) 734.

7. Cook, C. D. & Norcross, B. E., *J. Am. Chem. Soc.*, **78** (1956) 3797.
8. Okamoto, Y., Khojastech, M. & Hou, C. J., *J. Polym. Sci. Technol.*, **16** (1982) 111
9. Anderson, R., Arkles, B. & Larsen, G. L., *Silicon Compounds Review and Register.* Petrarch Systems, Bristol, PA, USA, 1987, p. 259.
10. Liu, H., Okamoto, Y., Skotheim, T. A., Pak, Y. S., Greenbaum, S. G. & Adamic, K. J., Polymer electrolytes with exclusively cationic conductivity. *Mat. Res. Soc. Symp. on Solid State Ionics*, Boston, Nov. 28–Dec. 3, 1988.

Perfluorosulphonimide Salts as Solute for Polymer Electrolytes

M. Armand

Laboratoire d'Ionique et d'Electrochimie du Solide de Grenoble, ENSEEG, BP 75, 38402 St Martin d'Hères Cedex, France

&

W. Gorecki, R. Andréani

Laboratoire de Spectrométrie Physique, Université Joseph Fourrier, 38401 St Martin d'Hères, France

ABSTRACT

The operation of polymer electrolytes using PEO and usual salts, $LiClO_4$ or $LiCF_3SO_3$ is limited to above 40–60°C due to the crystallinity of the complexes. New salts with extensive dispersion of the negative charge on the anion (favourable to complex formation), having as well a molecular design allowing fluxional flexibility (plasticizing effect) of the molecule have been prepared. The attachment of two electronegative CF_3SO_2-groups to a central nitrogen atom yields the stable $(CF_3SO_2)_2N^-$. The corresponding lithium salt easily forms with pristine PEO complexes having a wide amorphous domain and these solutions show excellent ionic conductivities in the range $\sigma \approx 5 \times 10^{-5} \, (\Omega \, cm)^{-1})$ at 25°C.

INTRODUCTION

The simplest polymer electrolytes are based on poly(ethylene oxide) (PEO). A significant advantage is the availability of PEO of various chain lengths. Complexes made from salts and this macromolecule have the best performances yet in terms of total ionic conductivity at temperatures where the system is entirely amorphous. Simple salts like $LiClO_4$ or $LiCF_3SO_3$ form eutectics with PEO, composition $O/Li \approx 8$, melting point $T_m \approx 40°C$ and $O/Li \approx 60$, $T_m \approx 60°C$ respectively. Thus, operation of PEO electrolytes with these salts is mainly limited to $T > T_m$.

Various attempts have been made to plasticize PEO–salt complexes.[1] Upon addition of polar molecules like PC or $\alpha-\omega$ methoxy PEO oligomers, the conductivity is improved while the proportion of crystalline phase at room temperature decreases. However such expedient brings out a higher risk of reaction or co-intercalation at the electrodes, a situation usually encountered with liquid electrolytes.

A more elegant approach consists in the design of new salts acting as plasticizers for the complexes.

The criteria for the choice of anions giving complexable salts are known and are those used for aprotic electrolytic solutions: extensive charge dispersion resulting in low surface charge and thus requiring little solvation energy; this also corresponds to a low lattice energy and easy competition from solvation enthalpy of the cation.

In addition, lithium battery operation requires a large electrochemical stability window, ideally from 0 to $+4$ V versus Li/Li^+. These two conditions are often quite contradictory since the electrons on large anions derived from heavy elements tend to be loosely bound in 'soft' orbitals and are easily oxidized (e.g. I^- or RS^-). Also the importance of acid–base reactions involving the solvent have been overlooked, especially with ethers and coordination salts derived from strong Lewis acids, like PF^{6-} of $AlCl^{4-}$. C—O bond breaking takes place, resulting in ring opening oligomerization for THF and dioxolane, chain scission and loss of mechanical properties for PEO-type polymers, as shown in Scheme 1. Another important criterion is safety, certainly not met by ClO_4^- or AsF_6^-. If perchlorate appears inert when associated with PEO

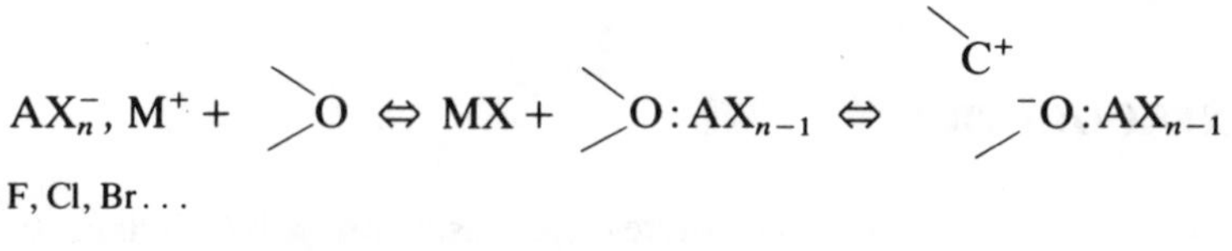

$$AX_n^-, M^+ + \ \rangle O \Leftrightarrow MX + \ \rangle O:AX_{n-1} \Leftrightarrow \overset{C^+}{\underset{}{\rangle}} {}^-O:AX_{n-1}$$

$X = F, Cl, Br \ldots$

Scheme 1. Degradation mechanism of polyethers with anions derived from Lewis acids.

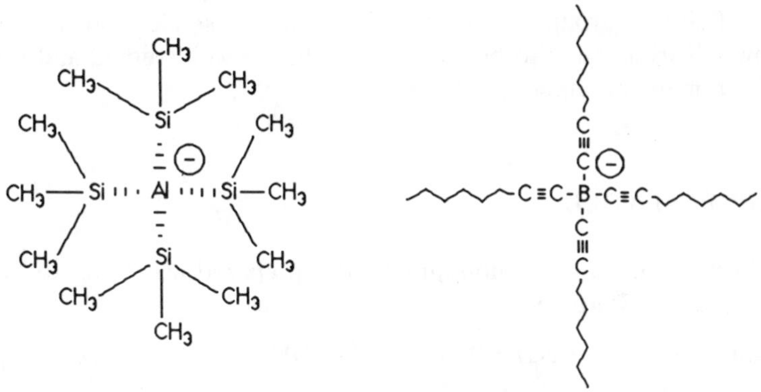

Figure 1. The chemical structure of tetrakis(trimethylsiloxy)alanate and tetrakis(alkynylborate) anion (C_8 shown).

in the solid state, the handling of large quantities of $LiClO_4$ in organic solutions during preparation of the complexes is a possible hazard.

Following these guidelines, anions of the type depicted in Fig. 1 were recently synthesized and tested with polyethers.[2]

Moderate enhancements in conductivity were reported and could be attributed to the low polarity of the moieties, trimethylsilyl or alkyl chains, unable, though very flexible, to modify the chain dynamics in the vicinity of the ions. Less polar PPO showed indeed much larger relative improvement than PEO.

Imide anions of the type $(X-N-X)^-$ where X is an electron withdrawing group present interesting characteristics:

- Covalent, non-coordination anion account for acid–base stability.
- Participation of two electronegative centres (X) to charge delocalization.
- Weaker interactions between A type elements (alkali metals) and nitrogen than with oxygen, though keeping a large electronegativity ($E = 3$, Pauling scale).
- Mechanical flexibility due to a flexible 'hinge' $X-N-X$, especially when $d - p$ 'π' bonding is involved. This is similar to Si—O—Si bond in polysiloxane which is responsible for their very low T_g.

We have chosen to study the compound

$$(CF_3SO_2-N-SO_2CF_3)^-$$

the CF_3SO_2— group being one of the strongest electron attractors known. This anion is to be compared with the well-studied and stable trifluoromethanesulphonate $CF_3SO_3^-$.

EXPERIMENTAL

Bis(trifluoromethanesulphone)imide was prepared according to Foropoulos & DesMarteau.[3]

$$CF_3SO_2F + 3NH_3 \rightarrow NH_4F + NH_4CF_3SO_2NH$$

$$\xrightarrow{HCl} CF_3SO_2NH_2 \xrightarrow{NaOH} NaCF_3SO_2NH$$

$$NaCF_3SO_2NH + \tfrac{1}{2}(Me_3Si)_2NH \rightarrow \tfrac{1}{2}NH_3 + Na(CF_3SO_2NSiMe_3)$$

$$Na(CF_3SO_2NSiMe_3) + CF_3SO_2F \rightarrow Me_3SiF + Na(CF_3SO_2)_2N$$

Free $(CF_3SO_2)_2NH$ is obtained from the sodium salt by distillation from anhydrous H_2SO_4 (b.p. 90°C under 2×10^{-3} Torr). The lithium salt is easily prepared from the carbonate and dried in vacuum at 120°C.

PEO complexes were prepared according to the now classical procedure from acetonitrile solution and evaporation of solvent in a closed loop argon circuit inside a dry box (<1 vpm H_2O, O_2). Conductivities were measured using impedance spectroscopy under dynamic vacuum.

RESULTS AND DISCUSSION

The lithium imide salt is a white hygroscopic solid, m.p. ≈220°C, exceedingly soluble in donor solvents and it easily forms complexes with PEO.

The DSC traces for compositions from O/Li = 3 to 30 are shown in Fig. 2 for the heating scans. All samples have been annealed at room temperature in the dry box for several weeks. The salt rich O/Li = 3 complex shows no event from 200 to 450 K and is completely amorphous; this is in complete contrast with all salts tested up to now, for which this composition corresponds to a high-melting (430 K) stoichiometric complex. Composition 6/1 on the other hand shows a

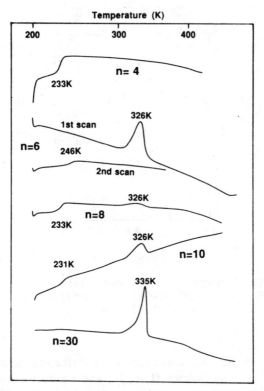

Figure 2. DSC traces (heating scans) recorded at 20°C/mn for P(EO)$_n$Li(CF$_3$SO$_2$)$_2$N; n = O/Li values shown on figure.

melting endotherm similar to the second stoichiometric (O/Li = 6) complex only found with large symmetrical anions (ClO$_4^-$, AsF$_6^-$). This stoichiometric compound forms an amorphous phase (eutectic) with PEO for compositions 8–10 O/Li. The formation of PEO crystallites takes place at higher dilutions. An important feature is the low T_g observed on these scans ($\approx$225 K,) with little or no variation with concentration, in contrast with other salts where $\partial T_g/\partial c \approx 30°C\,mol^{-1}$. This is a clear indication of the plasticizing effect of the salt on the polymer chains. A more precise investigation of the phase diagram is under way but is quite delicate as all compositions below O:Li = 20 are easily supercooled into a fully amorphous elastomer stable for several days.

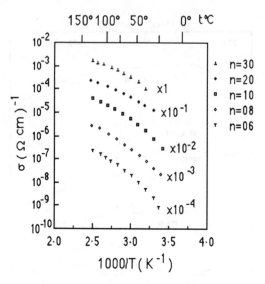

Figure 3. Conductivity (log scale) versus T^{-1} plots for $P(EO)_n Li(CF_3SO_2)_2N$; $n = O/Li$ values shown on figure. Successive curves are shifted by a decade.

The plots of the total conductivity, in Arrhenius coordinates, are shown for five compositions (Fig. 3). The conductivities show little variation with composition and for clarity the successive data have been shifted by a decade. The typical curvature of the plots indicates an amorphous phase regime (VTF of WLF) in the whole temperature range. The values for conductivity are excellent with $\sigma \approx 5 \times 10^{-5} (\Omega\,cm)^{-1}$ at 25°C, almost independent of composition, as shown by the roughly equidistant spacing of the curves.

The electrochemical stability window of these salts was also found satisfactory and compatible with battery operation. The perfluorosulphonimide salts thus represent a major improvement for polymer electrolytes, and possibly also for aprotic solutions.

ACKNOWLEDGEMENTS

This work was supported by Hydro-Québec within the ACEP project. We thank Mrs D. Foscallo for her help in sample preparation.

REFERENCES

1. Gray, F. M., In: *Polymer Electrolyte Reviews 1,* ed. J. R. MacCallum & C. A. Vincent. Elsevier Applied Science, London, 1987, pp. 139–72 and references therein.
2. Armand, M. & El Khadiri, F., Plasticizer salts for polymer electrolytes. *Lithium Battery Symposium,* ed. A. N. Dey. Electrochemical Society, Pennington, N.J., 88-1, 1987, pp. 502–8.
3. Foropoulos, J. & DesMarteau, D. D., *Inorganic Chemistry,* **23** (1984) 3720–6.

Polydioxolane as Solvating Host Polymer

G. Goulart, S. Sylla, J. Y. Sanchez & M. Armand

Laboratoire d'Ionique et d'Electrochimie du Solide de Grenoble, ENSEEG, BP 75, 38402 St Martin d'Hères Cedex, France

ABSTRACT

Poly(1,3-dioxolane) is a polyacetal with the repeat unit $(CH_2OCH_2CH_2O—)_n$. The polymer has been obtained by cationic ring-opening polymerization of the five-membered ring monomer using $\Phi COSbF_6$ as initiator. The material obtained is crystalline ($T_f = 55°C$) with a $M_w = 40\,300$ and polydispersity index (I) of 2·8 typical of this type of polymerization. Polydioxolane forms conductive complexes either with trifluoromethane sulphonate (Tf) or bis(trifluoromethanesulphone)imide (TFSI) lithium salts. While the former complexes are crystalline, DSC shows the existence of an amorphous phase centred around O/Li = 12 for LiTFSI. The conductivities obtained with this salt $[\sigma_{25°C} = 3 \times 10^{-6}\,(\Omega\,cm^{-1})]$ are quite high and comparable with those obtained with PEO.

INTRODUCTION

Polymer electrolytes represent a wide class of materials, however, most studies have dealt with poly(ethylene oxide) (PEO) and poly(propylene oxide) (PPO)-based materials. With the recognition that an elastomeric phase is needed for conductivity, the most successful strategies for improving the performances of polymer electrolytes came from the choices of suitable architectures for PEO segments in order to avoid the formation of crystalline phases. Such possibilities include crosslinked networks, comb polymers or copolymers.[1,2]

Replacement of ether oxygen by —NH— as in poly(aziridine) gives

99

materials which are less conductive, probably due to the lower polarity of the solvating group and to the existence of hydrogen bonds hindering diffusion.

On the other hand, polymers with a different oxygen spacing from PEO and PPO, i.e. —C—C—O—, show poorer solvating capabilities as poly(formal) $(CH_2O)_n$ and poly(oxetane) $(CH_2CH_2CH_2O)_n$ are reportedly unable to form salt complexes.[3] However, tests were made with either NaI and NaSCN as a comparison with standard PEO complexes known at that time. Li^+ being more strongly solvated than other alkali metal ions, one can expect that the tieline for complex formation (or the solubility limit) can be extended to less solvating host matrices. This is especially true with new anions forming salts with extremely low lattice energy and extensive solubilities.

Poly(1,3-dioxolane) PDXL:

$$(CH_2OCH_2CH_2O—)_n$$

was selected as an interesting candidate as the repeat unit still contains one —C—C—O— sequence. Also, the monomer (DXL) is extensively used in non-aqueous electrochemistry as it gives highly conductive solutions with lithium salts which also show good metal plating efficiency. The monomer has a tendency to polymerize spontaneously in the presence of salts derived from Lewis acids (Scheme 1).

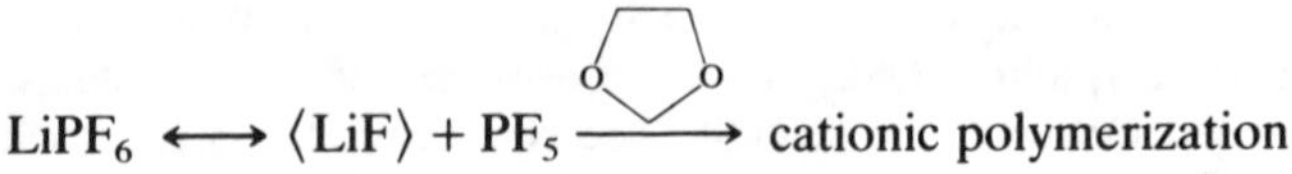

$$LiPF_6 \longleftrightarrow \langle LiF \rangle + PF_5 \xrightarrow{\quad} \text{cationic polymerization}$$

Scheme 1

Similarly, at positive potential (>3.8 V/Li:Li^+), the carboxonium cation formed upon oxidation initiates ring opening (Scheme 2). These phenomena were long underestimated when dioxolane was used as solvent for lithium batteries. Recently Foos & Erker[4] polymerized DXL in bulk in the presence of $LiAsF_6$ into a soft plastic material with appreciable conductivity at room temperature, but some monomer may still have been present (though the conductivity did not change after exposure to vacuum for 1 h). In addition, oligomers acting as

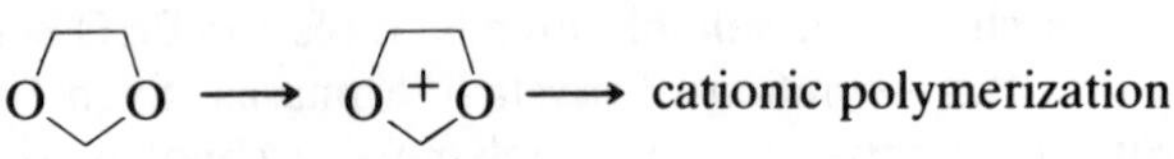

$$O\smile O \longrightarrow O\overset{+}{\smile}O \longrightarrow \text{cationic polymerization}$$

Scheme 2

plasticizer were certainly formed in appreciable amounts under these conditions.

EXPERIMENTAL

DXL (99 + %) was purified by double distillation, first with small addition of triflic acid to remove the stabilizer (Et_3N) and then over CaH_2. Polymerization was initiated by the benzoyl cation obtained from $\Phi COCl$ and $AgSbF_6$ ($2·75 \times 10^{-3}$ mol/litre) added to a cold ($-10°C$) solution of the monomer in dichloromethane and under dry argon. After 16 h, the polymerization was stopped with sodium ethoxide and the solvent and excess monomer evaporated. The solid residue was dissolved in THF and precipitated in pentane. The polymer was characterized by GPC in THF with Ultrastyragel 500, 10^3 and 10^4 Å columns.

Membranes were cast from an acetonitrile solution of preweighted amounts of polymer and salt corresponding to a given O/Li ratio and the solvent evaporated in a closed-loop system over molecular sieves in a dry box (<1 vpm water, oxygen). Conductivity measurements were performed under dynamic vacuum from room temperature to 100°C using impedance analysis.

RESULTS AND DISCUSSION

Polydioxolane (PDXL) is a crystalline polymer with a melting point of 55°C and a glass transition temperature of $-65°C$, comparable to that of PEO: these results could be expected as the C—C and C—O bonds have comparable energy barriers for rotation (~12 kJ). The polymer is soluble in many solvents including water, acetonitrile and THF.

After controversies over two decades, Szymanski et al.[5] recently established that cationic polymerization of cyclic acetals leads either to a majority of large cycles, the growth proceeding by ring expansion, or mostly to linear polyacetals according to the reaction conditions. With an initial DXL concentration lower than 0·8 M/litre the cycles predominate. On the other hand an increase far beyond this value favours the formation of linear chains. The growing species are mostly cyclic, in equilibrium with other active species involving carboxonium ions.

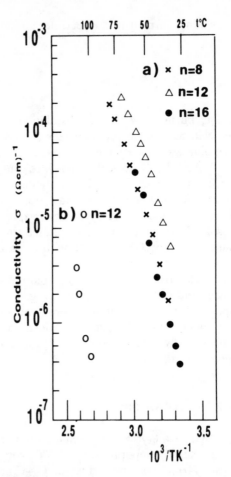

Figure 1. Arrhenius plot of the total conductivity for poly(dioxolane) versus T^{-1} for the salts: (a) LiTFSI, ×8/1; △12/1 and ●16/1. (b) TfLi ○12/1.

In order to obtain predominantly linear PDXL the polymerization has been carried out with a high DXL concentration: 5·4 M/litre. After precipitation in pentane, washing with the same solvent and drying, the polymer yield reaches 85%. The average molecular weights M_w and M_n were determined using GPC analysis with a polystyrene calibration curve. We find $M_w = 40\,300$; $M_n = 14\,400$. The polydispersity index $I = 2·8$ reveals a broad molecular weight distribution which is classical in the cationic polymerization of cyclic acetals.

The experimental $DP_n \approx 195$ is far from the theoretical value (1670), due probably to transfer reactions, but the goal was to prepare a large amount of polymer in order to carry out all the electrochemical studies with the same batch. Therefore we used a large amount of DXL, and that may explain the importance of transfer reactions.

The polymer electrolytes were prepared using two salts: lithium trifluoromethane sulphonate $LiCF_3SO_3$, (TfLi) and Bis(trifluoro-methanesulphone)imide lithium salt $Li(CF_3SO_2)_2N$ (LiTFSI).

The complexes PDXL/LiTFSI are amorphous in a wide range of salt concentrations, whereas they are always highly crystalline when using TfLi.

In Fig. 1, we report $\log \sigma$ versus T^{-1} for three concentrations of LiTFSI, and for one concentration of TfLi. A well defined maximum of conductivity is observed with LiTFSI, for the ratio O/Li = 12, while the values obtained with TfLi are at least an order of magnitude lower.

In Fig. 2 are reported the conductivity isotherms for three temperatures, 25°, 35°, 45°C, versus the ratio O/Li. The complex PDXL/TFSI corresponding to the ratio 12 reaches a conductivity of about 3×10^{-6} $(\Omega\ cm)^{-1}$ at room temperature, whereas the conductivity of the PDXL/TfLi is immeasurably low.

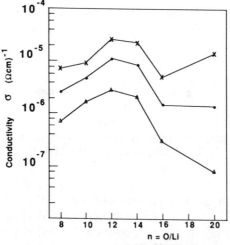

Figure 2. Conductivity isotherms (log scale) as a function of the salt composition for polydioxolane LiTFSI: △ 25°C; ● 35°C; × 45°C.

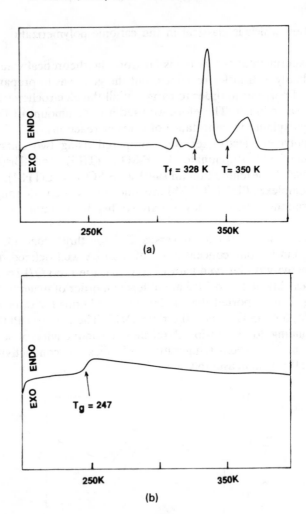

Figure 3. DSC traces recorded at 20°C/mn for: (a) polydioxolane-TfLi 12/1; (b) polydioxolane-TFSILi 12/1.

This result is in good agreement with the DSC recordings of these two samples (Fig. 3). The sample PDXL/LiTFSI appears completely amorphous and exhibits a glass transition $T_g = 247$ K ($-26°C$). On the other hand the sample PDXL/TfLi is very crystalline and the trace exhibits two main exothermic peaks. The first one at 328 K is probably the fusion peak of pure DXL, whereas the second at 350 K is tentatively attributed to the melting of a stoichiometric polymer/salt complex.

The promising conductivity values suggest further investigation of these complexes. The study will be extended to cyclic polydioxolane, and to copolymers of dioxolane and dioxepane. Also, electrochemical measurements on transference numbers and stability windows are required for final assessment.

REFERENCES

1. MacCallum, J. R. & Vincent, C. A. (eds), *Polymer Electrolyte Reviews 1* Elsevier Applied Science, London, 1987, pp. 1–22 and 68–138.
2. Armand, M. & Gauthier, M., Polymer electrolytes, the immobile solvent concept. In: *High Conductivity Solid Ionic Conductors*, ed. T. Takahashi. ASSI Costed World Scientific Publisher, Singapore, 1988, pp. 114–45.
3. Armand, M., Chabagno, J. M. & Duclot, M., In: *Fast Ion Transport in Solids*, ed. P. Vashishta. North Holland, New York, 1979, pp. 131–7.
4. Foos, J. S. & Erker, S. M., *J. Electrochem. Soc.*, **134** (1987) 1725.
5. Szymanski, R., Kubisa, P. & Penczek, S., *Macromolecules*, **16** (1983) 1000–8.

Crosslinked Comb-shaped Polymers Based on Ethylene Oxide Macromers

M. Andrei, L. Marchese, A. Roggero

Eniricérche SpA, Via Maritano 26, 20097 San Donato Milanese, Milan, Italy

&

S. Passerini, B. Scrosati

Dipartimento di Chimica, Università di Roma 'La Sapienza', Piazzale A. Moro 5, 00185 Rome, Italy

ABSTRACT

Crosslinked, comb-shaped polymers have been synthesized from vinyl-ether monomers containing variable number of ethylene oxide (EO) units with terminal methoxy or ethoxy groups.

Complexes of these polymers with lithium salts have been prepared and their properties investigated in terms of the role of the length of side chain in the monomer, the structure of the crosslinking agent, and EO/Li composition ratio.

INTRODUCTION

In recent years substantial work has been devoted to the study of the properties of polymeric materials which, in combination with suitable metal salts, could give electrolytic membranes of interest for the development of plastic-like batteries, sensors and optical displays.[1,2]

The most common of these membranes are the complexes between poly(ethylene oxide) (PEO) and metal (especially lithium) salts. The ionic conductivity of these membranes is reasonably good at temperatures around 80–100°C but it decays dramatically below 60°C due to the formation of crystalline phases characterized by a very low ionic mobility. Therefore, it appears of great technological importance to develop new polymer electrolytes having improved low-temperature electrical properties.

To achieve this goal we have attempted to synthesize modified polymer hosts having lower glass transition temperatures (T_g) than PEO, in order to obtain lithium salt complexes which could retain amorphicity (and thus high conductivity) down to room temperature.

Recently Armand[3] and Cowie *et al.*[4] have focused attention on comb-shaped polyethers in which the basic structure contained short and active ethylene oxide side chains attached to a supporting non-active polymer backbone. These polymers are indeed completely amorphous at room temperature. However, the complexes with lithium salts obtained with these simple homopolymers have very poor mechanical properties. To improve these, we have crosslinked the basic polymer host, directly into the polymerization medium, using a suitable agent belonging to the class of difunctional vinyl ethers.[5]

Preliminary electrochemical characterizations carried out in our laboratories[6] have indeed shown that the complexes between the crosslinked polyethers and lithium salts give rise to membranes having an acceptable mechanical stability accompanied by an improved room temperature conductivity. Similar conclusions have been recently reached by Cowie & Sadaghianizadeh.[7]

In this work we have further characterized the system and here we report results related to the role of the monomer, of the structure of the crosslinking agent and of the EO/Li composition ratio on the conductivity of crosslinked polyether-lithium perchlorate electrolytic membranes.

EXPERIMENTAL

Synthesis of the Monomers

The monomers having the structure:

$$CH_2{=}CHO{-}(CH_2CH_2O)_n{-}R \tag{1}$$

with $n = 2,\ 3,\ 4,\ 5,\ 7$ and R = Et, Me, were obtained by vinylic transetherification with mercuric acetate as catalyst and ethyl vinylether as vinylic agent.

In the case of the monomers with two and three ethylene oxide (EO) units commercially available starting materials have been used while in those with 4, 5 or 7 EO units the starting materials have been obtained by a classical Williamson synthesis. The former were purified by distillation and the latter by preparative HPLC.

All the products were obtained with satisfactory yields (about 80% in the case of the monomers with 2 and 3 EO units and 60% in the case of the monomers with 4, 5, 7 EO units) and with purity higher than 99%.

Synthesis of Crosslinking Agent

The crosslinking agents having the structure:

$$CH_2\!=\!CHO\!-\!(CH_2CH_2O)_m\!-\!CH\!=\!CH_2 \tag{2}$$

with $m = 2$, 3, 4, were obtained by vinylic transetherification with procedure similar to that described above, starting from commercial glycols.

Synthesis of the Polymers

All polymers were synthesized by cationic polymerization in chlorinated solvents at low temperature using etherate BF_3 as catalyst. The polymerization temperature was changed depending on the number of EO units in the monomer I (see Table 1). In fact, working at $-78°C$ with the higher monomers, the polymerization medium was not homogeneous in the concentration range used. The most difficult part in this synthesis was the definition of the optimal value of the molar ratio between the main monomer and the crosslinking agent, in order to obtain polymers which could combine a high molecular weight with a sufficient solubility to allow casting procedures for the realization of the electrolytic membranes. The best solution was found to be a polymerization process using 3% of crosslinking divinylether.

Under these conditions the polymerization is very fast and the conversion complete. The final polymers were isolated from the polymerization medium by the usual separation techniques. The structures of the polymers were confirmed by IR and NMR techniques

Table 1

$$CH_2\!=\!CHO\!-\!(CH_2CH_2O)_n\!-\!R \qquad \text{(I)}$$

$$CH_2\!=\!CHO\!-\!(CH_2CH_2O)_m\!-\!CH\!=\!CH_2 \qquad \text{(II)}$$

conc. (I) = 1 – 2 mol/litre
conc. (II) = 0·03 – 0·06 mol/litre

n	m	Polym. T. (°C)	T_g (°C)
2	2	−78	−70
	3		−68
	4		−70
3	3	−65	−68
	4		−70
4	2	−50	−69
	3		−70
	4		−73
5	2	−40	−68
	4		−71
7	4	−10	−64

that showed the complete absence of vinyl double bonds and consequently the absence of unreacted monomers. The glass transition temperatures, T_g, were also determined and their value reported in Table 1, together with some polymerization conditions.

Preparation of the Electrolytic Membranes

The electrolytic membranes, consisting of complexes between the above mentioned polymers and lithium salts (e.g. $LiClO_4$ and $LiBF_4$) were prepared by a casting technique, using purified acetonitrile as solvent. Transparent membranes, with a plastic appearance and a thickness between 50 and 150 μm were obtained.

Electrochemical Characterization

The conductivity of the selected electrolytic membranes were determined by frequency response analysis (Solartron, mod. 1250) using cells with blocking electrodes for ionic transport.

RESULTS AND DISCUSSION

The Polymer Hosts

Figure 1 shows the basic structure of the crosslinked polymers here considered. The variable parameters are the length of the crosslinking agent ($m = 2$, 3 and 4) and the length of the side chain ($n = 2$, 3, 4, 5 and 7).

Figure 2(a) illustrates a typical differential scanning calorimetry (DSC) trace of a polymer with $n = 4$ and $m = 4$. A sharp T_g (at about $-70°C$) without any evidence of melting is revealed by the trace, this indicating that the polymer is completely amorphous all over the range of temperature considered. A similar behaviour is also shown by the polymers having up to five units of ethylene oxide in the side chain ($n = 5$). However, when the side chain becomes longer, some crystalline regions appear, as revealed by the DSC trace of the $n = 7$, $m = 4$ polymer which shows, besides the expected low T_g at $-64°C$, a neat peak of melting at $-3°C$ (Fig. 2(b)).

The Polymer Host–Lithium Salt Electrolytic Membranes

Figure 3 shows the dependence of the conductivity of electrolytic membranes on the values of the side chain length (n) when the

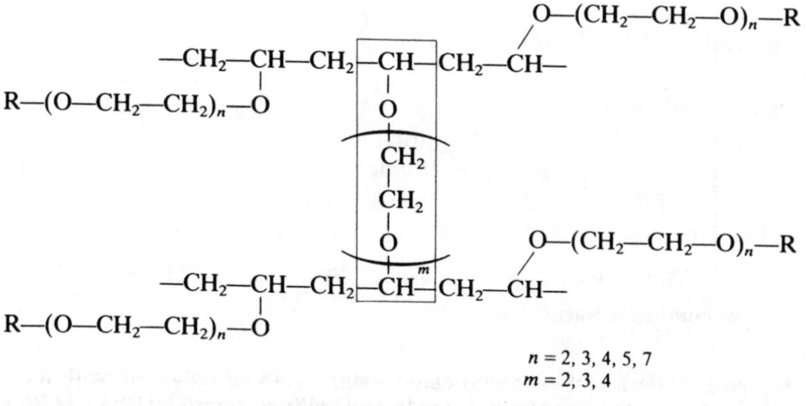

Figure 1. Basic structure of crosslinked polyvinylethers.

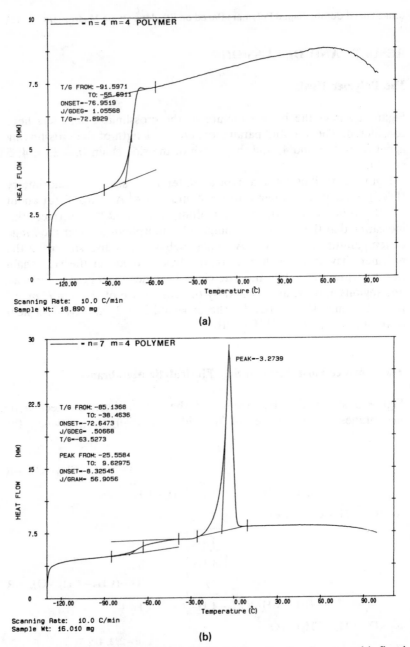

Figure 2. Differential scanning calorimeter traces of polymers with fixed value of crosslinking chain ($m = 4$) and with a length of the EO side chains (a) $n = 4$ and (b) $n = 7$.

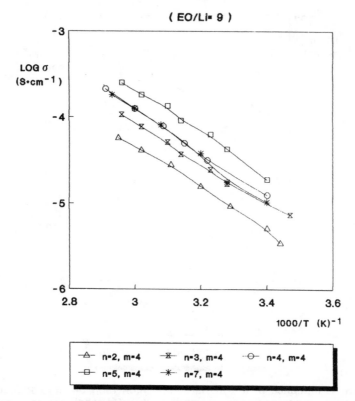

Figure 3. Influence of the length of the EO side chain on the conductivity of an electrolytic membrane having a EO/Li = 9 composition ratio.

crosslinking chain length ($m = 4$) and the composition ratio (EO/Li = 9) are kept fixed. It may be noticed that the conductivity increases as the number of EO units in the side chain increases from 2 to 5. This may be explained considering that longer chains are likely to have higher flexibility which in turn facilitates the ion mobility.[1] However, the trend is not followed when the side chain reaches 7 EO units, possibly because with such a length the polymer host acquires a certain degree of crystallinity (see Fig. 2(b)).

Figure 4 illustrates the dependence of the conductivity at 30°C and T_g of electrolytic membranes on the EO/Li composition ratio at the temperature of 30°C for a polymer with $n = 2$ and $m = 3$. It may be noticed that the maximum value is obtained for a composition of 12.

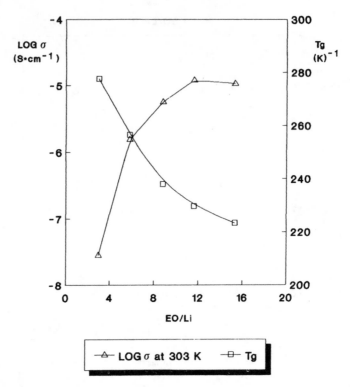

Figure 4. Influence of the EO/Li composition ratio on the conductivity (*A*) and the transition temperature (*T*$_g$) of electrolytic membranes.

The related membrane has a high conductivity (about 10^{-5} S cm^{-1}) still retaining a very low glass transition temperature ($T_g = -44°C$).

The favourable electrical properties of the EO/Li = 12 composition system are also confirmed by Fig. 5 which illustrates the thermal behaviour of the conductivity of membranes of the same polymer having different LiClO$_4$ content. Conductivity values of 10^{-5} and 10^{-4} S cm^{-1} at 30 and 60°C, respectively, are indeed shown by electrolytic membranes having a EO/Li composition ratio around 12.

Also in the case of the membranes obtained by complexing the $n = 2$, $m = 3$ polymer with LiBF$_4$, the conductivity reaches the maximum values, in the whole range of temperature, when the EO/Li ratio is 12 (Fig. 6).

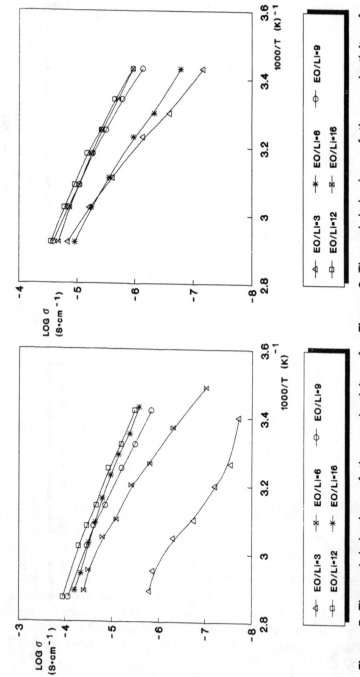

Figure 5. Thermal behaviour of the conductivity of electrolytic membranes at various EO/LiClO₄ composition ratios.

Figure 6. Thermal behaviour of the conductivity of electrolytic membranes at various EO/LiBF₄ composition ratios.

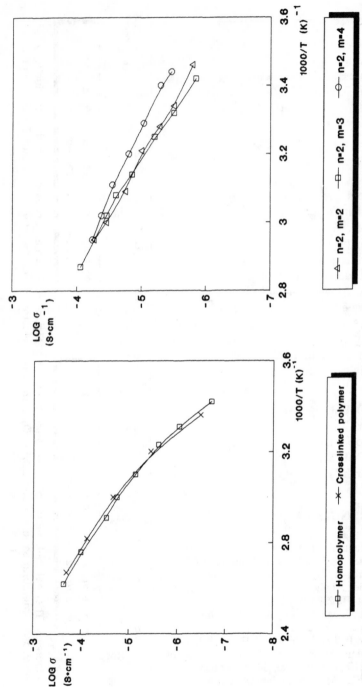

Figure 8. Influence of the length of the crosslinking agent on the conductivity of electrolytic membranes having an EO/Li = 9 composition ratio.

△— n=2, m=2   □— n=2, m=3   ○— n=2, m=4

Figure 7. Thermal behaviour of the conductivity of electrolytic membranes of homopolymers and cross-linked polymers complexed with LiClO$_4$ having an EO/Li = 5 composition ratio.

□— Homopolymer   ✕— Crosslinked polymer

Furthermore, Fig. 7 shows that the presence of the crosslinks does not depress the conductivity with respect to the analogue homo-polymer. Finally, Fig. 8 shows that the influence on the electrolyte conductivity of the length of the cosslinking agent is negligible.

CONCLUSIONS

The results reported in this work indicate that crosslinked poly-vinylethers are promising polymers for obtaining membranes with lithium salts which are easy to handle and have interesting electrical properties. When a composition of EO to Li around 12 is used, ionic conductivities of the order of $10^{-5}\,S\,cm^{-1}$ and $10^{-4}\,S\,cm^{-1}$ at room temperature and at 60°C, respectively, are obtained. These values are several orders of magnitude higher than those offered by electrolytes based on conventional PEO polymer hosts. Comparable results with similar membrane compositions have been independently obtained by Cowie *et al.*[4] and this may confirm that by properly selecting the nature of the crosslinked polymers, further successful steps may be achieved in obtaining electrolytic membranes having even higher ionic conductivity at room temperature.

ACKNOWLEDGEMENTS

This work has been carried out with the financial support of ENIRICERCHE. One of us (S.P.) is grateful to ENIRICERCHE for a Research Fellowship.

REFERENCES

1. MacCallum, J. & Vincent, C. A. (eds) *Polymer Electrolyte Reviews I.* Elsevier Applied Science, London, 1987.
2. Linford, R. G. (ed.) *Electrochemical Science and Technology of Polymer.* Elsevier Applied Science, London, 1987.
3. Patent FR 2523769, 1982.
4. Cowie, J. M., Martin, A. C. S. & Firth, A. M., *Brit. Polymer J.*, **20** (1988) 247.
5. Italian Patent application 22305 A/87, 1987.
6. Pantaloni, S., Passerini, S., Croce, F., Scrosati, B., Roggero, A. & Andrei, M., *Electrochim. Acta*, **34** (1989) 635.
7. Cowie, J. M. & Sadaghianizadeh, K., *Polymer*, **30** (1989) 509.

Polymer Solid Electrolytes Based on Poly(Ethylene Oxide)–Poly(Methyl Methacrylate) Blends and Ethylene Oxide Copolymers

K. Such,[a] Z. Floriańczyk,[b] W. Wieczorek[a] & J. Przyłuski[a]

[a] *Institute of Solid State Technology, Warsaw University of Technology, ul. Noakowskiego 3, 00-664 Warszawa, Poland*
[b] *Institute of Polymer Technology, Warsaw University of Technology, ul. Koszykowa 75, 00-662 Warszawa, Poland*

ABSTRACT

Polymer electrolytes comprising poly(ethylene oxide)/poly(methyl methacrylate) and NaI have been prepared. The maximum of room temperature conductivity of this type of materials is in the range 10^{-4}–10^{-5} S/cm. Polymer blends prepared by thermal polymerization of methyl methacrylate in the presence of poly(ethylene oxide) provided higher conductivities than those prepared by a solvent casting procedure. The superior properties of the former materials are connected with the presence of grafted copolymer. In order to improve mechanical properties of copolymer electrolytes, blends of poly(methyl methacrylate) and ethylene oxide–propylene oxide copolymer were prepared. Electrolytes were thermally stable up to 60–70°C.

INTRODUCTION

A number of methods of modification of solid electrolytes based on complexes of poly(ethylene oxide)(PEO) with alkali metal salts, leading to an improvement in their mechanical properties have been developed in the last few years. One of the approaches is to apply PEO blends with polymers of high T_g like poly(styrene),[1,2] poly(methacrylic acid)[3] or poly(methyl methacrylate) (PMMA).[4] Recently we have found[5] that some electrolytes obtained from blends of PEO with PMMA exhibited much higher room temperature conductivities than analogue materials based on PEO. The aim of this work is to study the effect of PEO/PMMA blends' composition and methods of preparation on the conductivity of their complexes with NaI. The blends were prepared by solvent casting of polymer mixtures or by thermal polymerization of methyl methacrylate (MMA) in the presence of PEO.

EXPERIMENTAL

Materials

PEO (M_w 5×10^6—Polysciences) and PMMA (M_w $1 \cdot 3 \times 10^5$—POCH) were used without further purification. NaI (reagent grade) was dried for 24 h at 100°C. MMA and solvents were purified by distillation in the presence of drying agents. All materials were stored and subsequent preparations were carried out under a nitrogen atmosphere.

Preparation of Polymer Blends

Solvent cast blends were obtained by mixing of the 5 wt % solutions of PEO in acetonitrile and PMMA in dichloromethane at room temperature followed by slow evaporation of solvents carried out in an evacuated desiccator. Thermal polymerization of MMA in the presence of PEO was carried out at 90°C in a non-solvent system or at boiling temperature of the solvent used for 48 h according to the procedure described elsewhere.[5] o-dichlorobenzene, chlorobenzene, acetonitrile and chloroform were used as the solvents. The composition of these materials was determined on the basis of the intensity of the IR peak at 1720 cm^{-1} characteristic for stretching vibrations of a carbonyl group in MMA monomeric units. Samples used in IR studies were prepared in the form of 5 wt % solutions in acetonitrile.

Isolation of the Grafted Copolymer

The blends obtained by polymerization of MMA in the presence of PEO were agitated in boiling methanol for 48 h in order to separate PMMA which is insoluble in this solvent. The methanol solution was treated with a large excess of water and the emulsion obtained was extracted several times with dichloromethane. The grafted copolymer was isolated after evaporation of the organic phase and purified by dissolution in methanol and precipitation with water.

Preparation of Polymeric Electrolytes

Complexes of NaI with polymers were prepared in a thin film configuration by a standard casting procedure. The molar ratio of PEO monomeric units to NaI was equal to 10.

Conductivity Measurements

Conductivity of polymer foils was evaluated from impedance data. Experiments were performed in the frequency range 5 Hz–500 kHz applying stainless steel blocking electrodes.

RESULTS AND DISCUSSION

Conductivities of the electrolytes obtained from blends prepared by solvent casting versus reciprocal temperature are shown in Fig. 1. With increasing PMMA concentration an accompanying decrease of conductivity in the portion of conducting materials can be observed. However, samples containing up to 20 wt% of PMMA exhibited consistently higher conductivity than pristine PEO–NaI electrolyte up to the melting temperature of the non-conducting PEO crystalline phase (about 65°C). This effect may be explained by lower crystallinity of PEO chains in blends resulting from the presence of non-crystallizable component.[6]

Figure 2 shows the variation of the conductivity over the temperature range 20–100°C for the materials comprising blends prepared by

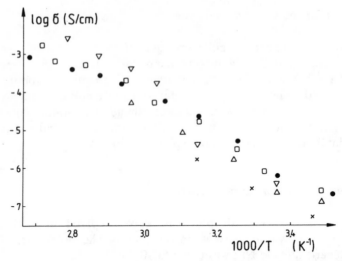

Figure 1. Bulk conductivity versus reciprocal temperature for the PEO–PMMA–NaI system. ▽, 0 wt% PMMA; ●, 10 wt% PMMA; □, 20 wt% PMMA; △, 30 wt% PMMA; ×, 50 wt% PMMA.

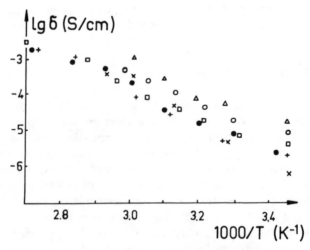

Figure 2. Bulk conductivity versus reciprocal temperature for blends prepared by thermal polymerization of MMA carried out without solvent. ⊘, 6 wt% PMMA; ○, 15 wt% PMMA; △, 20 wt% PMMA; □, 36 wt% PMMA; ×, 43 wt% PMMA; +, 50 wt% PMMA.

polymerization of MMA in the presence of PEO carried out without solvent. One can observe that electrolytes prepared in such a manner exhibited higher conductivities than solvent cast materials. The highest conductivity was measured for samples containing 20 wt% of PMMA in the blend and was equal to $2 \cdot 1 \times 10^{-5}$ S/cm at 20°C. It should be noticed that even for samples containing 50 wt% of PMMA, room temperature conductivity is higher than for pristine PEO–NaI electrolytes.

Blends prepared by thermal polymerization of MMA contained some PMMA—PEO grafted copolymers, which were isolated according to the procedure described in the experimental section. Materials exhibiting the highest conductivity value contained about 7 wt% of the grafted copolymer. The weight ratio of MMA to ethylene oxide monomeric units was equal to 12:88. The X-ray spectrum (Fig. 3) of this copolymer is characterized by diffraction maxima at $2\theta = 19 \cdot 7°$ and $2\theta = 23 \cdot 9°$, which are typical for the PEO crystalline phase, and

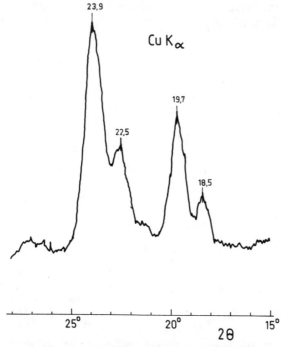

Figure 3. X-ray spectrum of the grafted PEO–PMMA copolymer.

by additional peaks at $2\theta = 18\cdot5°$ and $2\theta = 22\cdot5°$. The occurrence of these additional peaks may be explained by changes in the parameters of some unit cells of crystalline PEO caused by grafting PMMA to the PEO chain. The degree of crystallinity of PEO in a copolymer is about 53%, i.e. practically the same as in blends containing 12 wt% of PMMA. ^{1}H NMR studies showed that the tacticity of the PMMA chain in the grafted copolymer is practically the same as for the homopolymer isolated from a blend (mole fraction of racemic dyads was 0·21 and 0·22 respectively and triad distribution followed Bernoullian statistics).

The conductivity of the electrolyte prepared from the grafted phase was rather poor (Fig. 4). However, materials obtained by mixing of PMMA, PEO and the grafted copolymer, in the same concentrations as found from chemical analysis, exhibited considerably higher conductivity than all previously reported PEO–PMMA blends (Fig. 5). The role of the grafted phase on increasing ionic conductivity is not yet clear. The preliminary X-ray and SEM studies showed that in blends containing this phase PEO crystallites have a smaller size but the degree of crystallinity of blends characterized by the same

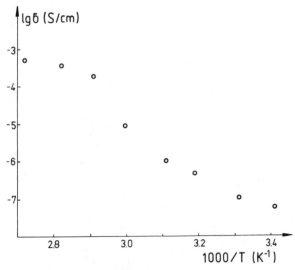

Figure 4. Bulk conductivity versus reciprocal temperature for PEO–PMMA grafted copolymer–NaI system.

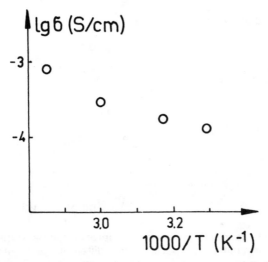

Figure 5. Bulk conductivity versus reciprocal temperature for PEO–PMMA grafted phase blend based electrolyte doped with NaI.

PEO/PMMA ratio practically does not depend on the method of preparation. One can expect that the effect of the grafted phase is attributed mainly to an increase of the segmental motion in the highly amorphous phase of polymeric electrolytes.

The formation of the grafted phase was observed also during thermal polymerization of MMA in an environment of a highly viscous suspension of PEO with o-dichlorobenzene and was not observed in the reactions carried out in chloroform or acetonitrile. The blends obtained in these last two solvents possessed properties similar to solvent cast blends described previously. The highest values of conductivities for electrolytes obtained from blends prepared by polymerization of MMA in the presence of PEO and various solvents are shown in Fig. 6.

All of the studied electrolyte films based on PEO/PMMA blends were mechanically stable at least to 70–80°C and their thermal stability increased with increasing PMMA concentration in the sample. The ionic conductivity of the above-mentioned systems is time stable. These observations suggested we prepare a blend of PMMA with ethylene oxide–propylene oxide based electrolytes described previously by us.[7] Copolymer electrolytes exhibit very high room tem-

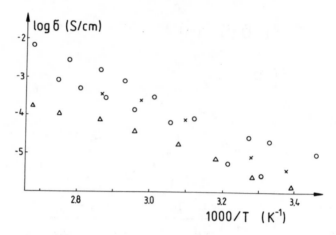

Figure 6. Bulk conductivity versus reciprocal temperature for PEO–PMMA–NaI systems prepared in different solvents. ○, o-dichlorobenzene; ×, chlorobenzene; △, acetonitrile; □, chloroform.

perature conductivity but their poor thermal stability considerably limits their usage. As can be seen from Fig. 7 by blending of the copolymer electrolyte with PMMA, materials thermally stable up to 60–70°C can be obtained without imparing room temperature conductivity.

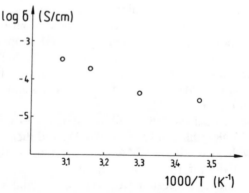

Figure 7. Bulk conductivity versus reciprocal temperature for EO–PO-based copolymer electrolyte doped with $LiBF_4$ and blended with PMMA.

CONCLUSIONS

1. Blending of PEO homopolymer with PMMA enables us to obtain polymeric electrolytes exhibiting higher room temperature conductivity and better mechanical stability than PEO-based electrolytes.
2. Electrolytes obtained from blends prepared by thermal polymerization of MMA in the presence of PEO have superior conductivities to solvent cast materials.
3. The formation of the grafted PEO/PMMA copolymer seems to be responsible for a significant increase in the conductivity of the studied systems.

ACKNOWLEDGEMENT

This work was financially supported by C.P.B.P 01.15 research programme.

REFERENCES

1. Steele, B. C. H. & Weston, J. E., *Solid State Ionics,* 2 (1981) 347.
2. Gray, F. M., MacCallum, J. R. & Vincent, C. A., *Solid State Ionics,* 18, 19 (1986) 252.
3. Tsuchida, E., Ohmo, H., Tsunami, K. & Kobayashi, N., *Solid State Ionics,* 11 (1983) 227.
4. Przyłuski, J., Such, K. & Wieczorek, W., *Materials Science Forum,* 42 (1989) 185.
5. Floriańczyk, Z., Wieczorek, W., Such, K. & Przyłuski, J., Patent Application No P-275812, 1988.
6. Gornick, F. & Mandekorn, L., *J. Appl. Phys.,* 33 (1982) 907.
7. Floriańczyk, Z., Krawiec, W., Wieczorek, W. & Przyłuski, J., *Angev. Makromol. Chem.* (in press).

Crosslinked Polyether Electrolytes with Alkali-Metal Salts for Solid-State Batteries: Relationship between Network Structure and Ionic Conductivity

Jean-François Le Nest & Alessandro Gandini

Ecole Française de Papeterie (INPG), BP 65, 38402 St Martin d'Hères, France

ABSTRACT

New data on the variation of ionic conductivity as a function of salt concentration are presented. They concern mostly PEO-based networks containing lithium perchlorate in which the crosslink density and the nature of the crosslink agent were varied. The results indicate that even at constant free-volume fraction the mobility of the ions is affected by the bulkiness and/or the rigidity of the crosslink structures. These findings have provided sound criteria for the optimization of conductivity.

INTRODUCTION

We have previously established that ionic transport in crosslinked polyether-based electrolytes follows remarkably well a free-volume dependence.[1] This major conclusion was reached on the basis of a thorough characterization of the materials and of the excellent concordance among a large set of data obtained with many different types of networks and salts from a wide array of independent probing techniques. These afforded specific volumes, glass-transition temperatures, equilibrium swelling in various liquids, dynamic mechanical parameters, nuclear magnetic relaxation features of atoms belonging to the polymer and to the salt, ionic conductivities, transport numbers and cyclic voltammetry diagrams.[1]

Of course such a comprehensive investigation also provided the means for a better insight into both the fundamental and the more applied aspects related to the field of 'polymer electrolytes', such as the understanding of interactions at the ion-chain level and the problems related to the applications of these materials. It is also worth reiterating that the original idea of using crosslinked structures proved particularly useful in minimizing crystallization of the polyether segments and in avoiding creep. Finally, this work gave ample evidence to the fact that the ionic conductivity is ensured for the most part by the anion, i.e. t^+ is typically only 0·2–0·3. In order to circumvent this drawback, ionomeric networks possessing anions covalently bound to the polymer chains were also prepared.[1]

This communication presents results concerning the dependence of ionic conductivity on the salt concentration (mostly lithium perchlorate) for a whole series of different polyether networks. The structure of these crosslinked materials was modified by changing the type of reticulating agent (aliphatic and aromatic isocyanates and cyclic hydrosiloxanes), the nature of the polyether segments and their length.

The aim of this investigation was to further our understanding of the mechanisms involved in charge transport on the basis of the added information obtained from new results concerning ionic conductivity and to reach sound criteria for the optimization of that parameter.

EXPERIMENTAL

The synthesis and conditioning of polyether-based networks crosslinked with isocyanates[1] and hydrocyclosiloxanes[2] have already been described. The determination of ionic conductivities[3] and transport numbers[4] was carried out following previously published procedures.

RESULTS AND DISCUSSION

The ionic conductivity in a solvating polymer electrolyte can be expressed as a two-term equation reflecting the separate contributions of anion (a) and cation (c).[5]

$$\sigma = Cq^2 \left[B_a \exp \left(-\frac{\gamma f_a^*}{f_g + \alpha(T - T_g)} - \frac{E_a + W/2\varepsilon}{RT} \right) \right.$$

$$\left. + B_c \exp \left(-\frac{\gamma f_c^*}{f_g + \alpha(T - T_g)} - \frac{E_c + W/2\varepsilon}{RT} \right) \right] \tag{1}$$

Within each of these terms one finds both the free-volume and the Arrhenius components. With respect to the equation in Ref. 5, the overall energies E are replaced here by the specific terms referring to the activation energy E and to the ionization energy of the salt $W/2\varepsilon$ related to the polarity of the medium. In the temperature range T_g to $T_g + 100$, the WLF contribution predominates and therefore the reduced conductivity for each ion becomes:

$$\log (\sigma_T/\sigma_{T_g}) = -\log a_T = C_1(T - T_g)/(C_2 + T - T_g) \tag{2}$$

This situation is characteristic of most of our previous work and of the present study.

Above about $T_g + 100$, the free-volume term begins to decrease and therefore the Arrhenius contribution gains progressively in importance until it becomes dominant. The appropriate expression for the reduced conductivity is then:

$$\log (\sigma_T/\sigma_{T_g}) = \frac{E + W/2\varepsilon}{2\cdot 3R} (1/T_g - 1/T) \tag{3}$$

Given the general feature of systems made up of a network and an added salt whereby the anions contribute to conductivity for at least 80%,[1,4] the term pertaining to the cations in the above equations can be neglected in the first approximation. Of course, when one deals with ionomers bearing anions fixed to the network, the situation is reversed to 100% cationic conductivity.[6]

Using eqn (2) applied only to anionic conductivity, sets of C_1 and C_2 values were obtained using a classical WLF approach.[5] The agreement found for a good number of systems differing in network structure and/or in the nature of the salt was taken as a clear indication of the validity of the above assumptions. For ionomers with mobile cations, C_1 and C_2 values were again found to fall within the above range, thus confirming that cations and anions move under an electric field following identical patterns. Within the scope of the present investigation, new networks were prepared which called upon hydrocyclosiloxanes as crosslink agents.[2] Conductivity measurements with $LiClO_4$ as added salt (Fig. 1) yielded C_1 and C_2 values of 11 and 50 K

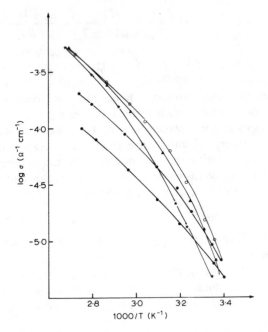

Figure 1. The dependence of ionic conductivity on the reciprocal temperature for PEO ($M_n = 1050$)-based networks bearing siloxane crosslinks[2] and containing $LiClO_4$ at different concentrations (mole dm^{-3}): ● 0.10, ✹ 0.20, ○ 0.45, ▲ 0.70, ★ 0.85.

respectively, i.e. once again within the close range reported previously.[5]

The fact that polyether-based networks bearing either urethane or siloxane crosslinks, but also both types of units in the same reticule display similar values of C_1 and C_2 from conductivity measurements below $T_g + 100$, seemed a strong indication that the determining factor in the transport properties of these materials is the glass transition temperature. This was moreover corroborated by results arising from other types of measurements, since the C_1 and C_2 values fell within the same small range.[1]

In principle, the above conclusion would also suggest that the specific structure of the crosslink units should not influence the conductivity in any way other than that already taken into account by its effect on the glass transition temperature, which is included in the calculations based on eqn (2).

The purpose of this study was to find a more accurate way to confirm (or disprove) this generalization. It was felt that reasoning only on the basis of C_1 and C_2 values was not sufficient to ensure the accuracy needed for such an important issue. In other words, it became essential to establish whether the small variation in the values of C_1 and C_2 were due to random fluctuations arising from experimental errors or to intrinsic features of the systems, i.e. whether specific trends relating the ionic conductivity to the network structure could be revealed. An analysis of the conductivity data as a function of salt concentration for a number of different networks seemed a more accurate way of answering this question, rather than looking for minor trends in the individual values of C_1 and C_2 possibly marred by statistical 'noise'.

The ionic conductivity measured in these salt–network systems arises from two distinct contributions: (i) the actual concentration of free ionic species (charge carriers) and (ii) their mobility. The extent of ionization is governed by the nature of the salt, the polarity (dielectric constant) of the polymer medium and by possible specific interactions between the dissolved salt molecules (or ions) and a particular site on the macromolecular chain (local solvating power). On the other hand, the mobility of the free ions depends on the temperature: if the system is in the free-volume regime (eqn (2)), the ease of charge transport will increase as the free-volume fraction increases, i.e. as the temperature is raised. The free-volume fraction is not however the only factor which determines the mobility of ions in polymer networks. Specific local structures can in principle constitute 'barriers' or 'obstacles' in the shortest statistical pathways available to the moving ions, quite apart from the classical hindrance arising from chain stiffness which affects the free-volume fraction at a given temperature.

Plotting the ionic conductivity at constant reduced temperature (i.e. at constant $T - T_g$, which corresponds to constant free-volume fraction) as a function of the added salt concentration for a given network implies that any change in σ can only arise from a corresponding change in ionic concentration. This in turn will depend in this specific context on the position of the dissociation equilibria as a function of salt concentration and of the actual temperature, which varies with it through the corresponding changes in T_g.[2,7] We constructed such plots in a log-log mode for a series of networks built with the same polyether structure, viz. ethylene oxide units, and with the same salt, viz. LiClO$_4$, as shown in Fig. 2.

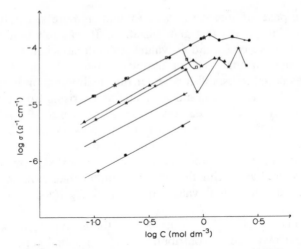

Figure 2. Log-Log plots of ionic conductivity versus LiCLO$_4$ concentration at $T - T_g = 100°C$, for a series of networks based on PEO of different M_n and on different crosslink agents: ✱, PEO400 + HC(C$_6$H$_4$NCO)$_3$; △, PEO600 + HC(C$_6$H$_4$NCO)$_3$; ★, PEO1000 + HC(C$_6$H$_4$NCO)$_3$; ▲, PEO1000 + OCN(CH$_2$)$_6$N[CONH(CH$_2$)$_6$NCO]$_2$; ●, PEO2000 + HC(C$_6$H$_4$NCO)$_3$; ☆, PEO4000 + HC(C$_6$H$_4$NCO)$_3$; □, PEO1000 + [Si(CH$_3$)(H)O]$_4$.

An analysis of each curve shows three major features: (i) straight linearity up to a critical salt concentration, (ii) slope of unity (±0.1) for the straight portion and (iii) the onset of a 'wavy' pattern above the critical concentration. Whereas the first two characteristics are displayed by all systems in Fig. 2, the appearance of minima and maxima at high salt contents are only visible with certain systems (see discussion below).

The unit slope of all these plots below the critical point indicates that the conductivity is directly proportional to the salt concentration. Given the high concentrations used this feature cannot be rationalized by a straightforward explanation based on complete dissociation throughout the range studied. This problem has already been tackled[5] in terms of a set of multiple equilibria, then reduced to three:

$$nMA \rightleftharpoons (MA)_n \qquad\qquad K_1 \qquad\qquad (4)$$

$$(MA)_n \rightleftharpoons M^+ + (MA)_{n-1}A^- \quad K_2 \qquad\qquad (5)$$

$$(MA)_n \rightleftharpoons A^- + (MA)_{n-1}M^+ \quad K_3 \qquad\qquad (6)$$

In the subsequent development one should work with activities, but lack of knowledge forced us to use concentrations at the inevitable expense of rigour. The resulting expressions for the free ions are then

$$[M^+] = (K_1 K_2 [MA]^n)^{1/2} \qquad [A^-] = (K_1 K_3 [MA]^n)^{1/2}$$

but in the first approximation based on transport number measurements, only the anion will be considered as a conducting entity. Since experimentally the conductivity is directly proportional to the initial salt concentration $[MA]_0$, it follows that

$$[A^-] = K'[MA]_0 \tag{7}$$

because at constant free-volume fraction the conductivity must be directly proportional to the concentration of anions as dominant conducting species.

This empirical relationship must now be compared with the expression of $[A^-]$ obtained above as a consequence of the postulated equilibria (4)–(6), i.e.

$$K_1 K_3 [MA]^n = K''[MA]_0^2 \tag{8}$$

where $K'' = K'^2$. It seems logical to postulate that the extent of ionic species arising from the salt is small compared with the undissociated entities, given the high concentrations involved in these systems. Moreover, at temperatures higher than about 35°C, which correspond to the lower end of the data reported in Fig. 2, the extent of aggregation of the salt must be small compared with the concentration of salt molecules in the form of ion pairs, written here as MA. This is corroborated by dynamic mechanical measurements which showed that on the rubbery plateau the modulus was independent of salt concentration for a given network studied above 35°C,[1] indicating the absence of ionic-type crosslinks arising from salt aggregation. It follows that the concentration of ion pairs $[MA]$ is close to that of the added salt $[MA]_0$.

Substituting this equality into eqn (8) gives

$$K_1 K_3 [MA]_0^n = K''[MA]_0^2 \tag{9}$$

The obvious conclusion that $n = 2$ is submitted to the condition that the product $K_1 K_3$ remains approximately constant within the temperature range scanned in each straight line. In other words, the temperature interval corresponding to the increase in T_g of a network as the salt concentration is increased[2,7] is sufficiently small not to affect that product by a detectable amount. This interval was typically 20–25°C. Within this small range it is reasonable to assume that the increase of K_3 with temperature should be approximately counter-balanced by a corresponding decrease of K_1.

In conclusion, the straight portion of the log-log plots shown in Fig. 2 reflects a situation characterized by the fact that the added salt is predominantly under the form of dipoles (ion pairs). The very high dielectric constants measured in such situations,[5] i.e. above about 35°C and with salt concentrations above 0·1 M (ε = several hundred units), constitute good supporting evidence for this model.

Results relating the ionic conductivity to salt concentration with a fixed free-volume fraction at temperatures below about 35°C were confined to rather low $LiClO_4$ concentrations because of the increase in T_g accompanying the introduction of the salt. Log-log plots similar to those shown in Fig. 2 were constructed within this limited concentration range at temperatures around 20°C. The slopes with the same type of networks used previously ranged between 0·55 and 0·65 instead of unity above 35°C. It appears that the temperature fall is accompanied by a substantial aggregation of the salt, as indeed corroborated by the additonal observation showing that the dielectric constant of the medium decreases very considerably (ε around 30). Moreover, the swelling characteristics of these systems at room temperature with non-polar liquids[8] clearly indicate that ionic aggregates are present and manifest themselves through chain coupling giving non-covalent crosslinks.[1] It is therefore plausible to postulate that the lowering of the temperature from above about 35°C to around 20°C and below is accompanied by a drastic change in the relative proportion of ion pairs and higher aggregates. The equilibrium (4) with the value of $n = 2$ found above must now be considered as being shifted heavily to the right hand side at low temperature to give predominantly quadrupoles. The decrease in the slopes mentioned for these lower temperatures is a result of this change. Indeed if one uses the set of equilibria (4)–(6) with the new situation, the concentration of A^- becomes directly proportional to the square root of the added salt concentration, which is close to the experimental findings giving slopes just above 0·5.

Whereas the slope of unity is a general feature of all networks based on polyethylene oxide (PEO) above 35°C, changing the nature of the polyether chain to polypropylene oxide (PPO) induces it to raise to 3 (log-log plots like that of Fig. 2).[5] A similar treatment of the above equilibria leads to a value of n as high as 6. This is not surprising considering the much lower solvating power of the PO units compared with that of EO units.[7] With block and graft copolymers containing PEO sequences plus polysiloxane and/or PPO chains, the slopes revert to 1·1–1·2 because the salt molecules will place themselves at the most solvating sites, i.e. the EO units, with extremely high selectivity. This phenomenology has been observed in a clear-cut manner in a study of the T_g of these materials.[2] Indeed, when the PEO sequences in these copolymers have high DPs, the slope becomes unity again.

Returning to the plots in Fig. 2, and more specifically to the waves appearing above a critical salt concentration, it seems likely that the saturation points correspond to a maximum organized filling of solvating sites (at well-defined salt-to-EO units ratios) and the dips to transitory microphase formation.[1,5] This phenomenon is also observed when studying proton magnetic relaxation in PEO chains as already mentioned;[1] a detailed account of this work will be submitted shortly.

The next feature which deserves close scrutiny has to do with a comparison of the various straight portions of the plots in Fig. 2. As already specified, the only structural differences among these systems are the length of the PEO chains and/or the nature of the crosslink sites.

If one compares the behaviour of a series of networks plus LiClO$_4$ in which only the DP of the PEO segments have been varied, the conductivity at reduced temperature is the higher, the longer the polyether chain up to about 45 EO units, but remains unchanged for longer segments, as shown in Figs 2 and 3.

If on the other hand the comparison is based on the structure of the crosslink sites with PEO chains of equal length, the trend is such that the more bulky and/or rigid the site, the lower the conductivity at reduced temperature, as shown in Fig. 2. Moreover, the highest conductivity at a given salt concentration, attained with siloxane crosslinks, is the same as that observed with the longest PEO chains in the previous comparison, indicating that the same limiting situation is reached.

In these analyses the data are compared for the same free-volume fractions, but inevitably for different temperatures. In both situations described above, the higher the conductivity measured, the lower the corresponding temperature at which the measurement was made. Thus, the observed dependencies cannot be attributed to a temperature

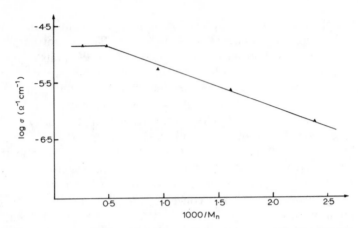

Figure 3. The variation of ionic conductivity at $T - T_g = 100°C$ as a function of the reciprocal M_n of the PEO chains for networks bearing aromatic urethane crosslinks[1] and containing $LiClO_4$ at a concentration of 0.1 mol dm^{-3}.

effect on the conductivity, since they follow the opposite trend. The origin of these trends is therefore not a temperature effect because the ionic mobility increases with increasing temperature at constant $T - T_g$. On the contrary, the effects observed are reduced in magnitude by the temperature variations and yet they remain clearly visible.

The only reasonable explanation for both sets of trends is one calling upon specific effects of the crosslink sites, and in particular their stiffness and/or bulkiness. Indeed the conductivity increases either when the concentration of crosslinks decreases or when their structure is more flexible and/or their size reduced. These qualitative correlations stop in both instances at an upper limit of conductivity specific to the given combination of polyether and salt (here PEO and lithium perchlorate). This implies that the detrimental effects of the crosslink units towards ionic movements become negligible if the DP of the PEO chain is higher than about 45 or if the crosslink structure is sufficiently supple, as with siloxanes.

The above interpretation supposes that although the comparisons are made among systems which have been normalized as much as possible (same salt, same polyether and same free volume fraction for all materials plus same DP of the PEO chains or same crosslink structure for each set), and although the effect of the measuring temperature goes against the observed trends, the nature of the crosslink sites can still affect the ionic transport by playing the role of a more or less important barrier. This role cannot be attributed to a decrease in the extent of ionization of the salt caused by the lack of solvating power of these sites (low dielectric constant or absence of complexation) because the siloxane structures do not affect the conductivity (see Fig. 2 with maximum values with these networks) and yet are known to be incapable of solvating $LiClO_4$.[2]

The most reasonable interpretation of this phenomenology considers that bulky or stiff sites constitute obstacles on the pathway of the ions migrating under the effect of an electric field. At sufficiently low concentration and/or low steric barrier, the effect of these sites is negligible because the statistical walk of the ions is not affected. Above critical values, ionic 'percolation' becomes perturbed and this results in a decrease in ionic mobility despite the fact that we are comparing systems at constant free-volume fraction. Obviously then, the more pronounced the 'obstacle' and/or the higher its concentration, the greater the conductivity drop, as indeed observed here.

This model turns out to be useful for explaining the variation in the values of transport numbers as a function of crosslink density for a given polyether (PEO), triisocyanate and salt ($LiClO_4$). It was found[1] that t^+ decreased as the crosslink density decreased. The improved mobility of the anions associated with a decrease in the concentration of crosslink units would indeed favour the contribution of these ions to the overall conductivity relative to that of the cations even further, because the low mobility of the latter charge carriers, complexed with the ether groups, would be largely insensitive to changes in the crosslink density.

The limiting upper conductivities obtained with either long PEO chains or 'soft' crosslink sites (cf. Fig. 2) seem to represent the maximum values attainable with the specific systems involving networks based on PEO and containing $LiClO_4$. In other words when the polyether chains and the salt are defined, the ionic conductivity will reach its maximum value at room temperature (i.e. at the most typical application temperature of solid-state batteries) with low crosslink densities, non-rigid and non-bulky crosslink structures and a salt concentration just below that giving microphase separation (waves in the diagrams of Fig. 2). A calculation carried out on a model system consisting of amorphous PEO of very high molecular weight containing an amount of $LiClO_4$ just below the saturation of one Li^+ per 12 EO units[1] leads to an expected conductivity of 4×10^{-5} ohm^{-1} cm^{-1} at 22°C. This is the maximum value which can be obtained for PEO-based networks containing $LiClO_4$. We attempted to improve this situation by adding to the systems various complexing agents for the cation in order to increase the extent of ionic dissociation. With either tetraglyme or the crown ether specific to Li^+ (1,4,7,10-tetraoxacyclododecane) added in equimolar amounts with respect to the cation, the conductivities were not enhanced, but rather decreased slightly. These results suggest that the complexing role of the polyether segments dominates even in the presence of these added agents. This conclusion is supported by the observation that t^+ was not affected by the presence of these substances.

One can envisage a number of ways to produce an increase in conductivity with respect to the maximum values discussed above:

—Polymers with T_g values lower than PEO but with similar solvating power (this excludes for example polysiloxanes).
—Higher bulk dielectric constants by the introduction of polar groups in the network structure, provided these groups do not interact with either ions from the salt:
—Different salts and more specifically new anions.

If the improvement of conductivity is a paramount factor for the application of these materials, the concurrent switch from a highly anionic to a predominantly cationic conductor is obviously of equal importance. The latter problem finds an obvious solution through the use of ionomers bearing anions bound to the polymer chains,[1,6,9–11] but this is unfortunately not accompanied by a corresponding solution of the former problem, i.e. the optimization of the conductivity.

REFERENCES

1. Le Nest, J. F., Gandini, A. & Cheradame, H., *Brit. Polym. J.*, **20** (1988) 253–68.
2. Le Nest, J. F. & Gandini, A., *Polym. Bull.*, **21** (1989) 347–51.
3. Killis, A., Le Nest, J. F., Cheradame, H. & Gandini, A., *Makromol. Chem.*, **183** (1982) 2835–45.
4. Leveque, M., Le Nest, J. F., Gandini, A. & Cheradame, H., *Makromol. Chem. Rapid Commun.*, **4** (1983) 497–502.

5. Le Nest, J. F., Cheradame, H. & Gandini, A., *Solid State Ionics*, **28–30** (1988) 1032–7.
6. Le Nest, J. F., Gandini, A., Cheradame, H. & Cohen-Addad, J. P., *Polym. Comm.*, **28** (1987) 302–5.
7. Le Nest, J. F., Gandini, A., Cheradame, H. & Cohen-Addad, J. P., *Macromolecules*, **21** (1988) 1117–20.
8. Le Nest, J. F., Defendini, F., Gandini, A., Cheradame, H. & Cohen-Addad, J. P., *J. Power Sources*, **20** (1987) 339–44.
9. Ganapathiappan, S., Chen, K. & Shriver, D. F., *Macromolecules*, **21** (1988) 2299–301.
10. Tsuchida, E., Kobayashi, N. & Ono, H., *Macromolecules*, **21** (1988) 96–100.
11. Zhou, G., Khan, I. M. & Smid, J., *Polym. Comm.*, **30** (1989) 52–5.

Transport Properties of Poly(Ethylene Oxide)–Siloxane Networks Containing Lithium Perchlorate

L. Lestel,[a] S. Boileau[b] & H. Cheradame[a]

[a] *Laboratoire de Chimie Macromoléculaire et Papetière, École Française de Papeterie (INPG), BP 65, 38402 St Martin d'Hères Cedex, France*
[b] *Collège de France, Laboratoire de Chimie Macromoléculaire, Associé au CNRS: UA 24, 11 Place Marcelin Berthelot, 75231 Paris Cedex 05, France*

ABSTRACT

Networks were synthesized by reacting diallyloxy telechelic poly(ethylene oxide) with tetramethylcyclotetrasiloxane (D4H) in the presence of a hydrosilylation catalyst. These materials were filled with lithium perchlorate, and their conductivity was measured as a function of temperature and salt concentration. Their glass transition temperature was also determined, which allowed study of their behavior at constant reduced temperature, i.e. at constant $T - T_g$. These materials exhibited an unexpected behavior since the conductivity increased as the square root of the salt concentration, at constant $T - T_g$, rather than at the first power. Apparently, the salt behaves like a weak electrolyte, which shows, by comparison with the poly(ethylene oxide)–urethane networks, that the crosslinks have an influence on the charge carrier generation.

INTRODUCTION

Our laboratory has been involved for a long time in research on ionically conducting polyether-based networks containing alkali salts,

143

because improved mechanical properties, and higher conductivity due to lower crystallinity, were expected. In order to increase segmental chain motions, allowing ion transport through polymer matrix, more flexible chains and lower glass transition temperatures have been much sought after. Few works deal with poly(ethylene oxide) crosslinked with siloxane compounds.[1,2] Some properties of poly(ethylene oxide) networks having tetrafunctional crosslinks involving Si–C bonds, less sensitive to hydrolysis than the Si–O–C linkages previously used,[3] and filled with lithium perchlorate are discussed in this paper.

EXPERIMENTAL

Networks were prepared by hydrosilylation in toluene of diallyloxy telechelic poly(ethylene oxide) (PEO) of different molecular weights (1000 or 2000) with tetramethyltetrahydrocyclotetrasiloxane (D4H).[4] The tetrafunctional crosslinks have the following structure:

$$
\begin{array}{c}
\quad\quad\quad\quad\quad CH_3 \quad CH_3 \\
\quad\quad\quad\quad\quad | \quad\quad | \\
PEO-CH_2-CH_2-CH_2-Si-O-Si-CH_2-CH_2-CH_2-PEO \\
\quad\quad\quad\quad\quad | \quad\quad | \\
\quad\quad\quad\quad\quad O \quad\quad O \\
\quad\quad\quad\quad\quad | \quad\quad | \\
PEO-CH_2-CH_2-CH_2-Si-O-Si-CH_2-CH_2-CH_2-PEO \\
\quad\quad\quad\quad\quad | \quad\quad | \\
\quad\quad\quad\quad\quad CH_3 \quad CH_3
\end{array}
$$

The synthesis as well as the conductivity measurements and some related properties of the networks filled with lithium perchlorate have been published in detail elsewhere.[4]

RESULTS AND DISCUSSION

The dependence of the logarithm of the conductivity on the reciprocal temperature, at different lithium perchlorate concentrations for PEO 1000 and PEO 2000 based networks, is shown in Fig. 1. The conductivities are higher than those obtained in the case of PEO networks crosslinked either with aromatic urethanes[5] or with polydimethylsiloxane.[1] They are of the same order of magnitude as those obtained in the case of PEO networks crosslinked by aliphatic urethanes,[6] and of poly(siloxane-g-polyether).[2,6,7] Conductivity values of about $10^{-5}\,S\,cm^{-1}$ and $3 \times 10^{-5}\,S\,cm^{-1}$ have been found at 25°C for PEO 1000 based networks ($\%\,Li/O = 3\cdot7$) and for PEO 2000

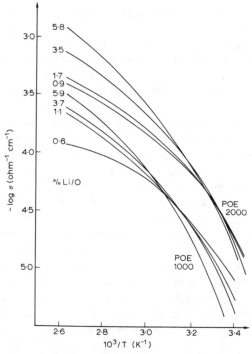

Figure 1. Arrhenius plot of the logarithm of the conductivity for networks based on PEO 1000 or 2000 for different LiClO$_4$ contents. On the figure the contents are expressed in 10^2 molar fraction Li/O, O being the molar content of ethereal type oxygen atoms per volume unit.

($\%$ Li/O = 3·5) respectively. These values are slightly lower than those observed in the case of PEO–PPO–PEO block copolymers crosslinked by aliphatic urethanes (7×10^{-5} S cm^{-1} at room temperature).[6] The materials of the present study are among the best ionic conductors based on lithium salts ever met in the domain of solvent free organic electrolytes.

Figure 1 deserves a second comment. Indeed, it could be assumed that the curves corresponding to the same network cross at the same point. It is clear that, within the range of salt concentrations investigated in the present paper, the conductivity is decreasing with increasing salt concentration at low temperature, while the reverse is true at high temperature. Thus, there is a zone of temperatures for which the conductivity is approximately independent of the salt concentration. In this zone, the decrease of mobility is quasi-counterbalanced by an increase in charge carriers concentration.

However, there is no reason why the curves should cross exactly at the same point.

The conductivity data have been analysed in terms of the WLF equation[8] according to a procedure already described in our previous work.[4] It has been found that the values of the C_1 and C_2 parameters are close to the values found in the case of networks crosslinked by urethane functions ($C_1 = 10$ and $C_2 = 40$ K, when the reference temperature is the glass transition temperature), only at high salt concentrations. The low values of T_g (between $-60°C$ and $-50°C$) at low salt concentration entail that at high temperature the material is not at a temperature falling within the domain for which a free volume behaviour is predominant, according to Ferry.[9] The interpretation of this finding is quite simple: above this range, in which the free volume is the predominant parameter, a rise in temperature does not bring about the expected increase of segmental mobility corresponding to the free-volume law. In the case of amorphous networks, based on urethane crosslinks, the T_g is often too high to allow a direct observation of this situation, due to the obvious limitation of the temperature range in which the conductivity could be measured. It is worth recalling here that the dominant role of the glass transition temperature observed for the materials of the present study is in

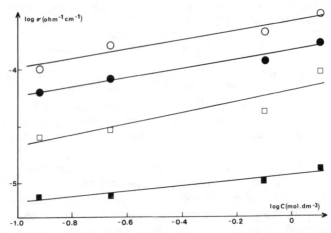

Figure 2. Logarithm of the conductivity of PEO 1000 networks cross-linked by siloxane linkages versus salt concentration, at constant reduced temperature $(T - T_g)$. ■, $T - T_g = 80°C$; □, $T - T_g = 100°C$; ●, $T - T_g = 125°C$; ○, $T - T_g = 150°C$.

agreement with all the previous studies carried out on ionically conducting polymers.[10–13]

According to a technique used in our laboratory for a long time, the conductivity measurements were analysed at constant reduced temperature, i.e. at constant $T - T_g$. The results are shown in Figs 2 and 3 for networks based on PEO 1000 and PEO 2000 respectively. Despite some scattering, it can be deduced from these two figures that the logarithm of the conductivity increases linearly with salt concentration at given $T - T_g$.

It is worth recalling here that the conductivity is the product of the mobility of the charge carriers into the number of the same charge carriers. Since the conductivity at constant reduced temperature actually means conductivity at constant mobility, the plots shown in Figs 2 and 3 reflect the variations of the number of charge carriers with salt concentration, i.e. the variations of the salt ionization. This type of analysis has been found to be extremely useful in understanding the effect of the influence of the network structure on the charge carrier generation.

The fact that the chemical nature of the crosslinks could play a role in the ionization process has not been recognized until now.[4] It has

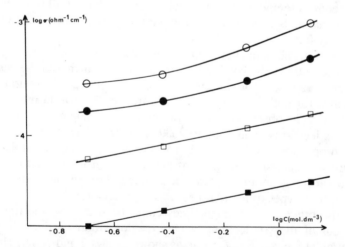

Figure 3. Logarithm of the conductivity of PEO 2000 networks cross-linked by siloxane linkages versus salt concentration, at constant reduced temperature $(T - T_g)$. ■, $T - T_g = 80°C$; □, $T - T_g = 100°C$; ●, $T - T_g = 125°C$; ○, $T - T_g = 150°C$.

Table 1. Conductivity at Constant Reduced Temperature for Membranes (PEO) Synthesized with Two Different Crosslinking Systems[a]

Reduced temp.	Conductivity (S cm^{-1}) PEO 1000 polyurethane (Desm R)	Conductivity (S cm^{-1}) PEO 1000 siloxane
$T - T_g = 100°C$	$1·6 \times 10^{-5}$	$3·7 \times 10^{-5}$
$T - T_g = 125°C$	$5·2 \times 10^{-5}$	$9·3 \times 10^{-5}$
	PEO 2000 polyurethane (Desm R)	PEO 2000 siloxane
$T - T_g = 100°C$	$4·2 \times 10^{-5}$	$7·7 \times 10^{-5}$
$T - T_g = 125°C$	$1·2 \times 10^{-4}$	$1·9 \times 10^{-4}$

[a] Salt content: 0·32 mole dm^{-3} lithium perchlorate.

been possible to shed more light on this problem by varying the nature of the crosslinks, as in the present study. In this context conductivity data on membranes obtained by hydroslilylation of diallyloxy telechelic PEO with D4H are useful. The most relevant results are shown in Table 1 and are displayed under the form of conductivity at the same reduced temperature for materials crosslinked by siloxane bonds and for materials crosslinked by the 'classical' isocyanates used in our laboratory for many years.[11–12]

Analysis of the data shows that, at a given $T - T_g$ and for the same salt concentration, the conductivity is significantly higher in the case of siloxane crosslinks than in the case of the corresponding urethane crosslinks, for both lengths of PEO. Moreover, a twofold increase in conductivity is observed when the molecular weight of the PEO segments increases from 1000 to 2000 for both values of $T - T_g$. This direct proportionality is probably specific for the range of molecular weights investigated in the present paper, since it is believed that the charge carrier generation does not proceed through the same equilibria in both types of membranes, as seen thereafter.

On the other hand, straight lines could be drawn by plotting the logarithm of the conductivity versus the logarithm of the salt concentration at a given $T - T_g$ value as shown in Figs 2 and 3. The slopes of these lines are close to 0·5, which means that the conductivity increases linearly with the squre root of the salt concentration. This behaviour is quite different from that observed for PEO networks crosslinked with urethanes for which a slope equal to 1 has been found.[12]

It is worth recalling here that we explain the conductivity behaviour of the urethane based PEO networks by the following set of equilibria:

$$(AB)_2 \rightleftharpoons 2(AB) \tag{1}$$

$$(AB)_2 \rightleftharpoons (AB)B^+ + A^- \tag{2}$$

$$(AB)B^+ \rightleftharpoons (AB) + B^+ \tag{3}$$

in which the inorganic salt AB generates the charge carriers through equilibria (2) and (3).[14] The slope equal to one found in this case has been attributed to the presence of quadrupoles playing a predominant role in the charge carrier generation. In the case of siloxane based PEO networks, the curvatures of the plots found for high $LiClO_4$ concentrations, at high $T - T_g$ (Fig. 3), could be due to the presence of some quadrupoles formed by precipitation from dipoles (salt molecules). When a slope of around 0·5 is observed, one can assume that the salt is present in this solid electrolyte under the predominant form of dipoles, and behaves like a weak electrolyte. Above a critical temperature, which depends on the network structure and salt concentration, quadrupoles are formed resulting in an ionization process which becomes predominant and is similar to that which is observed for urethane crosslinks and described above.

In conclusion, the behaviour of the siloxane based PEO networks is different from that of the PEO networks crosslinked with urethanes. More flexible crosslinks such as cyclosiloxanes lead to materials which present higher conductivity values than less flexible ones such as urethanes. However, the beneficial influence of the siloxane crosslinks is not restricted to the mobility of charge carriers, but also deals with the ionization process since at constant salt concentration and reduced temperature the conductivity is higher. Further work on PEO networks with crosslinks of different nature and flexibility is in progress.

REFERENCES

1. Wintersgill, M. C., Fontanella, J. J., Smith, M. K., Greenbaum, S. G., Adamic, K. J. & Andeen, C. G., *Polymer,* **28** (1987) 633.
2. Fish, D., Khan, I. M. & Smid, J., *Makromol. Chem., Rapid Commun.,* **7** (1986) 115.

3. Plumb, J. B. & Atherton, J. H., In: *Block Copolymers,* ed. D. C. Allport & W. H. Janes. Elsevier Applied Science, London, 1973.
4. Lestel, L., Thèse Docteur Ingénieur, Grenoble, 1987; Lestel, L., Cheradame, H. & Boileau, S., *Polymer* (in press).
5. Cheradame, H., Gandini, A., Killis, A. & Le Nest, J. F., *J. Power Sources,* **9** (1983) 389.
6. Le Nest, J. F., Gandini, A. & Cheradame, H., *Brit. Polym. J.,* **20** (1988) 253.
7. Bouridah, A., Dalard, F., Deroo, D., Cheradame, H. & Le Nest, J. F., *Solid State Ionics,* **15** (1985) 233.
8. Williams, M. L., Landel, R. F. & Ferry, J. D., *J. Am. Chem. Soc.,* **77** (1955) 3701.
9. Ferry, J. D., *Viscoelastic Properties of Polymers.* John Wiley, New York, 1980.
10. Cheradame, H., Souquet, J. L. & Latour, J. M., *Mat. Res. Bull.,* **15** (1980) 1173.
11. Cheradame, H., *IUPAC Macromolecules,* ed. H. Benoit & P. Rempp. Pergamon Press, Oxford, 1982, p. 251.
12. Cheradame, H. & Lenest, J. F., In: *Polymer Electrolyte Reviews 1,* ed. J. R. MacCallum & C. A. Vincent. Elsevier Applied Science, 1987, p. 103.
13. Ratner, M. A. & Shriver, D. F., *Chem. Revs.,* **88** (1988) 109.
14. Cheradame, H. & Niddam-Mercier, P., *Farad. Disc. Chem. Soc.,* **88** (1989) 77.

The Conductivity Behaviour of Gamma-irradiated PEO–LiX Electrolytes. II

R. J. Neat

Solid State Chemistry Group, MDD, Harwell Laboratory, Oxfordshire OX11 0RA, UK

&

E. Kronfli, K. V. Lovell

Applied Physics and Electro Optics Group, RMCS Cranfield Institute of Technology, Shrivenham, Swindon, Wiltshire SN6 8LA, UK

ABSTRACT

In an attempt to enhance the ambient temperature ionic conductivity of PEO–LiX electrolytes, aqueous solutions of PEO–LiX have been gamma-irradiated at room temperature. As a result of cross-links formed between the randomised chains, the polymer gels and forms an insoluble film. The paper discusses the effect of preparation dose on the resulting PEO–LiX electrolyte film/gels. Exposure of a PEO–LiCF$_3$SO$_3$ ([EO units]/[Li] = 9) aqueous solution to 8·0 Mrad of gamma radiation produces a rubbery, totally amorphous electrolyte film with an ionic conductivity of $\sim 1 \cdot 0 \times 10^{-5}$ S cm^{-1} at 25°C. The use of an electron beam is shown to reduce the treatment time from several days to a few seconds.

INTRODUCTION

Polymer electrolytes based on lithium salts (LiX) and poly(ethylene oxide) (PEO) have proved ideal materials for all solid-state advanced

151

lithium batteries.[1,2] Operating at temperatures of 100–140°C, these batteries have many potential applications where thermal management is an acceptable additional problem, e.g. vehicle traction and general space power. However, for the majority of consumer applications, the high operating temperature is clearly unacceptable. In order to adapt this promising technology to 'room' temperature operation (10–15°C), a polymer electrolyte is required that can support a high level of ionic conductivity ($\sim 10^{-4}\,S\,cm^{-1}$) over this temperature range. In the search for such an electrolyte many approaches have been adopted, both at Harwell and elsewhere; these include synthesis of novel polymers and modification of the original PEO–LiX materials identified by Armand.[3] In this paper we wish to report the results of one of these approaches.

In terms of large-scale production of polymer batteries, the family of polymer electrolytes PEO–LiX has many desirable properties; they are easily processed, relatively inexpensive and readily available. In order to preserve these properties, we attempted to modify several PEO–LiX systems with a view to preventing the crystallisation process which dramatically reduces the ionic conductivity at room temperature. One of several methods available to retard crystallisation in polymeric systems is the introduction of cross-links into the amorphous form. Charlesby[4] has shown that gamma-irradiation can be used for this purpose, and MacCallum et al.[5] have previously applied this method to PEO–LiX electrolyte systems. In a follow-up of this work, we have reported a study of the effect of gamma irradiation on the ionic conductivity of both PEO–LiCF$_3$SO$_3$ ($x = 9$) where $x = [EO]/[Li]$ and PEO–LiClO$_4$ ($x = 20$).[6] In both cases the irradiations were performed at 78°C, a temperature at which both electrolytes are substantially amorphous.[6] We postulated that irradiation would introduce cross-links into the amorphous form and hence prevent crystallisation of the amorphous conducting polymer on cooling to room temperature. In this way, ionic conductivities of $\sim 10^{-5}\,S\,cm^{-1}$ for PEO–LiCF$_3$SO$_3$ ($x = 9$) and $10^{-4}\,S\,cm^{-1}$ for PEO–LiClO$_4$ ($x = 20$) were predicted by extrapolation of the 70–120°C conductivity data to 25°C. However, subsequent analysis using DSC, optical microscopy and variable temperature ionic conductivity indicated that no significant retardation of the amorphous to crystalline transition had occurred over a wide range of radiation dose (1–12 Mrad), and hence no improvement in the room temperature conductivity was achieved.

Doses in excess of 12 Mrad caused mechanical breakdown and rendered films unusable.

As an explanation for the above results, it was postulated that the long-range order of the polymer chains in these crystalline solvent-cast electrolytes is only moderately affected by heating to 78°C. Thus although amorphous, the PEO chains remain partially ordered. Introducing cross-links into this structure has little effect, since polymer chain repositioning is relatively easy and requires only a small activation energy. To be effective the cross-links must be introduced into a totally randomised amorphous form, where repositioning for crystallisation would require large chain movement.

In this paper we report a method of producing homogeneous, totally amorphous PEO–LiCF$_3$SO$_3$ and PEO–LiClO$_4$ electrolytes via gamma irradiation of aqueous solutions of both PEO–LiCF$_3$SO$_3$ ($x = 9$) and PEO–LiClO$_4$ ($x = 20$). The introduction of cross-links into the viscous polymer solution is very effective since the polymer chains are randomly positioned. This causes the polymer to gel and form an insoluble film. The paper will discuss the effect of the irradiation dose on the ionic conductivity of the resultant PEO–LiCF$_3$SO$_3$ ($x = 9$) and PEO–LiClO$_4$ ($x = 20$) gelled electrolyte films, suggest alternative solvents and briefly report the use of alternative cross-linking methods.

EXPERIMENTAL

Film Preparation

Electrolyte solutions of PEO–LiCF$_3$SO$_3$ ($x = 9$) and PEO–LiClO$_4$ ($x = 20$) were prepared from appropriate amounts of PEO, LiCF$_3$SO$_3$ or LiClO$_4$ and distilled water to yield a 2·5 wt% concentration based on PEO. The mixtures were mechanically stirred for 24 h to create a homogeneous solution. Aliquots of solution were placed in plastic containers (10 cm diameter) and gamma-irradiated using a ^{60}Co source at room temperature and in the presence of oxygen. A dose rate of 0·1 Mrad h^{-1} (1·0 kiloGray h^{-1}) was used for a range of exposure times between 10 and 90 h. Samples were thus irradiated to total doses of 1–9 Mrad (10–90 kiloGray). After irradiation the samples were allowed to dry in the open laboratory at room temperature before final drying in a vacuum oven. The resulting films (100–200 μm) were stored in sealed polyethylene bags and all subsequent handling performed in a lithium standard dry room (RH 0·5%).

AC Conductivity Measurements

The ionic conductivity of the films was determined between 25 and 90°C using the complex impedance method. Discs of the films were sandwiched between two polished stainless steel electrodes which were spring-loaded to maintain good interfacial contact. The electrode assembly was mounted in a Büchi furnace with the temperature control performed by a BBC microcomputer. The impedance of the electrode assembly was measured by a Solartron 1174 FRA coupled with a 1186 ECI, and data collected every 5°C between 25°C and 90°C on both the heating and cooling cycles.

At low temperatures the impedance data, when plotted in the complex plane, consist of a high frequency semi-circle and a low frequency sloping spike. At higher temperatures the impedance plots show only the spike. The exact physical meaning of the semi-circle remains unclear, but the authors are confident that the sloping spike is produced by the electrode/electrolyte interface. Thus extrapolation of the upper part of this spike to the real axis yields the bulk polymer resistance. This method has been used to determine all conductivity data reported here.

DSC Analysis

Samples of each material were sealed in DSC cans and their thermal characteristics evaluated between 25 and 100°C. No glass transition temperatures were measured but endotherms and exotherms are reported both in terms of peak maximum and peak area. Experiments were performed on a Perkin-Elmer DSC1B at a programmed heating/cooling rate of 4°C min^{-1}.

RESULTS AND DISCUSSION

Exposure of PEO–LiX aqueous solutions to gamma irradiation causes the polymer to gel as cross-links are formed. This process produces a rubbery high molecular mass material which is insoluble in the aqueous medium. The result is a thin gelled polymer film (the film thickness depends on the concentration of the solution and the amount of solution placed in the plastic container), and an aqueous liquor which is allowed to evaporate slowly thus returning any dissolved salt to the polymer matrix. Electrolyte films produced by this method have excellent mechanical properties: they are rubbery and tolerate a high degree of mechanical deformation.

PEO–LiCF$_3$SO$_3$ ($x = 9$) Gel Electrolytes

Figure 1 shows the cooling conductivity data for a typical solvent cast film of PEO–LiCF$_3$SO$_3$ ($x = 9$) electrolyte plotted in the Arrhenius-like format (log σ versus $1/T$). On cooling from 90 to 55°C, the data exhibit a linear decrease in conductivity, followed by a significantly larger, and nonlinear, decrease between 55 and 45°C. This larger decrease is caused by the conversion of a significant proportion of amorphous polymer (the conducting medium) into crystalline polymer. The data from 45 to 25°C resume the linear behaviour but with an increased activation energy. This overall pattern is typical of the many

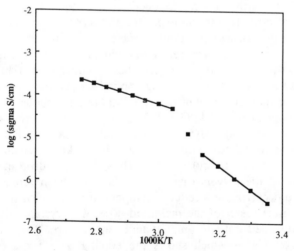

Figure 1. Variation of the ionic conductivity with temperature for a typical solvent-cast PEO–LiCF$_3$SO$_3$ ($x = 9$) electrolyte.

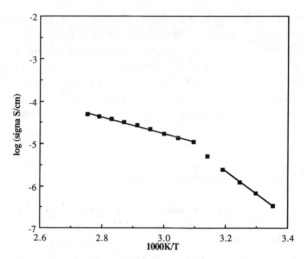

Figure 2. Variation of the ionic conductivity with temperature for a PEO–LiCF$_3$SO$_3$ ($x = 9$) gel after a 1·0 Mrad exposure to gamma radiation.

PEO–LiX electrolytes which exhibit a strong tendency to undergo rapid recrystallation on cooling from elevated temperatures.

Figure 2 shows the cooling conductivity data for a PEO–LiCF$_3$SO$_3$ ($x = 9$) gel electrolyte after an exposure of 1·0 Mrad. Comparing these data to the solvent-cast equivalent, it is clear that a modification to the crystallisation event has occurred. Although still present, the nonlinear decrease between 55 and 45°C has been reduced, and hence the conductivity levels of the remaining cooling points are higher. These data indicate that the amount of crystallisable amorphous polymer in the film has been reduced. Figure 3 shows the equivalent data for a 2·0 Mrad exposed electrolyte gel. The cooling data profile for this material consists of two straight lines with a knee (change to an increased activation energy) at 40°C. Thus the large drop in conductivity found in the solvent-cast cooling data has been replaced by a simple knee occurring at a reduced temperature. This clearly indicates that, as a direct result of the increased dose, the crystallisation event has been retarded both in amount and temperature. This pattern is confirmed by Fig. 4 which shows the cooling data for a 6·0 Mrad exposed PEO–LiCF$_3$SO$_3$ ($x = 9$) electrolyte. All these data lie on a slight curve rather than a straight line (an indication of an increased

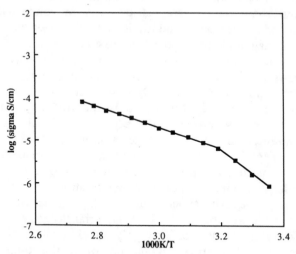

Figure 3. Variation of the ionic conductivity with temperature for a PEO–LiCF$_3$SO$_3$ ($x = 9$) gel after a 2·0 Mrad exposure to gamma radiation.

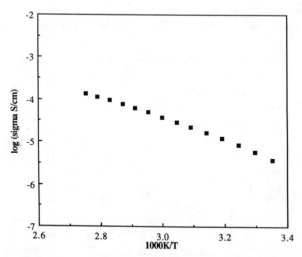

Figure 4. Variation of the ionic conductivity with temperature for a PEO–LiCF$_3$SO$_3$ ($x = 9$) gel after a 6·0 Mrad exposure to gamma radiation.

glass transition temperature and the amorphous nature of the electrolyte) and show only a gradual decrease in conductivity. The resulting room temperature conductivity has thus been increased by 1·5 orders of magnitude over the solvent-cast equivalent.

The effect of the radiation dose on the ionic conductivity discussed above, shows that this method of cross-linking is very effective at creating amorphous PEO–LiCF$_3$SO$_3$ ($x = 9$) electrolytes. Increasing the radiation dose from 1·0 Mrad to 6·0 Mrad gradually reduces the amount of amorphous polymer that is capable of crystallising, until at 6·0 Mrad, no recrystallisation event occurs during the time scale of the conductivity experiment. However, examination of time dependent impedance plots at 25°C (see Fig. 5), indicates that the 6·0 Mrad sample undergoes time-dependent recrystallisation outside the time frame of the conductivity experiment, i.e. the conductivity value for 25°C is not stable with time. This is an important consideration when reporting conductivity data on new polymer electrolytes. The above time-dependent behaviour is eliminated when the total dose is increased to 8·0 Mrad. This film has a set of conductivity values almost identical to the 6·0 Mrad electrolyte but moreover, exhibits no change in impedance over a 24 h time period at 25°C.

Thus the technique of gamma-irradiating the aqueous polymer electrolyte solution, rather than the amorphous solvent-cast polymer

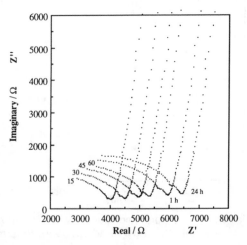

Figure 5. Time-dependent AC impedance data collected at 25°C on a PEO–LiCF$_3$SO$_3$ ($x = 9$) gel after a 6·0 Mrad exposure to gamma radiation.

electrolyte film, produces totally amorphous, mechanically strong films with ionic conductivities of $\sim 10^5$ S cm^{-1} at 25°C.

PEO–LiClO$_4$ ($x = 20$) Gel Electrolytes

Although the above study of PEO–LiCF$_3$SO$_3$ ($x = 9$) gel electrolytes achieved a successful result, the room temperature ionic conductivity obtained is too low for many practical, high rate, applications. There are, however, many additional PEO–LiX electrolyes which as solvent-cast films have higher overall levels of ionic conductivity. An example is PEO–LiClO$_4$ ($x = 20$) which undergoes a similar drop in conductivity on cooling, but from a higher level. Thus if the recrystallisation event in PEO–LiClO$_4$ ($x = 20$) could be retarded in a similar manner, a $\sigma_{25°C}$ level of $\sim 10^{-4}$ S cm^{-1} is predicted. Thus a similar study of the effect of gamma irradiation on the conductivity of PEO–LiClO$_4$ ($x = 20$) gel electrolytes was carried out.

The results are in broad agreement with the PEO–LiCF$_3$SO$_3$ study, although in general a higher dose was required to produce the same effect. This indicates that the type and amount of lithium salt may have an effect on the cross-linking process. Figure 6 shows the

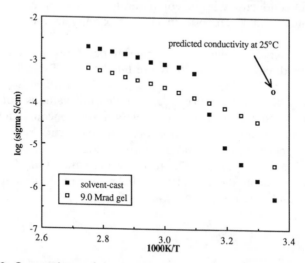

Figure 6. Comparison of the ionic conductivity of a PEO–LiClO$_4$ ($x = 20$) solvent-cast electrolyte with the equivalent gel electrolyte after 9·0 Mrad of gamma irradiation.

Table 1. Ionic Conductivity Values at Various Temperatures for a Series of PEO–LiX Electrolyte Films

Electrolyte type	$\sigma_{90°C}$ (S cm^{-1})	$\sigma_{60°C}$ (S cm^{-1})	$\sigma_{30°C}$ (S cm^{-1})	$\sigma_{25°C}$ (S cm^{-1})
PEO–LiCF$_3$SO$_3$ ($x = 9$) (solvent-cast)	1.99×10^{-4}	6.37×10^{-5}	6.09×10^{-7}	3.05×10^{-7}
PEO–LiCF$_3$SO$_3$ ($x = 9$) (8.0 Mrad gel)	3.24×10^{-4}	8.69×10^{-5}	2.38×10^{-5}	7.43×10^{-6}
PEO–LiClO$_4$ ($x = 20$) (solvent-cast)	1.80×10^{-3}	7.73×10^{-4}	9.02×10^{-6}	5.40×10^{-7}
PEO–LiClO$_4$ ($x = 20$) (9.0 Mrad gel)	6.37×10^{-4}	2.08×10^{-4}	3.54×10^{-5}	3.35×10^{-6}

conductivity data for a solvent-cast PEO–LiClO$_4$ ($x = 20$) film on the cooling cycle and indicates the predicted conductivity level expected given full recrystallisation retardation. Figure 6 also shows the conductivity data for a PEO–LiClO$_4$ ($x = 20$) electrolyte gel after an exposure of 9.0 Mrad. Even at this dose, the clear drop in conductivity between 30°C and 25°C indicates the film still contains non-cross-linked amorphous material. In addition, there is a significant drop in the overall conductivity levels compared to the solvent-cast film. This is associated with a decrease in chain flexibility in the cross-linked matrix.

The conductivity values at several temperatures for the various electrolytes discussed in this study are tabulated in Table 1. The best conductor at 30°C is the PEO–LiClO$_4$ ($x = 20$) gel which has an ionic conductivity of 3.45×10^{-4} S cm^{-1}. Although this value is not as high as the predicted value, it is nevertheless a very encouraging result. However, the minor crystallisation event which occurs between 30°C and 25°C renders this material a poorer conductor at 25°C than the PEO–LiCF$_3$SO$_3$ ($x = 9$) gel. The fact that the amount and/or type of lithium salt changes the cross-linking behaviour of these materials is interesting and requires further investigation.

DSC Analysis

Differential scanning calorimetry (DSC) analysis was performed on all gelled electrolyte samples. Surprisingly, all samples of gelled PEO–

Table 2. DSC Analysis of Gamma-Irradiated PEO–LiClO$_4$ ($x = 20$)
Electrolyte Gels.

Dose (Mrad)	Sample (wt/mg)	Endotherm		Exotherm	
		Peak (°C)	Peak area wt (cm^{-2} g)	Peak (°C)	Peak area wt (cm^{-2} g)
1·0	11·60	48·0	1·147	31·3	0·853
2·0	12·90	40·0	1·093	—	—
3·0	11·70	38·0	0·966	—	—
4·0	15·80	38·0	0·335	—	—
5·0	14·00	—	—	—	—
6·0	10·89	—	—	—	—
7·0	15·10	—	—	—	—
8·0	15·03	—	—	—	—
9·0	12·06	—	—	—	—

LiCF$_3$SO$_3$ ($x = 9$) electrolyte produced no thermal events in the DSC
experiment; the sensitivity of the heat flow measurement was identical
to that used during analysis of the solvent-cast electrolytes.[5] This is not
consistent with the conductivity data which clearly show both melting
and recrystallisation events in the low dose samples. Table 2 shows the
DSC results for the PEO–LiClO$_4$ ($x = 20$) gels. These data, although
showing both melting and recrystallisation events, are again inconsis-
tent with the conductivity data for this series of electrolytes. However,
the peak positions and sizes show a clear decrease with the increase in
cross-linking density. These results show that the melting/
crystallisation of small amounts of crystalline material, which are too
small to be detected by the DSC experiment, have pronounced effects
on the ionic conductivity.

Alternative Solvents

One of the major drawbacks of the gelling technique described above
is the use of water as the solvent. Although the resulting electrolytes
are vigorously dried before testing, the possibility of small amounts of
bound water is a major concern when lithium batteries are constructed. Thus a series of alternative gelling solvents was investigated;
these included propylene carbonate, benzene, acetonitrile, dichloro-
ethane and tetrahydrofuran.

In all cases it was not possible to gel the solutions, and a large viscosity loss was observed. This indicates that chain scission rather than cross-linking is the dominant process, and is explained by the poor hydrogen radical reactivity of most aprotic solvents.

Alternative Cross-Linking Methods

The use of gamma irradiation is only one of many methods available to introduce cross-links into polymers. Although the use of a ^{60}Co source has produced encouraging results, it has several drawbacks. These include logistical problems of sample mounting, source configuration and general operator inconvenience. In addition, and perhaps more importantly, treatment times are very long, e.g. 80 h for the optimum PEO–LiCF$_3$SO$_3$ $(x = 9)$ material. Two alternative cross-linking methods were therefore investigated: ultraviolet and electron beam radiation.

Several experiments were performed using a 2·5 kW medium pressure UV lamp. The aqueous electrolyte solutions, with additional cross-linking initiators, were placed 30 cm from the lamp, and the solution exposed for various times. No gelling was observed with any combination of time and initiator and the conductivity of the resulting films was very similar to normal solvent-cast samples. Further work is required, particularly on improved initiators, if electrolyte gelling is to be successfully induced by the lower levels of energy available from an UV source.

Electron beam irradiation provides a large amount of energy in a concentrated region. Several aliquots of PEO–LiClO$_4$ $(x = 20)$ aqueous solution were irradiated for various numbers of passes under an electron beam. Each pass is equivalent to 1·25 Mrad of gamma irradiation and takes ~1 s. Exposure of the aqueous solutions to the electron beam induced the polymer to gel in the same way as the ^{60}Co source. Figure 7 shows the cooling conductivity of a 5·0 Mrad (4 passes; ~4 s) exposed electrolyte. Again the recrystallisation event has been significantly modified producing improved room temperature conductivities. This is a very encouraging result since the electron beam process requires only 4 s compared to 50 h for the equivalent dose of gamma irradiation.

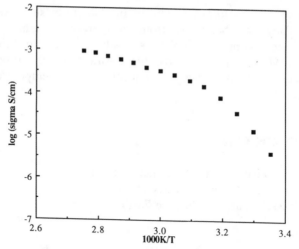

Figure 7. Variation of the ionic conductivity with temperature for a PEO–LiClO$_4$ ($x = 20$) gel after a 5·0 Mrad exposure to an electron beam.

CONCLUSIONS

Gamma irradiation of aqueous solutions of PEO–LiCF$_3$SO$_3$ ($x = 9$) and PEO–LiClO$_4$ ($x = 20$) electrolytes has been shown to produce rubbery polymer electrolyte films which are mechanically strong and substantially amorphous at room temperatures. The ionic conductivity of the PEO–LiCF$_3$SO$_3$ ($x = 9$) film/gel after 8·0 Mrad of exposure, for example, shows an improvement over the standard solvent-cast material and is $\sim 1{\cdot}0 \times 10^{-5}$ S cm^{-1} at 25°C. The effect of increasing the radiation dose from 1·0 to 8·0 Mrad is to retard, and eventually eliminate, the formation of crystalline material in the electrolyte films. However, similar exposure of PEO–LiClO$_4$ ($x = 20$) solutions does not eliminate all crystalline material, and thus room temperature conductivity values are lower than predicted.

An investigation of several alternative gelling solvents to water has shown that chain scission rather than cross-linking occurs in most common aprotic solvents.

The use of electron beam irradiation dramatically reduces the treatment time from several days to a few seconds, though UV irradiation has, so far, proved unsuccessful.

The electrolyte materials produced in this study show significant promise. They can be made from inexpensive, readily available materials using a process which is easily adapted to large scale production. Moreover, they have the desirable combination of excellent mechanical properties and improved room temperature conductivity levels.

ACKNOWLEDGEMENTS

The authors wish to acknowledge the financial support of the Department of Trade and Industry and the members of the Harwell-based Solid-State Battery Working Party.

REFERENCES

1. Hooper, A. & North, J. M., *Solid State Ionics*, **9, 10** (1983) 1161.
2. Hooper, A., Belanger, A. & Gauthier, M., Polymer electrolyte lithium batteries. In: *Electrochemical Science and Technology of Polymers, Vol. 2*, ed. R. G. Linford. Elsevier Applied Science, London, 1990.
3. Armand, M. B., Chabagno, J. M. & Duclot, M. J. In: *Fast Ion Transport in Solids*, ed. P. Vashishta, J. N. Mundy & g. K. Shenoy. North-Holland, Amsterdam, 1979, p. 131.
4. Charlesby, A., *Atomic Radiation and Polymers*. Pergamon Press, Oxford, 1959.
5. MacCallum, J. R., Smith, M. J. & Vincent, C. A., *Solid State Ionics*, **11** (1984) 307.
6. Kronfli, E., Lovell, K. V., Hooper, A. & Neat, R. J., *British Polymer Journal*, **20** (1988) 275.

Solid State Voltammetry of Ferrocene in Network Poly(Ethylene Oxide) Electrolytes

Masayoshi Watanabe, Hiroyuki Shibuya, Kohei Sanui & Naoya Ogata

Department of Chemistry, Sophia University, 7-1 Kioi-cho, Chiyoda-ku, Tokyo 102, Japan

ABSTRACT

The methodology of solid state voltammetry has been established and its usefulness and limitation in examining transport properties of polymer electrolytes are demonstrated by using ferrocene dissolved in network poly(ethylene oxide) (PEO) electrolytes. The molecular solute, ferrocene, undergoes a reversible one-electron redox reaction even in the absence of any low-molecular-weight solvent. The diffusion coefficient of ferrocene in the network PEO electrolytes, determined by solid state voltammetry, is $10^{-7} \, \mathrm{cm^2 \, s^{-1}}$ at 100°C and approaches $10^{-9} \, \mathrm{cm^2 \, s^{-1}}$ at 40°C.

INTRODUCTION

Polymer electrolytes[1-3] are ion-conducting solid solutions of electrolyte salts in ion-coordinating polymers, typically, poly(ethylene oxide)(PEO). The discovery[4] of ion-conducting polymer electrolytes has been followed by rigorous studies on the transport property of ions in polymers, on syntheses of new polymer hosts suitable for forming highly conductive polymer electrolytes, and on the electrochemical stability and the interfacial charge transfer properties in contact with certain kinds of electrode materials.[5,6] One of the motivations of the active studies has been undoubtedly the potential applicability of polymer electrolytes as solid electrolytes to a variety of electrochemical devices.

Additonally, an interesting property of polymer electrolytes is that electroactive molecules can be dissolved in and diffuse through the polymeric phases. Using this property as well as their ion-conducting property allows polymer electrolytes to be used as solid state solvents for electrochemical reaction of the molecules dissolved in them, just as fluid electrolyte solutions are used for this purpose.[7] The studies on this point of view can provide information on mass-transport and heterogeneous/homogeneous electron-transfer properties of the incorporated molecules in the polymeric phases. This information, especially on mass-transport properties, in turn, can be exploited as a probe for ion transport and segmental dynamics in the polymer electrolytes. More importantly, the development of the study will promisingly open new scope for solid state chemistry.

There have been several reports[7-12] on the electrochemical reaction of electroactive molecules dissolved in polymer electrolytes, which is explored by a voltammetric method; so-called solid state voltammetry. This paper describes the electrochemical response of ferrocene dissolved in network PEO electrolytes by means of solid state voltammetry. In the previous studies,[9-12] linear PEO electrolytes are mainly used as solid solvents for electroactive solutes. Due to the highly crystalline nature of the linear PEO electrolytes, electrochemical response, reflecting mass transport property, is influenced and complicated by the existence of the crystalline phase and also by the phase transition. An advantage in using network PEO electrolytes for voltammetric study lies in their completely amorphous nature. This study has been aimed at establishing the methodology of solid state voltammetry and demonstrates its usefulness and limitation in examining the transport property of ferrocene in the network PEO electrolytes.

EXPERIMENTAL

Network PEO electrolytes were prepared by a cross-linking reaction of PEO triol ($M_n = 3000$) with toluene-2,4-diisocyanate, followed by doping with LiClO$_4$ by an immersion method. Precise details of the preparation method are described elsewhere.[13] Concentration of LiClO$_4$ in the PEO network polymers was controlled at [LiClO$_4$]/[EO unit] $= 0.02-0.04$ in order to optimise the ionic conductivity of the network PEO electrolytes.[13,14]

Ferrocene was dissolved in the network polymer electrolytes as follows. Chloroform solutions of ferrocene (0·5–10 mM) were dropped onto disk samples (13 mm dia., c. 0·5 mm thick) of the network PEO electrolytes, placed on a Teflon (polytetrafluoroethylene, PTFE) sheet, by using a microsyringe. The disk samples absorbed the solutions and swelled. In order to obtain homogeneously swelled samples, the concentration of ferrocene and the amount of absorbed solution were adjusted so as to make the weight swelling ratios of the samples about 1–2. The samples swelled by the chloroform solutions of ferrocene were dried under reduced pressure at room temperature for 18 h.

Figure 1 shows the structure of an air-tight cell used for solid state voltammetry. A working electrode (3 mm dia. Pt disk) and a quasi-reference electrode (a small piece of Ag wire) were placed closely (distance $\simeq 1$ mm) on one side of a disk sample, and a counter electrode (13 mm dia. Pt disk) was placed on the other side of the sample (see inset of Fig. 1). The contact between the electrodes and

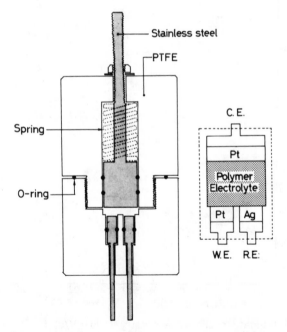

Figure 1. Structure of electrochemical cell used for solid state voltammetry.

the sample were ensured by pressure applied by a spring loaded in the cell. The network PEO electrolytes with dissolved ferrocene were packed and sealed in the cells inside an Ar-filled glove box and subjected to voltammetric measurements at benchtop. A conventional potentiostat and functional generator were used for the potential control. In this experiment, no *iR* compensation for the electrode potential was carried out.

RESULTS AND DISCUSSION

In order to obtain well-defined and reproducible solid state voltammo-grams, proper design of solid state cells and electrode configuration was necessary. Especially, to make good and steady contact between

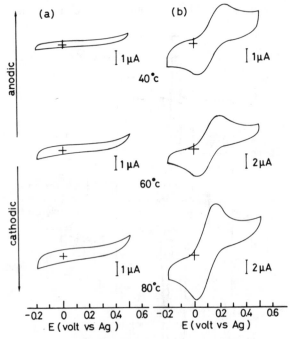

Figure 2. Solid-state cyclic voltammetry (100 mV/s) for network PEO electrolytes at various temperatures: (a) in the absence of ferrocene; (b) in the presence of ferrocene (3 mM).

the electrode and the sample was a basic but not so easily achievable requirement. Solid state voltammetry using the cell shown in Fig. 1 gave reasonable results. Figure 2 shows solid state voltammetric responses of ferrocene at three different temperatures together with background scans. In the background scans (Fig. 2(a)), no appreciable background current was observed in this potential range. The voltammograms in the presence of ferrocene (Fig. 2(b)) gave a much higher current level, compared with that of the background scan. Consequently, these responses can be attributed to the Faradaic current of the redox reaction of ferrocene. Figure 3 shows cyclic voltammograms of ferrocene at various scan rates and 80°C. The profile of voltammograms having current peaks and diffusion tails is characteristic of the depletion of the redox species near the electrode surface. The anodic and cathodic peak currents were the same in absolute magnitude at each scan rate. The peak separation was lower than 100 mV at the lower scan rate. These results show that ferrocene molecules incorporated in the polymer electrolyte undergo a reversible one-electron redox reaction even in the absence of any low-molecular-weight solvent.

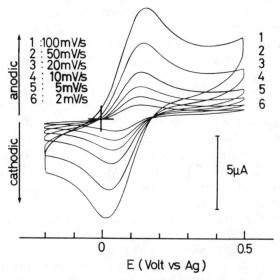

Figure 3. Solid-state cyclic voltammetry (80°C) of ferrocene (3 mM) in network PEO electrolyte at various scan rates.

As can be easily supposed from the lower conductivity of polymer electrolytes compared to that of ordinary electrolyte solutions and from the large change in the conductivity depending on such variables as temperature and salt concentration, the iR drop is a big problem in making quantitative measurements. This problem is typically seen in Figs 2 and 3. The peak separation became larger and the voltammograms were distorted at lower temperature (Fig. 2), due to the increase of the iR drop which results from the large decrease of ionic conductivity of the polymer electrolytes with decreasing temperature. Even at a high temperature, the peak separation slightly increased with increase in the scan rate (Fig. 3). The effective way to get rid of this iR drop problem is to use microelectrodes,[15] as has been introduced in this field of study.[9–12]

Figure 4 shows the temperature dependence of the anodic peak current at $2\,mV\,s^{-1}$ obtained by cyclic voltammetry of the network PEO electrolytes with dissolved ferrocene at various concentrations. The ionic conductivity of the network PEO electrolyte is also shown for comparison. As is well known, the positively curved profile of the Arrhenius plot of the ionic conductivity is due to the likewise-changing ionic diffusivity which is cooperative with the local segmental motion.[14] The temperature dependence of anodic peak current was moderate compared with that of the ionic conductivity, but it also seemed to show positively curved profiles. In the inset of Fig. 4 is shown a typical relationship between anodic peak current and square root of scan rate, found in cyclic voltammetry. Its good linearity together with the positively curved profiles of the Arrhenius plots of the anodic peak currents indicate that the redox reaction of ferrocene in the network PEO electrolyte is a diffusion-controlled process. If the linear diffusion theory[16] is applicable to the present system, the slope of the relation in the inset of Fig. 4 gives the diffusion coefficient of ferrocene. Figure 5 shows the temperature dependence of the ferrocene diffusion coefficient in the network PEO electrolytes. Since the thickness of the diffusion layer during our experimental time scale, estimated from the diffusivity in Fig. 5, is far smaller than the electrode size and also far smaller than the film thickness, the linear diffusion theory is consequently applicable. The diffusivity of ferrocene nearly reaches $10^{-7}\,cm^2\,s^{-1}$ at 100°C and decreases down to the order of 10^{-8}–$10^{-9}\,cm^2\,s^{-1}$ at 40°C. These values are consistent with our recent diffusion data[17] obtained by using microelectrodes and comparable to the ionic diffusion coefficient in linear PEO electrolytes above their melting points.[18]

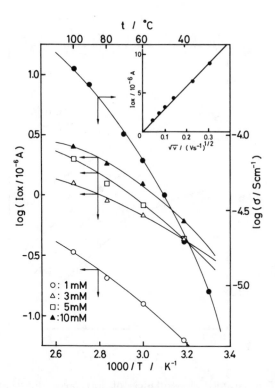

Figure 4. Oxidation peak current (left axis) of ferrocene in network PEO electrolytes found in cyclic voltammetry (2 mV/s) and ionic conductivity (right axis) of network PEO electrolyte as a function of reciprocal absolute temperature; inset: relation between oxidation peak current of ferrocene (3·4 mM) and square root of sweep rate at 80°C.

The ferrocene diffusivity at 10 mM was distinctly lower than that at the other concentrations. This decrease resulted from the levelling off of the anodic peak current as a function of ferrocene concentration above *c*. 5 mM, though this relation was linear up to that concentration. One possible reason is the stiffening effect of ferrocene toward the network PEO electrolytes. Actually, we observed the decrease of ionic conductivity of the network PEO electrolytes with increasing ferrocene concentration, whereas we could not detect any appreciable increase of the glass transition temperature due to the dissolution of ferrocene. Another reason may come from the possibility that the real concentration of ferrocene is lower than the apparent concentration

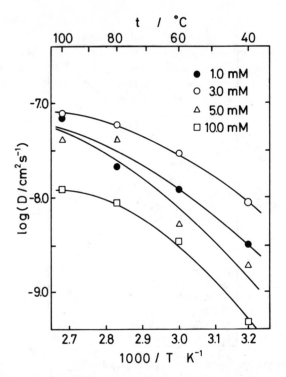

Figure 5. Temperature dependence of ferrocene diffusion coefficient calculated from the results of solid state voltammetry.

and this tendency becomes severe with increase in concentration, due to the sublimation problem of ferrocene. However, we have no conclusive reason for this phenomenon up to now.

CONCLUSION

It has been shown that solid state voltammetry can be performed by a combination of the methodology demonstrated in this paper with the network PEO electrolytes. This style of solid state chemistry provides a lot of fundamental information on transport property in polymers. However, numerous problems remain unclarified in terms of mass-transport and homogeneous and heterogeneous electron-transfer phenomena. For example, how salt concentration affects the diffusivity of electroactive solutes is still not clear. Furthermore, when the concentration of the electroactive solutes is high, the possibility of homogeneous electron-transfer between the solutes may be appreciable due to slower physical diffusion of the solutes than that in fluid solutions.[12] To clarify these problems may also contribute to the development of new materials and devices.

REFERENCES

1. Ratner, M. A. & Shriver, D. F., Ion transport in solvent-free polymers. *Chem. Rev.*, **88** (1988) 109–24.
2. Vincent, C. A., Polymer electrolytes. *Prog. Solid St. Chem.*, **17** (1987) 145–261.
3. Armand, M. B., Polymer electrolytes, *Ann. Rev. Mater. Sci.*, **16** (1986) 245–61.
4. Wright, P. V., Electrical conductivity in ionic complexes of poly(ethylene oxide). *Brit. Polym. J.*, **7** (1975) 319–27.
5. MacCallum, J. R. & Vincent, C. A. (eds), *Polymer Electrolyte Reviews 1.* Elsevier Applied Science, London, 1987.
6. *Brit. Polym. J.*, **20** (1988) 171–297 (papers presented at First International Symposium on Polymer Electrolytes).
7. Watanabe, M. & Ogata, N., Ionic conductivity of polymer electrolytes and future applications. *Brit. Polym. J.*, **20** (1988) 181–92.
8. Watanabe, M., Shibuya, H., Sanui, K. & Ogata, N., Polymer electrolytes as reaction media. Redox reaction of ferrocene. *Polym. Prepr. Jpn* **36** (1987) 2501–4.
9. Reed, R. A., Geng, L. & Murray, R. W., Solid state voltammetry of electroactive solutes in poly(ethylene oxide) polymer films on microelectrodes. *J. Electroanal. Chem.*, **208** (1986) 185–93.
10. Geng, L., Reed, R. A., Longmire, M. & Murray, R. W., Solid state linear sweep voltammetry. A probe of diffusion in thin films of polymer ion conductors on microdisk electrodes. *J. Phys. Chem.*, **91** (1987) 2908–14.
11. Geng, L., Longmire, M. L., Reed, R. A., Parcher, J. F., Barbour, C. J. & Murray, R. W., Solid-state voltammetric measurement of plasticization transport enhancement in ionically conducting poly(ethylene oxide) films, *Chem. Mater.*, **1** (1989) 58–63.
12. Geng, L., Reed, R. A., Kim, M.-H., Wooster, T. T., Oliver, B. N., Egekeze, J., Kennedy, R. T., Jorgenson, J. W., Parcher, J. F. & Murray, R. W., Chemical phenomena in solid-state voltammetry in polymer solvents. *J. Am. Chem. Soc.*, **111** (1989) 1614–19.
13. Watanabe, M., Nagano, S., Sanui, K. & Ogata, N., Ionic conductivity of network polymers from poly(ethylene oxide) containing lithium perchlorate. *Polym. J.*, **11** (1986) 809–17.

14. Watanabe, M., Itoh, M., Sanui, K. & Ogata, N., Carrier transport and generation processes in polymer electrolytes based on poly(ethylene oxide) networks. *Macromolecules*, **20** (1987) 569–73.
15. Fleischmann, M., Pons, S., Rolison, D. R. &˙ Schmidt, P. P. (eds), *Ultramicroelectrodes*. Datatech Systems, Morganton, N.C., 1987.
16. Bard, A. J. & Faulkner, L. F., *Electrochemical Methods*, John Wiley, New York, 1980, p. 218.
17. Watanabe, M., Longmire, M. L. & Murray, R. W., A study of ferrocene diffusion dynamics in network poly(ethylene oxide) polymer electrolyte by solid state voltammetry. *J. Phys. Chem.*, **94** (1990) 2614–19.
18. Bridges, C., Chadwick, A. V. & Worboys, M. R., Radiotracer self-diffusion measurements in poly(ethylene oxide) and poly(propylene oxide) electrolytes. *Brit Polym. J.*, **20** (1988) 207–11.

Ionic Conductivity in Sulphonate End-capped Poly(Ethylene Oxide)

T. Hamaide, C. Carré & A. Guyot

Laboratoire des Matériaux Organiques, CNRS, BP 24, 69390 Vernaison, France

ABSTRACT

In an attempt to overcome the incompatibility problems due to various chemical structures bearing grafted ions, sulphonate end-capped poly(oxyethylene) has been synthesized by reaction of the propane sulphone with PEO alcoholates. Solid polymer electrolytes (SPE) and solid-state batteries have been prepared and characterised by impedance measurements. The best electrochemical performances have been obtained by using these telechelic PEO in conjunction with lithium triflate. The analysis of the impedance spectra clearly shows that these performances are due to a decrease of both resistance and relaxation time in the bulk electrolyte.

INTRODUCTION

It is well known that the conductivity in solid polymer electrolytes is mainly achieved by the transport of the ions located in the amorphous phase and a lot of work has been focused in this area in order to improve this key-metastable phase. Nevertheless, the enhancement of the amorphous domains is not the only point to take into account and the problems connected with the different transference numbers have to be faced, namely the electrode polarisation[1] and the dynamic space charges[2] which may greatly limit the current density.

As suggested by Armand,[1] a possible solution to this problem would be either to immobilise the motion of one or both ions by grafting it to the polymer backbone or to use a rather large anion, so that the transference number of its counterion would then be equal to one.

Therefore, polymer backbones bearing either carboxylato or sulphonato groups have been synthesised: Bannister *et al.*[3] reported polymers with anchored alkylsulphonic and perfluorocarboxylic acids. These strong acid salts were also chosen in order to reduce the degree of ion-pairing. Tsuchida made use of poly(lithium methacrylate-co-oligo(oxyethylene)methacrylate[4,5] and blends of PEO oligomers with lithium-containing perfluoro copolymers.[6] Cheradame[7,8] included phosphate ions in polyether networks. Hardy & Shriver[9] described poly(styrene-sulphonate)-based electrolytes plasticised with polyethyleneglycol. Recently, Zhou *et al.*[10] described poly(siloxanes) with pendant oligo-oxyethylene chains and sulphonate groups. On the other hand, polycationic supports with only one mobile anion were reported by Watanabe *et al.*[11] and Hardy & Shriver.[9,12]

Nevertheless, many of these structures display high T_g, so that the addition of low molecular weight polymers (PEO oligomers) or molecules (acetamide, propylene carbonate) is required as plasticisers in order to reach an appreciable level of conductivity. Moreover, incompatibility and therefore non-mixing problems may occur, owing to the great difference in chemical structure between PEO and ion-bearing polymers, which are other heterogeneity factors. Comb-polymers were also prepared in an attempt to separate the grafted anion from the backbone by long PEO side chains, but this way does not prevent the non-mixing of the main chains and therefore may be another heterogeneity factor.

This paper deals with the grafting of sulphonate groups at both ends of short PEO chains by deactivation of PEO alcoholates with propane sulphone.[13] This structure was chosen in an attempt to reduce the incompatibility problems which may be induced by using other kinds of polymer backbones. In addition, the sulphonate anion is a very poor nucleophilic agent, so that the grafted anion is not able to open another propane sulphone. Another advantage of these telechelic POE is to bring by themselves the ether oxygen atoms required for the cation complexation, so that the monomer to salt ratio may be expected to be kept around a constant value, without a dilution effect.

These telechelic PEO were added to an electrolytic medium, the amorphous domains of which had been highly enhanced by using a semi-interpenetrated network comprised of a high molecular weight PEO, an elastomer and a styrenic macromonomer of PEO (SEO).[14–16]

$$SEO : CH_2\!\!=\!\!CH\text{-}\phi\text{-}CH_2\text{-}O\text{---}(CH_2CH_2O)_n CH_3$$

A Li/SPE/MnO$_2$ solid-state battery was also constructed. Both electrolytes and battery were characterised by the usual a.c. measurement procedures. The results were discussed in terms of transfer of the ionic species inside the bulk electrolyte.

EXPERIMENTAL SECTION

(a) Synthesis of the Telechelic POE

Sulphonate end-capped poly(oxyethylenes) are synthesised by reacting the propane sulphone with the potassium or lithium alcoholates in THF. The experiments are carried out under an argon atmosphere in previously flamed and argon-purged flasks equipped with rubber septums. THF is distilled on the sodium-benzophenone complex. POE ($M_w = 1500$) is dried by azeotropic benzene distillation, lyophilised and then dissolved in THF. Propane sulphone is distilled under vacuum just prior to use and transferred to the reaction flask by means of flamed stainless-steel capillaries.

The PEO is first reacted with the alkali dihydronaphthylide, previously obtained by reaction of lithium or potassium with naphthalene. The resulting alcoholates are then deactivated with the propane sulphone, according to the ring-opening reaction described in Scheme 1.

The nature of the cation bound to the sulphonate groups is determined by the counterion (Li$^+$ or K$^+$) of the naphthylide anion

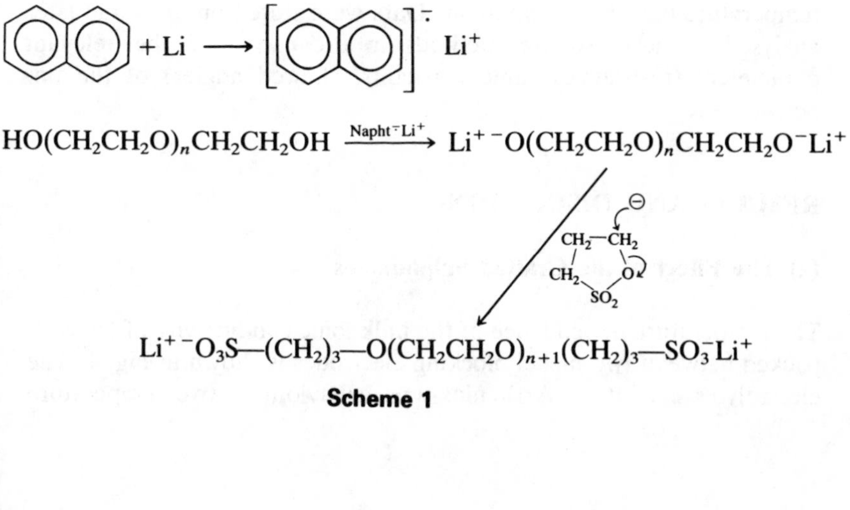

$$\text{[naphthalene]} + \text{Li} \longrightarrow \text{[naphthalene]}^{\bar{\ }}\ \text{Li}^+$$

$$\text{HO(CH}_2\text{CH}_2\text{O)}_n\text{CH}_2\text{CH}_2\text{OH} \xrightarrow{\text{Napht}^{\bar{\ }}\text{Li}^+} \text{Li}^+\ ^-\text{O(CH}_2\text{CH}_2\text{O)}_n\text{CH}_2\text{CH}_2\text{O}^-\text{Li}^+$$

$$\text{Li}^+\ ^-\text{O}_3\text{S}\text{—(CH}_2)_3\text{—O(CH}_2\text{CH}_2\text{O)}_{n+1}\text{(CH}_2)_3\text{—SO}_3^-\text{Li}^+$$

Scheme 1

Table 1. Composition (% in weight) of SPE with PEO bearing sulphonate end groups. PEO/SEO = 40/60 in weight. Elastomer = 10·5%

SPE No.	1	2	3	4	5
PEO + SEO	100	100	100	100	100
PEO(SO₃Li)₂		30		15	
PEO(SO₃K)₂			30		
CF₃SO₃Li	30			15	15
[OE]/[M⁺]	11	77	79	20·8	21·5

radical. Due to its marked ionic character, the polymer is recovered as a powder by evaporation of the solvent. It cannot be dissolved by aromatic solvents, so that naphthalene residues can be easily removed by washing. The polymer is finally dried under vacuum. The sulpho-end groups are characterised by elemental analysis and by IR spectroscopy.

(b) Characterisation of the Solid Polymer Electrolytes

The electrolytes were obtained and characterised according to procedures already described.[15,16] The composition of the electrolytes are reported in Table 1. The impedance measurements were carried out using a Transfer Function Analyser connected to a 1186 Electrochimie Interface Schlumberger. The experiments were monitored from a microcomputer. The samples are allowed to stay for 1 h at each temperature before measurement. Data were stored on disks and later analysed by least square methods in order to find the relevant parameters (resistances, time constants, skewed angles) of the impedance spectra.

RESULTS AND DISCUSSION

(a) The Effect of the Grafted Sulphonates

The temperature dependence of the bulk ionic conductivity of the SPE packed between two nickel blocking electrodes is shown in Fig. 1. The electrolytes exhibit an Arrhenius-type behaviour in two temperature

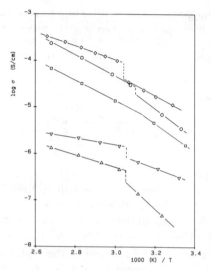

Figure 1. Arrhenius plots of the bulk conductivity for solid electrolytes with grafted sulphonato-end groups, inserted between two nickel electrodes. SPE No. 1 (○); 2 (△); 3 (▽); 4 (◇); 5 (□). The compositions are given in Table 1.

ranges, separated by a discontinuity which has already been described and usually attributed to the recrystallisation of the PEO amorphous phase.[15,17]

At low temperatures, the substitution of the triflate anion by the grafted sulphonate (SPE No. 2, 3) does not change in an appreciable way the level of conductivity: the conductivity is similar in magnitude to that of PEO-triflate,[17] although there are no fluorine atoms in the grafted sulphonate moiety, contrary to the triflate anion, so that a higher degree of ion-pairing has to be expected. Around the same performances were also reported with the grafted sulphonates on a poly(siloxane) backbone prior to adding tetraethyleneglycol.[10]

Higher performances are noticed when using the potassium cation instead of the lithium cation. Hibma attributed this effect to the larger cationic radius which limits the degree of crystallisation.[18] The same results have been described by Rietman et al.[19] who ascribed this effect to a more pronounced complexation of the small Li$^+$ cation by the PEO ether oxygens. The same effect was reported by Zhou et al. and was associated with the weakening of the cation–sulphonate interaction.[20]

Table 2. Activation energy (eV) of the ionic conductivity for low and high temperature ranges. Columns 6 and 7 refer to electrolytes containing lithium and potassium trifluorosulphonates[19] respectively, for which the monomer to salt ratio is equal to 10

SPE No.	1	2	3	4	5	6	7
Low temp.	0·63	0·79	0·32	0·43	0·50	1·32	1·78
High temp.	0·45	0·26	0·16	0·32	0·43	0·41	0·45

At low temperature, the activation energies E_a (Table 2) are smaller than those reported for electrolytes containing lithium and potassium triflate.[19] This effect has been attributed to the macromonomer, which increases the amorphous domain and promotes the ion transport through the bulk electrolyte.[15,16] This is also supported by the activation energies which are observed in the high temperature range for the SPE No. 1, 5, 6, 7: above the melting zone, the same E_a value around 0·43 eV is retrieved, that means that the macromonomer no longer promotes the ion transport. The great difference between the SPE No. 2 (POE–SO$_3$Li: $E_a = 0·70$ eV) and No. 3 (POE–SO$_3$K: $E_a = 0·32$ eV) could be related to various crystallisation and morphological states induced by the different cations, and also to aggregation phenomena leading to clusters which are always encountered with ionic polymers.[22] These clusters may also explain the very poor conductivity above the PEO melting zone. Adding a more polar component will certainly break these associations and yield better results, as well as the optimisation of the monomer to salt ratio.

In the same low temperature range, the impedance spectra reveal another high frequency semicircle which has been shown to be correlated with the ionic transport processes occurring inside the bulk electrolyte.[16,21] The related time constants $\tau = (2\pi f_c)^{-1}$ are displayed in Fig. 2. In the framework of the ionic transport, τ has to be interpreted as the inverse of the rate constant of the exchange processes between the complexation sites, and therefore may be correlated with the level of conductivity: the highest τ, the lowest σ.

(b) The Blends of Free and Grafted Anions

The conductivity plots related to the ESP Nos 4 and 5 show that adding lithium triflate to the grafted lithium sulphonate causes better

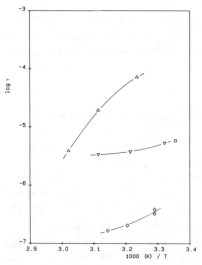

Figure 2. Temperature dependence of the time constants related to the high frequency process (inside the bulk electrolyte). Same symbols as in Fig. 1.

performances than only triflate or grafted sulphonate: the conductivity level is enhanced by half a decade with respect to that displayed by electrolytes which contain triflate. This increase cannot be related to a lithium concentration effect since the monomer to salt ratio was kept at the same value, around 21. The impedance spectra analysis show that it is connected to a lower value of the high frequency relaxation time, which corresponds to a higher rate constant of the transfer process. In the same way, the high frequency resistance is drastically decreased from 631 kΩ (SPE No. 2) to 1 kΩ (SPE No. 4) at room temperature. The activation energy in the low temperature range is also decreased with respect to those displayed by the SPE Nos 2 and 5.

This effect could be simply explained by looking at an additional anionic conductivity due to the trifluorosulphonate anion. It has also been noted previously that the telechelic oligomers change from a viscous liquid state to the solid state upon reaction with propane sulphone. This fact suggests ionomeric behaviour[22] where the lithium counterions are clustering. Then, one also may assume that there are two kinds of lithium populations. The first one would be constituted by 'free' lithium cations, while the other one would include all the clustered cations. In addition, clusters may be in various sizes. Exchange between both populations may occur and be important in the transport process. Under the electrical field, the free lithium salt, more mobile than the lithium in the clusters, would act as some bridge between the clusters, so that the current will pass more easily in the electrolyte.

(c) The Li/SPE/MnO₂ Solid-State Battery

Some Li/SPE/MnO$_2$ batteries with the SPE No. 4 have been constructed and are now under study by a.c. measurements. The measurements were carried out in a galvanostatic mode with a bias voltage of 3·25 V. Figures 3(a) and 3(b) display the complex impedance spectra of such a cell at 23°C and 115°C. The low temperature impedance spectrum shows two semicircles followed by a low frequency straight line. As described above, the high frequency semicircle (f_c = 41 kHz) involves relaxation processes which are due to ionic transport in the bulk electrolyte. The low frequency semicircle (f_c = 52 Hz) is associated with processes occurring at the electrode interfaces. Using a V$_6$O$_{13}$ insertion electrode, Weston & Steele[23] attribute this semicircle to an interfacial impedance between the electrolyte and the V$_6$O$_{13}$ electrode. At low temperature, the MnO$_2$ electrolyte displays a relaxation semicircle around 59 Hz.[16] However, the relaxation processes occurring at the Li–electrolyte interface may also appear in the same frequency range ($f_c \approx$ 100 Hz).[15] Therefore, R_1 is associated with the electrolyte resistance and ($R_2 - R_1$) with an interfacial transfer resistance. The straight line is ascribed to the diffusion processes of Li$^+$ inside the cathodic porous material.

At high temperature, both high and low frequency semicircles are overlapping and the first one can no longer be easily analysed. The analysis of these impedance spectra are currently under study and results will be published in a later paper.

CONCLUSIONS

The reaction of short PEO chains with the propane sulphone has been used in order to get sulphonate end-capped poly(oxyethylene). These telechelic polymers are suitable materials for the elaboration of solid polymer electrolytes displaying a single-ion conduction. Nevertheless, it has been found that the best electrochemical performances were

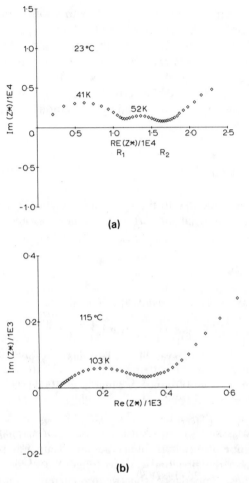

Figure 3. Impedance spectra of a Li/SPE No. 4/MnO$_2$ solid-state battery.
(a) 23°C; (b) 115°C.

obtained by using these grafted sulphonates in conjunction with the usual lithium trifluorosulphonate. The analysis of the ion transport in the bulk electrolyte clearly shows that both conduction resistance and time constant are decreased in this latter type of electrolyte. Finally, a Li/SPE/MnO$_2$ solid-state cell has been constructed, whose characteristics are currently under study.

Better conductivity may be expected with another monomer to salt ratio and also with another PEO chain-length between both sulphonate-end groups. It is worth underlining the question of clustering: these clusters may also be seen as physical crosslinking points and a balance must be found between the 'free' Li^+ required for the ionic conductivity and the 'crosslinking' Li^+. With that aim, the use of polar components will certainly be helpful. All these points are now under investigation.

ACKNOWLEDGEMENTS

The authors are indebted to the Compagnie Générale d'Electricité and to the CCE for financial support and technical assistance.

REFERENCES

1. Armand, M., *Solid State Ionics*, **9, 10** (1983) 745.
2. MacDonald, J. R., *J. Chem. Phys.*, **61** (1974) 3977.
3. Bannister, D. J., Davies, G. R., Ward, I. M. & McIntyre, J. E., *Polymer*, **25** (1984) 129.
4. Kobayashi, N., Uchimaya, M. & Tsuchida, E., *Solid State Ionics*, **17** (1985) 307.
5. Tsuchida, E., Kobayashi, N. & Ohno, H., *Macromolecules*, **21** (1988) 96.
6. Shigehara, K., Kobayashi, N. & Tsuchida, E., *Solid State Ionics*, **14** (1984) 85.
7. Gandini, A., Le Nest, J. F., Levêque, M. & Cheradame, H., In: *Integration of Fundamental Polymer Science and Technology*, ed. L. A. Kleintjens & P. J. Lemstra. Elsevier, Amsterdam, 1986, p. 250.
8. Le Nest, J. F., Gandini, A., Cheradame, H. & Cohen-Addad, J. P., *Polym. Commun.*, **28** (1987) 302.
9. Hardy, L. C. & Shriver, D. F., *J. Am. Chem. Soc.*, **107** (1985) 3823.
10. Zhou, G., Khan, I. M. & Smid, J., *Polym. Commun.*, **30** (1989) 52.
11. Watanabe, M., Nagaoka, K., Kamba, M. & Shinohara, I., *Polym. J.*, **14** (1982) 877.
12. Hardy, L. C. & Shriver, D. F., *Macromolecules*, **17** (1984) 975.
13. Sepulchre, M., Paulus, G. & Jérôme, R., *Makromol. Chem.*, **184** (1983) 1847.
14. Le Méhauté, A., Crépy, G., Marcellin, G., Hamaide, T. & Guyot, A., *Polym. Bull.*, **14** (1985) 233.
15. Carré, C., Hamaide, T. & Guyot, A., *Brit. Polym. J.*, **20** (1988) 269.
16. Hamaide, T., Le Méhauté, A., Crépy, G., Marcellin, G. & Guyot, A., *J. Electrochem. Soc.*, **136** (1989) 3152.

17. Robitaille, C. D. & Fauteux, D., *J. Electrochem. Soc. Electrochem. Sci. Technol.*, **133** (1986) 315.
18. Hibma, T., *Solid State Ionics,* **9, 10** (1983) 1101.
19. Rietman, E. A., Kaplan, M. L. & Cava, R. J., *Solid State Ionics*, **25** (1987) 41.
20. Zhou, G., Khan, I. M. & Smid, J., 2nd Int. Symp. on Polymer Electrolytes. Extended Abstracts. Siena, 14–16 June, 1989.
21. Weston, J. E. & Steele, B. C. H., *Solid State Ionics*, **7** (1982) 75, 81.
22. (a) Russel, T. P., Jérome, R., Charlier, P. & Foucart, M., *Macromolecules*, **21** (1988) 1709. (b) Fetters, L. J., Graessley, W., Hadjichristidis, N., Kiss, A. D., Pearson, D. S. & Younghouse, L. B., *Macromolecules*, **21** (1988) 1644.
23. Weston, J. E. & Steele, B. C. H., *Solid State Ionics,* **9, 10** (1983) 391.

Elastomeric Ionic Conductor from Poly(Oxyethylene) Grafted Poly(Siloxane) Doped with Lithium Perchlorate

Shinzo Kohjiya, Tsuneo Kawabata, Kazuyuki Maeda, Shinzo Yamashita

Department of Material Science, Kyoto Institute of Technology, Matsugasaki, Kyoto 606, Japan

&

Yutaka Shibata

Research and Development Center, Sumitomo Electric Industries, Shimaya, Osaka 554, Japan

ABSTRACT

As a matrix for polymer solid electrolytes poly(siloxane)-g-poly(oxyethylene) was synthesised. The copolymer was prepared by hydrosilylation of macro-monomers, i.e. poly(oxyethylene) carrying a vinyl group. Poly(methyl hydrogen siloxane)s were used as silylation agents. The copolymer was doped with lithium perchlorate, and a very good ionic conductivity ($2\cdot1 \times 10^{-4}\,\mathrm{S\,cm^{-1}}$ at 30°C) was observed.

The temperature dependency of the conductivity followed the WLF type equation, and not the Arrhenius type. A battery made from this ionic conductor showed good discharge characteristics at room temperature.

INTRODUCTION

Polymeric solid electrolytes[1] (PSE) have been paid much attention due to the importance of them for electronic devices, especially chemical batteries. PSE is expected to be of light weight and easily formed into thin films. We can fabricate thin and light-weight batteries of high performances, once we get an excellent PSE having high conductivity and good durability as well as feasibility to industrial processes for production.

Elastomeric solid electrolytes have a few more advnatages over inorganic and polymeric ones, the glass transition temperatures (T_g) of which are higher than room temperature, in addition to the advantages of solid electrolytes in general. Suitable elasticity can result in flat, thin and flexible solid electrolytes, which would be very useful material to manufacture thin and flexible paper-like batteries and electrochromic displays. Suitable elasticity can also give excellent contact between an electrolytic layer and an electrode in chemical batteries. The authors had been working on elastomeric materials, and made an estimation that an elastomer could be the best choice as a matrix for ion conduction, because the rubbery state is actually a liquid from a rheological view point even though it is classified as solid from its appearance.[2] In other words, an elastomer is a polymeric 'solvent', and a higher conductivity of ions is expected in elastomeric matrices than in the glassy or crystalline states of polymers. Because all the ions are much larger than the electron which is a carrier in electronic conductors, mobility in the matrix has to be considered first.

In this paper poly(siloxane) is chosen as a matrix for ionic conductors. Poly(siloxane)s are known to be one of the best elastomers due to their lowest T_g among polymers and their excellent thermal and chemical stabilities.[3] However, silicone elastomers do not accommodate ions because they are relatively non-polar. Many fillers and other ingredients have been compounded into silicone elastomers, but they are simply dispersed, not dissolved into the rubbery matrix. Therefore, some modification is necessary in order to use poly(siloxane) as an electrolytic matrix.

Nagaoka *et al.*[4] reported the results on the synthesis of poly(siloxane-co-oxyethylene) where both siloxane units and oxyethylene units formed the polymeric main chains. However, the conductivity of their composites with lithium perchlorate was not very encouraging, and the polymer matrix was subject to hydrolysis due to the presence of silicon–oxygen–carbon bonds. After consideration of molecular designs of PSE matrix, we chose a chemical modification of poly(siloxane) by poly(oxyethylene) (PEO) to produce a graft copolymer whose main chain is composed of poly(siloxane) and the side chains are poly(oxyethylene)s; the hydrophilic PEO side chains afford solubility of ions though the main chain is poly(siloxane), and the flexible main chain, i.e. a polymeric liquid, affords a good mobility to carrier ions. Actually, there appeared several reports[5] on the conductivity of poly(oxyethylene) grafted poly(siloxane)s. The grafting is essential to prevent the phase separation of PEO segments from poly(siloxane) matrix. However, so far published results seem to have been on the graft copolymers, the grafting points of which were of silicon–oxygen–carbon bonds. The Si—O—C bonds are well known to be susceptible to moisture and result in bond scission.[6] In order to obtain more stable materials, we attempted to synthesise graft copolymers consisting of silicon–carbon bonds. This objective was realised by hydrosilylation of PEO macromonomers using poly(methyl hydrogen siloxane)s as silylation agents.[7] The synthetic route is shown in Fig. 1.

EXPERIMENTAL

Materials

Following the synthetic route of PEO grafted poly(siloxane) shown in Fig. 1, the macromonomers were prepared from polyethylene glycol or its monomethyl ether and methyl acrylate using sulphuric acid as an ester exchange catalyst.[7] The hydrosilylations were carried out in toluene, catalysed by hexachloroplatinic acid (the Speier's catalyst).[6–8]

Sample Preparation

Toluene solution of the graft copolymer was mixed with lithium perchlorate dissolved in tetrahydrofuran, and cast onto a Teflon mould to prepare an elastomeric solid electrolyte film which was subject to drying at 80°C under vacuum for 4 days.

Analytical Measurements

Proton NMR was measured on a Varian T-60A NMR Spectrometer in deuterated chloroform. Infrared spectra were obtained on a Hitachi

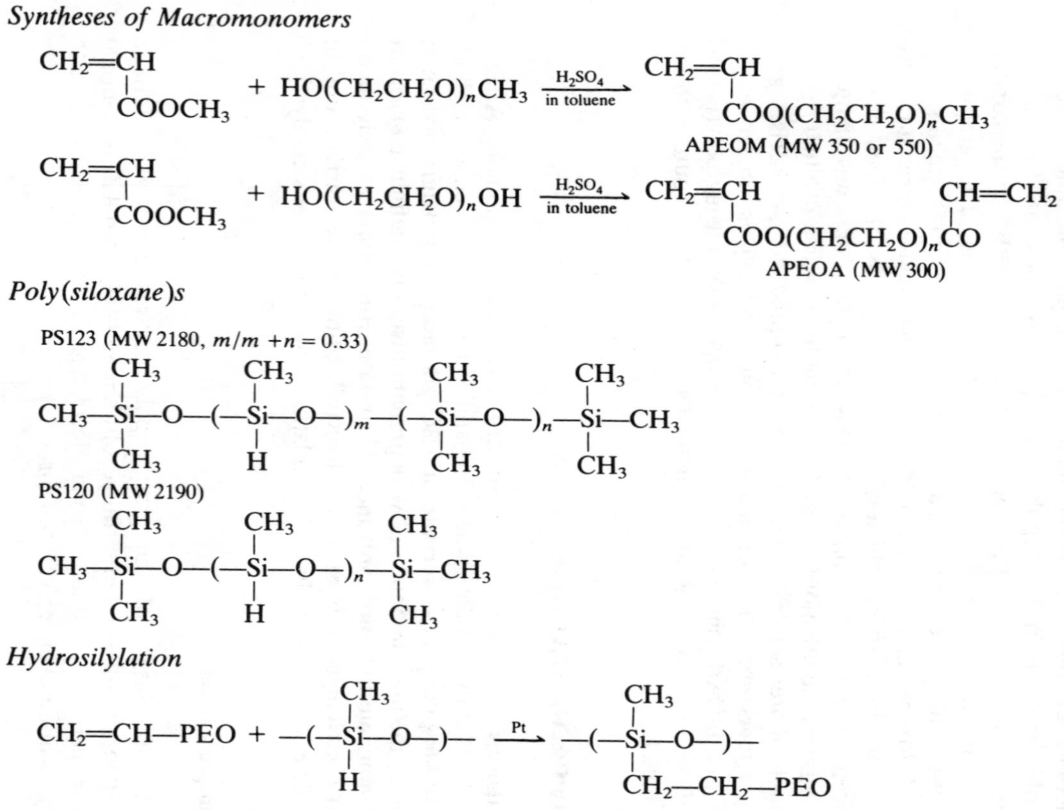

Syntheses of Macromonomers

$$CH_2=CH \atop |COOCH_3} + HO(CH_2CH_2O)_nCH_3 \xrightarrow[\text{in toluene}]{H_2SO_4} {CH_2=CH \atop |COO(CH_2CH_2O)_nCH_3}$$

APEOM (MW 350 or 550)

$$CH_2=CH \atop |COOCH_3} + HO(CH_2CH_2O)_nOH \xrightarrow[\text{in toluene}]{H_2SO_4} {CH=CH_2 \atop |COO(CH_2CH_2O)_nCO \atop |CH_2=CH}$$

APEOA (MW 300)

Poly(siloxane)s

PS123 (MW 2180, $m/m+n = 0.33$)

$$CH_3-Si-O-(-Si-O-)_m-(-Si-O-)_n-Si-CH_3$$

PS120 (MW 2190)

$$CH_3-Si-O-(-Si-O-)_n-Si-CH_3$$

Hydrosilylation

$$CH_2=CH-PEO + -(-Si-O-)- \xrightarrow{Pt} -(-Si-O-)- \atop |CH_2-CH_2-PEO}$$

Figure 1. Synthetic route to poly(oxyethylene) grafted poly(siloxane)s.

215 Grating IR Spectrophotometer. The sample film was subject to differential scanning calorimetry (DSC) measurements using a Seiko DSC 20. The temperature rise was adjusted to 10°C/min.

Electrical Conductivity

Polymer films doped with lithium perchlorate were treated in a dry box under a pure nitrogen stream to make up the conductivity cells using platinum electrodes. Electrical conductivity was evaluated by the complex impedance method using Yokogawa–Hewlett Packard LF Impedance Analyzer 41920A. The cell was placed in a constant-temperature bath, and the complex impedance was measured after thermal equilibrium was reached. The frequency was between 5 Hz and 1000 kHz.

RESULTS AND DISCUSSION

Figure 2 shows proton NMR spectra of two macromonomers, APEOA 300 and APOEM 350, where 300 and 350 are the molecular weights of PEO segment, A is acrylate and M is a methyl ether terminal. After the hydrosilylation of the macromonomer, the olefinic protons (*c.* 6 ppm) disappeared. Also, Si–H protons (4·7 ppm) in poly(methyl hydrogen siloxane)s were not detected. Infrared spectra indicated the

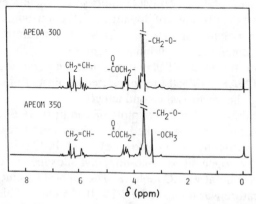

Figure 2. Proton NMR spectra of the macromonomers.

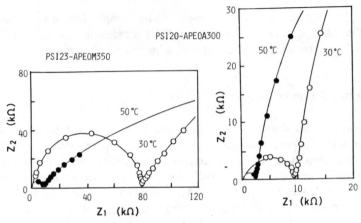

Figure 3. The Cole–Cole plots.

disappearance of Si–H $(2220\,\text{cm}^{-1})$ and vinyl groups (1690, $1670\,\text{cm}^{-1}$). These results suggest almost quantitative hydrosilylation reactions under the present conditions.

Figure 3 shows the Cole–Cole plots for PS120-APEOA300 and PS123-APEOM350. AC impedance was measured in the frequency range of 5–1000 Hz using Pt, i.e. blocking electrodes. The sample thickness was between 0·4 and 0·6 mm.

The optimal amount of $LiClO_4$ was found to be fairly low as suggested from the results in Fig. 4. Here, conductivity was plotted against $LiClO_4$ concentration relative to oxyethylene units. The presence of hydrophilic oxyethylene units surely facilitates dissociation of the salt, but the presence of dopant $LiClO_4$ contributes to increasing the T_g of the polymer matrix to a higher temperature. From these effects, the optimal value of lithium perchlorate content was observed.

To illustrate the effect of glass-transition temperature, the conductivity of PS120APEOM550s were plotted against T_g in Fig. 5. The higher the T_g, the lower is the conductivity.

Figure 6 shows the Arrhenius plots of conductivity of five samples. APEOA samples were crosslinked, the others were linear. PS120 is a homo poly(methyl hydrogen siloxane), and PS123 is a copolymer poly(methyl hydrogen siloxane-co-dimethyl siloxane) whose hydrogen siloxane units content was 33 mole %. The highest conductivity around room temperature was found to be $2·1 \times 10^{-4}\,\text{S cm}^{-1}$. This value is one of the highest conductivities so far reported on PSEs.[5]

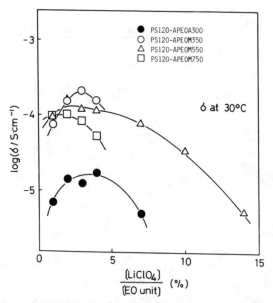

Figure 4. Relationship between conductivity at 30°C and [LiClO$_4$] relative to oxyethylene units in the matrix.

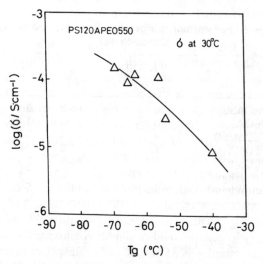

Figure 5. Relationship between glass-transition temperature and conductivity at 30°C for PS120-APEOM550.

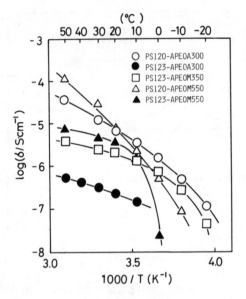

Figure 6. Temperature dependency of the conductivities.

Table 1. The WLF Parameters from the Temperature Dependency of Conductivity

Sample	$\dfrac{[\text{LiClO}_4]}{[\text{EO unit}]}$	T_g (°C)	C_1	C_2 (°C)
PS120-APEOA300	3	−64·8	5·1	70·6
PS123-APEOA300	3	−70·5	1·4	60·8
PS123-APEOM350	3	−77·4	2·7	24·7
PS120-APEOM550	7	−54·0	9·6	114·3
PS123-APEOM550	3	−66·8	2·7	5·1
Poly(oxypropylene)	0·042	−1·0	5·5	46·1
Poly(oxypropylene)	0·076	11	5·2	31·1

The WLF equation:

$$\log \frac{\sigma(T)}{\sigma(T_g)} = \frac{C_1(T - T_g)}{C_2 + (T - T_g)}$$

The values for poly(oxypropylene) are taken from Ref. 9 and [LiClO$_4$] was relative to oxypropylene units.

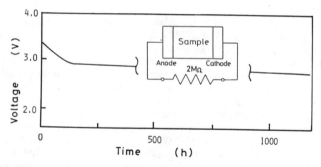

Figure 7. Discharge versus time curve of the battery.

The temperature dependence of conductivity in Fig. 6 suggests that the dependency was not of Arrhenius type but WLF type. The curves in the figure could be fully simulated by the WLF equation, indicating that the free volume is important to determine the mobility of carrier ions. Both the conductivity and T_g were influenced by PEO content in the polymer matrix and the amount of doped lithium perchlorate. These results are in conformity with the WLF type temperature dependency. Table 1 shows the C_1 and C_2 values obtained from the simulation of the experimental results shown in Fig. 6 together with those for poly(oxypropylene).[9]

Figure 7 shows the discharge characteristics of the battery composed of the present elastomeric electrolyte and metallic lithium (cathode) and fluorinated carbon (anode) electrodes. This is only a preliminary experiment, but the characteristics are found to be fairly good.

REFERENCES

1. MacCallum, J. R. & Vincent, C. A., *Polymer Electrolyte Reviews.* Elsevier Applied Science, London, 1978.
2. Lodge, A. S., *Elastic Liquid.* Academic Press, London, 1960.
3. Lynch, W., *Handbook of Silicone Rubber Fabrication.* Van Nostrand Reinhold, New York, 1978.
4. Nagaoka, K., Naruse, H., Shinohara, I. & Watanabe, M., High ionic conductivity in poly(dimethyl siloxane-co-ethylene oxide) dissolving lithium perchlorate. *J. Polym. Sci.: Polym. Lett. Ed.*, **22** (1984) 659–63.
5. (a) Bannister, D. J., Doyle, M. & Macfarlane, D. R., A water-soluble siloxane: poly(ethylene glycol) comb polymer. *J. Polym. Sci.: Polym. Lett.*

Ed., **23** (1985) 365–7. (b) Bouridah, A., Dalard, F., Deroo, D., Chera-dame, H. & Le Nest, J. F., Poly(dimethylsiloxane)-poly(ethylene oxide) based polyurethane networks used as electrolytes in lithium electrochemical solid state batteries. *Solid State Ionics*, **15** (1985) 233–40. (c) Fish, D., Khan, I. M. & Smid, J., Conductivity of solid complexes of lithium perchlorate with poly((-methoxyhexa(oxyethylene)ethoxy)methyl siloxane). *Makromol. Chem. Rapid Commun.*, **6** (1986) 115–20. (d) Hall, P. G., Davies, G. R., McIntyre, J. E., Ward, I. M., Bannister, D. J. & Le Brocq, K. M. F., Ion conductivity in polysiloxane comb polymers with ethylene-glycol teeth. *Polymer*, **27** (1986) 98–100. (e) Watanabe, M., Nagaoka, S., Sanui, K. & Ogata, N., Structure–conductivity relationship in polymer electrolytes formed by network polymers from poly(dimethylsiloxane-g-poly(ethylene oxide)) and lithium perchlorate. *J. Power Sources*, **20** (1987) 327–32.

6. Noll, W., *Chemistry and Technology of Silicone*. Academic Press, New York, 1968, p. 83.
7. Kohjiya, S., Maeda, K. & Yamashita, S., Polymer solid electrolytes from polyethylene oxide grafted poly(siloxane). *Polymer Preprints, Japan*, **35** (1986) 778 (in Japanese).
8. Kohjiya, S., Maeda, K., Yamashita, S. & Shibata, Y., Chemical modification of silicone elastomers for optoelectrical applications. Paper No. 20 presented at the Rubber Division 134th Meeting, American Chemical Society. Cincinnati, OH, October, 1988.
9. Watanabe, M., Sanui, K., Ogata, N., Kobayashi, T. & Ohtaki, Z., Ionic conductivity and mobility in network polymers from poly(propylene oxide) containing lithium perchlorate. *J. Appl. Phys.*, **57** (1985) 123–8.

Ambient Temperature Conductivity of Immobilised Liquid Membrane Electrolyte

K. Koseki, K. Saeki, T. Itoh

Corporate Research & Development Laboratory, Tonen Corporation, Ohi-machi Iruma-gun, Saitama Pref. 354, Japan

Cao Qi Juan

Institute of Physics, Beijing, People's Republic of China

&

O. Yamamoto

Department of Chemistry, Faculty of Engineering, Mie University, Kamihama, Tsu, 514, Japan

ABSTRACT

A new concept was introduced to design a polymer electrolyte having an appreciable ionic conductivity at an ambient temperature. Incorporation of liquid polyoxyalkenes complexed with $LiClO_4$ in a microporous polyethylene membrane having a mean pore radius of $0 \cdot 01 \, \mu m$ was carried out by taking advantage of capillary condensation forces. Surface tension and contact angle measurements suggested that the polymer electrolyte was immobilised in the pore structure. A gas permeability test confirmed the result. The immobilised liquid membrane electrolyte showed an ionic conductivity higher than $10^{-5} \, S \, cm^{-1}$ and an effective resistivity less than $100 \, ohm \, cm^2$.

INTRODUCTION

Considerable studies have been done on polymer-based solid electrolyte from viewpoints of both fundamental and applied sciences.[1-10] High ionic conducting solid polymers have been reported reaching around $10^{-5}\,S\,cm^{-1}$ at ambient temperature. However further improvement is desirable for applications such as batteries and electrochromic devices. In connection with the membrane studies, attempts were made in our laboratories to prepare a membrane electrolyte having a high ionic conductivity substantially in a solid state. We have proposed applying the concept of immobilisation where a polymeric liquid electrolyte is incorporated into the micropores of a membrane to make a quasi-solid state electrolyte.[11] In the present study, a low molecular weight liquid PEO/PPO complexed with $LiClO_4$ was immobilised in a microporous polyethylene membrane having a thickness of $10\,\mu m$ to prepare the immobilised liquid membrane electrolyte (ILME). The surface chemistry of the immobilisation and the comparison of ionic conductivity between the ILME and the polymer liquid electrolyte are described.

EXPERIMENTAL

Poly(oxyethylene, oxypropylene) glycol monoethers with average molecular weights of about 240 and 540 (PEO/PPO240 and 540, Sanyo Kasei), polyoxypropylene glycol monoethers with average molecular weights of about 340 (PPO340, Sanyo Kasei) and $LiClO_4$ (Aldrich) were used to prepare the polymer-based electrolytes. The porous polyethylene membranes having a thickness of $10\,\mu m$ were prepared from polyethylene. The preparation procedures were described elsewhere.[12] According to the preparation procedure, a polyethylene membrane having a mean pore radius around $0\cdot01\,\mu m$ and a porosity of about 80% was obtained. Immobilisation was carried out by immersing the porous polyethylene membranes in the polymer-based electrolyte solutions. In order to confirm uniform immobilisation in the pore structure of the membrane, volumetric gas permeation measurements were carried out. Oxygen or nitrogen gas was applied at a gauge pressure of $10\,kg\,cm^{-2}$, to one side of a circular ILME sample having a diameter of $47\,mm$, and then the volume of gas permeated through ILME was measured. Contact angle and surface tension measurements were made according to the Wilhelmy method and the liquid drop method (CA-Z and CBVP-A3, Kyowa Kaimenkagaku), respectively. The a.c. conductivity measurements were carried out over a frequency range of 0·1 Hz to 100 kHz using FRA (S-5720C, NF Electronic Instruments) and potentiostat/galvanostat (2000, Toho Giken). The ILME samples for the conductivity measurements were circular pieces having a diameter of 15 mm pressed between a pair of polished stainless steel electrodes. Experiments at ambient temperature have been done in a glove-box via hermetic signal terminals under an argon atmosphere at a 10 ppm moisture level.

RESULTS AND DISCUSSION

Immobilisation of a polymer electrolyte into micropores of a poly-ethylene membrane is greatly affected both by the properties of the polymer electrolyte solution and by the properties of the pores, such as mean pore radius, porosity, pore shape and pore size distribution. The force which holds a polymer electrolyte solution within a capillary is approximately represented by the Young–Laplace equation,

$$P = 2\gamma \cos \theta / r \qquad (1)$$

where P is the capillary condensation pressure of the liquid in the pore, γ is the surface tension, θ is the contact angle, and r is the radius of the cylindrical pore. Table 1 shows the properties of the immobi-lised liquid membrane described here. The high P value means that the system of the microporous polyethylene membrane and the PEO/PPO/LiClO$_4$ solution is immobile.

The gas permeation data confirmed that the polymer electrolyte solution was held strongly in the membrane pores as expected by eqn (1).

Table 1. Surface tension and contact angle data and capillary condensation forces calculated according to eqn (1)

Polymer	γ(dyne/cm)	θ(°)	P(kg/cm^2)
PEO/PPO240	32·1	19·3	61·4
PEO/PPO540	34·6	22·4	65·3
PPO340	30·1	17·0	58·7

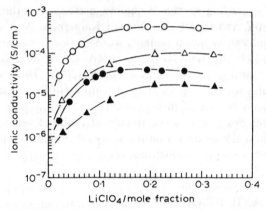

Figure 1(a). Ionic conductivity as a function of concentration for LiClO₄ in PEO/PPO at 25°C. ○, PEO/PPO240 solution; △, PEO/PPO540 solution; ●, PEO/PPO240 ILME; ▲, PEO/PPO540 ILME.

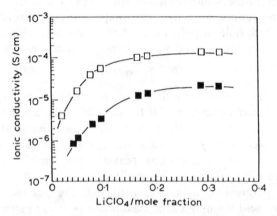

Figure 1(b). Ionic conductivity as a function of concentration for LiClO₄ in PPO at 25°C. □, PPO340 solution; ■, PPO340 ILME.

Relations between ionic conductivity and concentration of LiClO₄ in PEO/PPO are shown in Fig. 1.

As expected from the conductivity data for the solutions of PEO/PPO or PPO and LiClO₄, the higher ILME conductivity was obtained using the lower molecular weight PEO/PPO or PPO. The results on the PPO340 solution were consistent with the previous results obtained by M. Watanabe *et al.* for PPO solutions having molecular weights 420–2880.[9] The ionic conductivity of ILME increased with the $LiClO_4$ salt concentration and reached $4 \times 10^{-5} \, S \, cm^{-1}$. This increase in the ionic conductivity can be attributed to the increase in the carrier ion concentration in polymer electrolyte. At higher $LiClO_4$ concentrations the ionic conductivity reached a plateau, suggesting that the deceleration in the ionic movement resulting from the viscosity increase began to dominate. The importance of mobility factor in the ILME system is also confirmed by the present results which show that the ionic conductivity can be increased with lower molecular weight polymer electrolyte solution. It is important to note that the change in the ionic conductivity of ILME as a function of concentration for $LiClO_4$ in PEO/PPO or PPO was similar to that of the polymer electrolyte solution, approximately one order of magnitude lower level than that of the polymer solution. The similarity in the curve might suggest that an ILME behaves like a liquid electrolyte in the ion transport. The lower ionic conductivity of ILME was considered to be attributed to the geometrical blocking of solvated ion diffusion by the polyethylene network rather than the interaction between ion carriers and polyethylene matrix. Further studies would be needed in this case. In conclusion, by the use of the ILME concept it would be possible to design a polymer electrolyte which makes solid-like handling possible and gives a liquid-like ionic conductivity.

REFERENCES

1. Armand, M. B., Chabagno, J. M. & Duclot, M. T. In: *Fast Ion Transport in Solids*, ed. P. Vashishta, J. N. Mundy & G. K. Shennoy. North-Holland, Amsterdam, 1979, p. 131.
2. Shriver, D. F. & Farrington, G. C., Solid ionic conductors. *Chem. Engng News* (May 20, 1985) 42–57.
3. Watanabe, M., Nagano, S. Sanui, K. & Ogata, N., Ion conduction mechanism in network polymers from poly(ethylene oxide) and poly(propylene oxide) containing lithium perchlorate. *Solid State Ionics*, **18, 19** (1986) 338.
4. Scrosati, B. In: *Solid State Ionic Devices*, ed. B. V. R. Chowdari & S. Radhakrishana. World Scientific, Singapore, 1988, p. 341.
5. Vassort, G., Gauthier, M., Harvey, P. E., Brochu, F. & Armand, M. B., ACEP ambient temperature lithium batteries: improved performance, 172nd Meeting of the Electrochemical Society, Honolulu, Hawaii, October 1987.

6. Wright, P. V., Electrical conductivity in ionic complexes of poly(ethylene oxide). *Brit. Polym. J.*, **7** (1975) 319.
7. Payne, D. R. & Wright, P. V., Morphology and ionic conductivity of some lithium ion complexes with poly(ethylene oxide). *Polymer*, **23** (1982) 690–3.
8. Watanabe, M., Sanui, K., Ogata, N., Kobayashi, J. & Ohtaki, Z., Ionic conductivity and mobility in network polymers from poly(propylene oxide) containing lithium perchlorate. *J. Appl. Phys.*, **57** (1985) 123–8.
9. Watanabe, M., Ikeda, J. & Shinohara, I., Effect of molecular weight of polymeric solvent on ion conductive behavior in poly(propylene oxide) solution of LiClO$_4$. *Polym. J.*, **15** (1983) 65–9.
10. Blonsky, P. M., Shriver, D. F., Austin, P. & Allcock, H. R., Polyphosphazene solid electrolytes. *J. Am. Chem. Soc.*, **106** (1984) 6854–5.
11. Itoh, T., Saeki, K., Kohno, K. & Koseki, K., Immobilization of polymeric electrolytes in a microporous membrane. *J. Electrochem. Soc.*, **136** (1989) 3551–2.
12. Kohno, K., Mori, S., Miyasaka, K. & Tabuchi, J., Immobilized electrolyte membrane. US Pat. 86-4588633, 13 May, 1986; 86-4620955, 4 Nov. 1986.

Water, Heat and Spherulites: Results for Divalent Films

Malcolm D. Glasse, Roger G. Linford & Walkiria S. Schlindwein

Leicester Polytechnic, Leicester LE1 9BH, UK

ABSTRACT

Water cast films of PEO with zinc halides have been prepared at a range of stoichiometries. They form electrolytes with acceptable conductivity and mechanical properties, but their morphology differs from similar films cast from acetonitrile/methanol mixtures. This difference can be removed by heating to 180°C, confirming that 'melt-cast' films are free of solvent effects. The presence of high melting spherulites is influenced not only by the state of dryness but also by the choice of anion and by the heat treatment used to dry the films. This has implications for the conventional procedures used to produce water-free films. The determination of water content of polymer electrolytes is briefly discussed.

INTRODUCTION

The detection, control and influence of water in polymeric electrolyte films have been the subject of much debate. The presence of water can be considered undesirable for several reasons: it carries the risk of introducing proton or hydroxide ion conduction; it adds hydrated ions to the range of possible species present (which already includes ion pairs and triple ions); and it complicates our understanding of the role of the interaction between conducting cations and ether oxygen atoms.

From a practical point of view, the objections to water are few. It can be argued that neither the complete absence of water, nor a knowledge of the precise water content is necessary in electrolytes for battery or device application. All that is required is reproducibility, which might be obtained by standardizing the humidity during preparation.

For a device manufacturer, there are certain advantages in producing polymer electrolyte films with water as a casting solvent. It is a cheap, easily available, safely disposable, non-toxic solvent which can be used even under confined conditions. Poly(ethylene oxide) (PEO) is water soluble, unlike most polymers, and water is also a suitable solvent for a wide range of inorganic salts. This means that it is viable to consider water as an alternative to the more usual non-aqueous solvents such as acetonitrile, methanol or ethanol.

Electrode–electrolyte compatibility is necessary for batteries and related devices and so lithium-based systems cannot be prepared using wet films. Moisture-free atmospheric conditions are also necessary. These restrictions do not apply to some other technologically interesting electrode metals such as zinc or lead. The divalent polymer electrolyte films that are compatible with these electrodes have already been shown to display a range of properties that are interesting in their own right.[1]

We therefore decided to explore the feasibility of preparing divalent polymer electrolyte films under laboratory atmosphere conditions, using water as a casting solvent. We were able to produce water cast films and we report on their properties below. We then studied the properties of films that had been dried by heating and evacuation under several different regimes, and compared their behaviour with films cast from more conventional solvents, as described in later sections. The precise determination of water content is not easy, and some of the difficulties are considered in the final part of the discussion.

Drying Procedures

Two classes of standard practice seem to have emerged. One involves the use of the driest possible ingredients and preparation conditions, or at least standardized conditions. The other involves drying the electrolyte films after they have been cast.

We have previously pointed out[2] that drying procedures are likely to change the properties of PEO electrolytes. Most of the solutions of salts in PEO which form interesting electrolytes also contain at least one crystalline phase. This phase can be seen as spherulites which often melt, or start to melt, just above room temperature. Any drying procedure which involves heating will inevitably introduce at least partial melting. The extent of this melting and any variation in the subsequent cooling can affect the recrystallization process and change the amount and form of the residual amorphous phase in which the conduction occurs.

As a consequence, there is a major dilemma for all those concerned with the effect of preparation variables upon the properties of crystalline polymeric electrolytes. If a drying procedure is not applied after casting, then any adventitious water or residual solvent could affect the film properties. If drying is carried out after casting, then the drying procedure itself, or the subsequent cooling can also alter properties, including conductivity.

Spherulites

As is well known, spherulites are three-dimensional, predominantly crystalline entities. The crystalline regions are lamellar in form, the lamellae being made up of polymer chains folded in a regular way which possesses long-range order.[3] Their structure can be studied by diffraction techniques. The lamellae radiate out from a central nucleating point. Amorphous material always fills the regions between the lamellae within a spherulite, and the spherulites are sometimes embedded in an amorphous matrix. These inter- and intraspherulite amorphous regions may or may not be the same in composition and other physical properties.

Spherulites may or may not contain salt within their crystalline regions. The PEO-like spherulites described in this work melt at about 65°C; their crystalline regions are formed from the polymer alone. They appear in dilute films. It seems that as they grow from the amorphous 'melt', salt is expelled into the remaining amorphous matrix, thus increasing its salt content and raising its T_g. Eventually, the chain mobility is so restricted that further crystalline growth is impeded.[4] Alternatively, crystalline 'complexes' involving polymer and salt in a fixed stoichiometry can form and these are usually salt-rich and high melting;[5] their melting point depends on the salt used, and is typically 120–180°C.

The existence of spherulites is easily detected by differential scanning calorimetry (DSC) and/or variable temperature polarizing microscopy (VTPM). In this work, some spherulites were observed which had melting points intermediate between those of the high-melting and the PEO spherulites. Because it is difficult to be confident that polymer electrolyte films are truly at equilibrium, we prefer not to describe intermediate or other spherulites in phase diagram terms. Instead, we use descriptions, based on experimental observations, of three types of spherulites: *PEO-like, intermediate* and *high melting.*

The morphology of the polymer electrolytes studied here is complicated and inhomogeneous. Different aliquots of film contain an amorphous matrix together with one or more spherulite type, mixed in different proportions. This morphology, and hence the properties, also vary with thermal history and preparation conditions. These factors make it particularly desirable to use several techniques simultaneously to study the same aliquot of film. In this work, we have been able to carry out VTPM and conductivity studies at the same time on the same region of a sample, but equipment restrictions have forced us to carry out DSC studies on different aliquots of the same sample.

Choice of System

This work is part of a continuing study of the properties of divalent polymer electrolyte films. For water-cast films, we started with zinc salts for several reasons. Zinc is a convenient battery anode material: Zn is now generally agreed to be transportable, although the nature of the mobile species is not yet established:[1] and there are some interesting and unresolved questions on the presence and absence of high-melting spherulites.[6]

Because of our interest in studying spherulitic systems, we intitially chose to examine $PEO_n:ZnX_2$ halide systems at a stoichiometry of $n = 8$. This high concentration had been found to lead to the formation of high-melting spherulites. Results are presented for $PEO_8:ZnBr_2$ and $PEO_8:ZnI_2$. In this work, we have examined a range of drying conditions and found that water-cast $PEO_8:ZnCl_2$ decomposes at 180°C. We are therefore, switching our attention to the $n = 12$ materials and in this paper we report the first stage of our investigation, namely $PEO_{12}:ZnCl_2$. The chloride is the most hygroscopic of the zinc salts under study and we felt that it would, therefore, be the most interesting candidate for a water-cast system.

We have commenced a parallel study of Ca systems; the initial results show the feasibility of preparing water-cast Ca polymer electrolytes. The DSC, VTPM and conductivity results contain similarities to those reported here for Zn systems; details will be published elsewhere.

EXPERIMENTAL

Films were prepared using PEO (relative molar mass 4×10^6) supplied by BDH which had been dried for 3 days at 50°C under vacuum before use. The inorganic salts, ZnI_2, $ZnBr_2$ and $ZnCl_2$ were also supplied by BDH, and were dried for 3 days at 140°C under vacuum. All the chemicals were stored in a glove box for further weighing. Triple distilled water was used as a solvent. Films of $PEO_n : ZnX_2$ (with n from 6 to 30 and $X = I$, Br or Cl) were prepared by first dissolving the stoichiometric amount of the salt in 100 ml of water before the addition of 1·0 g of PEO. The solution was stirred for 48 h and the final viscous solution was then cast in a glass ring on a polyethylene sheet in the open laboratory, but protected from dust. The solvent was allowed to evaporate for at least 2 weeks at room temperature.

The films obtained were typically 70–200 μm thick. Each film was divided into two halves, one of them being dried for 7 days at 50°C under vacuum (series D_{50}), the other being left as an undried film, i.e. in equilibrium with atmospheric moisture (series U). A subset of the series D_{50} films was subsequently heated at 140°C for 24 h and allowed to cool under vacuum to 25°C, taking 2 h (series D_{140}). The samples D_{50} and D_{140} were transferred to the dry glove box for handling. A comparison between the three series of samples has been made, drawing attention to the importance of thermal history effects in the formation of high-melting complexes. The various treatments are summarized in Table 1.

Films cast from water might be expected to become comparable with those cast from other solvents once those films have been thoroughly dried and melted. As a basis for comparison with conventional cast films, the two solvents technique[7] was used to prepare films from 75/25 v/v acetonitrile/methanol high purity solvents (Aldrich HPLC grade) in completely dry conditions. These films were given the same drying treatment as series D_{50}, resulting in another series of films, D_{50A}. The main comparison between water- and acetonitrile-cast films was made after heating D_{50} and D_{50A} samples at 180° for 2 h

Table 1. Treatment of series of samples under vacuum

Series reference	Previous treatment	Temp (°C)	Time	Cooling time (h)
U	As cast from H_2O	(No treatment)		
D_{50}	As cast from H_2O	50	7 days	
D_{50A}	As cast from $CH_3CN/MeOH$	50	7 days	
D_{140}	As D_{50}	140	24 h	2
D_{140A}	As D_{50A}	140	24 h	2
D_{180}	As D_{50}	180	2 h	3
D_{180A}	As D_{50A}	180	2 h	3

under vacuum followed by natural cooling to 25°C, which took 3 h in the vacuum oven (series D_{180} and D_{180A}). To ensure dryness during transfer, dry argon was used to fill the vacuum oven and the samples were quickly transferred to a small desiccator, which was then evacuated for 30 min before being transferred to the glove box.

Different techniques have been used to study the properties of these samples. Differential scanning calorimetry (DSC) was used to measure their thermal properties, using a Perkin–Elmer DSC-4 Differential Scanning Calorimeter. The variable temperature polarizing microscopy (VTPM) technique[5] was used to examine the spherulitic morphology of these polymeric electrolytes. The micrographs were obtained using an Olympus BHS polarizing microscope with an Olympus PM-10ADS automatic photographic system. The a.c. impedance measurements were obtained by using a Solartron Frequency Response Analyser (FRA) using a frequency range of 65·6 kHz–10 Hz.

FTIR studies were carried out on a Nicolet model 5 DXC spectrometer at room temperature. The samples were sealed between dried KBr plates in a sub-ppm water glove box using PTFE tape. They were transferred to a small vacuum desiccator and rapidly transported to the spectrometer. Ten scans were carried out in a period of 30 s.

Calibration was carried out by exposing the films to laboratory air for specified periods, and measuring the mass and the subsequent FTIR spectrum for each exposure period.

RESULTS AND DISCUSSION

The results shown in this paper are presented in three sections: Series U: as cast from water; Series D_{50} and D_{140}, dried under vacuum; and

Series D_{50A}, D_{140A}, D_{180} and D_{180A}, to permit a comparison of solvent effects.

Undried Films: Series U: As Cast from Water

PEO_n:ZnX_2 films ($n = 6$, 8, 10, 12, 15, 30 and X = I, Br, Cl) were cast from water and analysed by VTPM, DSC and a.c. impedance. In our opinion, all samples should ideally be studied at the same time after preparation. This is because the history of a sample affects its spherulitic morphology[2,5,8] since the growth of these crystalline regions is time and temperature dependent. In practice this is difficult to arrange but we have tried to base our comparisons on samples of similar age: typically 2 weeks after evaporation seemed to be complete. As an example, when the DSC analysis of PEO_6:$ZnCl_2$ was done immediately after complete solvent evaporation, a low-melting peak was found at 45°C, and this peak shifted to 55°C after 2 weeks.

Figure 1 shows the DSCs of PEO_n:ZnI_2, PEO_n:$ZnBr_2$ and PEO_n:$ZnCl_2$ samples from Series U. DSC results on PEO_n:ZnI_2 (Fig. 1(a)) can be interpreted most easily by dividing the samples in two categories: 'concentrated', $n = 6$, 8, 10 and 'dilute', $n = 12$, 15. The

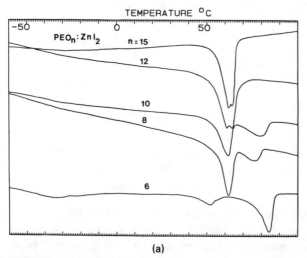

(a)

Figure 1. DSC traces of PEO_n:ZnX_2, $n = 6$, 8, 10, 12, 15. (a) X = I; (b) X = Br; (c) X = Cl. Figure 1(b) with $n = 6$ contains a scale change below 0°C to amplify the glass transition.

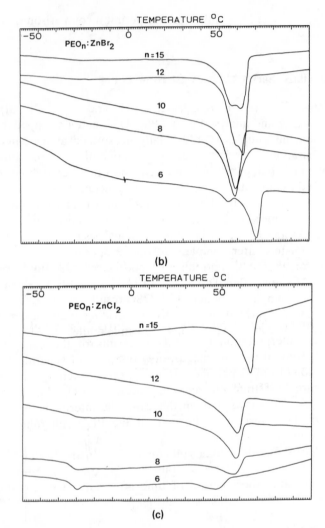

(b)

(c)

Figure 1—*contd.*

concentrated samples give rigid, almost brittle films containing intermediate melting material (75–85°C). Large (1 cm diameter) spherulites, which cover the whole of the film, can be seen. The dilute ones
show a melting temperature close to that of pure PEO (60–65°C).
Figure 2 makes this clear; very dilute ($n = 30$) samples of iodide (Fig.
2(a)) and the other halides (Figs 2(b, c)) have melting events which

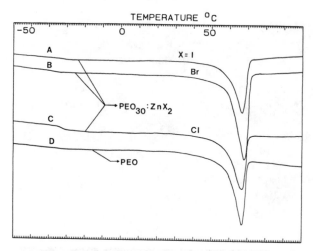

Figure 2. DSC traces of $PEO_{30}:ZnX_2$. (A) X = I; (B) X = Br; (C) X = Cl; (D) pure PEO for comparison.

closely resemble that of pure PEO (Fig. 2(d)). The VTPM analysis assisted the interpretation of the DSC results of the concentrated samples. It showed the presence of only one type of spherulite which melted steadily from its periphery to its core over a temperature range of 20°. Figure 3 illustrates the melting process of $PEO_8:ZnI_2$ at 20, 70, 75 and 82°C. This shows that, in this case, the intermediate material detected by DSC resides at the core of the PEO spherulite and is perhaps its nucleating centre.

Figure 1(b) shows the DSC results for $PEO_n:ZnBr_2$. In this case the presence of an intermediate crystalline phase is only detected for $n = 6$, if the same short period of time after solvent evaporation is considered. However, after one further month the crystallinity increases for $n = 8$ and 10. VTPM studies of the melting process in $PEO_6:ZnBr_2$ show large spherulites, as for the corresponding iodide, which melt from the periphery. The small melting event at about 55°C could not be detected by VTPM.

DSC of $PEO_n:ZnCl_2$ samples (Fig. 1(c)) does not show intermediate material to be present, and this absence is confirmed by VTPM. The values of the glass transition temperature, T_g, for all U samples are between −30 and −40°C (Fig. 1(c)) and no changes were observed when different concentrations of salt were used.

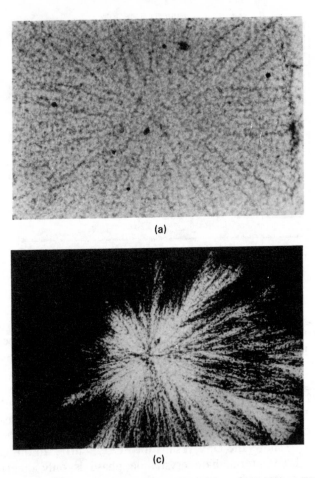

Figure 3. VTPM photographs of PEO$_8$:ZnI$_2$. (a) at 20°C; (b) at 70°C; (c) at 75°C; (d) at 82°C.

No high-melting spherulites were detected by VTPM or DSC for any of the undried halide films. Results of conductivity versus composition are presented in Fig. 4. The presence of high crystallinity in PEO$_n$:ZnI$_2$ concentrated samples ($n = 6$, 8, 10) could be the reason for the low conductivity values below their melting temperatures. In general the conductivity values at room temperature increase where n is greater than 10. At high temperatures, e.g. 120°C, the U samples are totally amorphous and the conductivity values are much the same

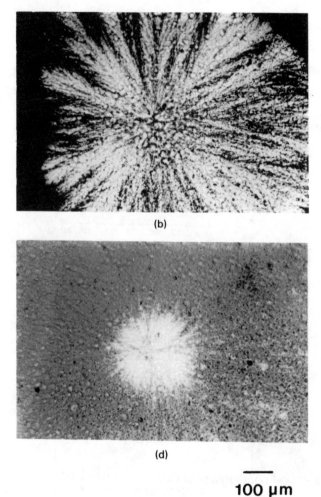

(b)

(d)

100 µm

Figure 3—*contd.*

for any concentration (Fig. 4(a)). For the $PEO_n : ZnBr_2$ samples the concentration has little influence upon the conductivity values (Fig. 4(b)). In the case of $PEO_n : ZnCl_2$ samples (Fig. 4(c)), higher conductivity values are found perhaps because no intermediate melting material is present. Below 40°C the conductivity of the halides increases in the order: $ZnI_2 < ZnBr_2 < ZnCl_2$.

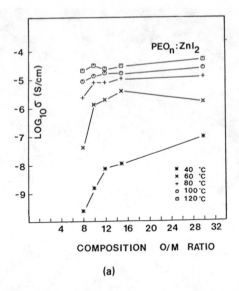

(a)

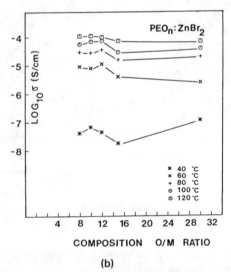

(b)

Figure 4. Log (conductivity) against composition for undried films of $PEO_n : ZnX_2$, $n = 8, 10, 12, 15, 30$ exposed to laboratory air for one month. (a) X = I; (b) X = Br; (c) X = Cl.

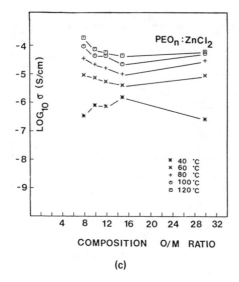

(c)

Figure 4—*contd.*

Dried Water-Cast Films: Series D_{50}, D_{140}: Dried under Vacuum

Initially it was thought that drying samples at 50°C for 7 days would not modify the crystallinity of the sample. For $PEO_8:ZnCl_2$ however, even after low temperature drying below the PEO melting point (Series D_{50}), there is evidence of enhanced PEO-like spherulites and a small peak around 155°C (Fig. 5(a)). Upon subsequent annealing at 140°C (Series D_{140}), this high-melting peak gets larger (Fig. 5(b)). VTPM confirms the presence of both PEO-like and high-melting spherulites for both drying regimes.

The analogous bromide and iodide samples showed essentially the same DSC behaviour. Drying at 50°C produces large DSC peaks at 65°C, accompanied by small intermediate melting peaks. These occur at higher temperatures (about 100°C) than for the corresponding undried samples. Annealing at 140°C increases the size of the higher temperature DSC peak, and raises its melting temperature to around 150°C. This suggests that the intermediate melting material seen in the undried sample is really high-melting spherulitic material which has a melting range depressed by the presence of water. For the bromide and iodide, the increase in amount of high-melting material is accompanied by a decrease in the size of the PEO peak.

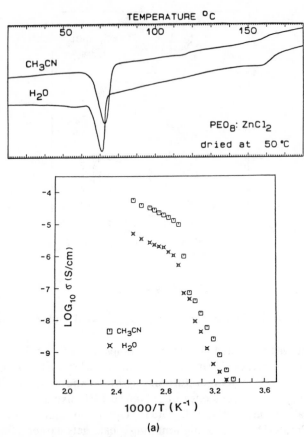

Figure 5. DSC traces (upper figures) and log (conductivity) versus $1/T$ plots (lower figures) for $PEO_8:ZnCl_2$ films cast and dried in different ways (see text). (a) Dried at 50°C for 7 days under vacuum; (b) dried as (a), then at 140°C for 24 h.

Comparison of Solvent Effects

A comparison was carried out of the DSC and conductivity results of $PEO_8:ZnCl_2$ described above with those obtained from films cast from the acetonitrile/methanol mixture and subsequently dried in the same ways as for the water cast films (Series D_{50A} and D_{140A}). Similar comparisons for bromide and iodide films dried at 50°C and 140°C are being carried out.

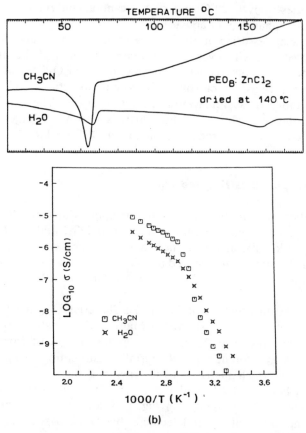

Figure 5—*contd.*

It can be seen from Figs 5(a) and 5(b) that the high-melting material is relatively more evident for water-cast films dried in either way. For both casting solvents, the higher temperature drying regime produces more high melting material. This is because high temperature drying inevitably results in annealing.

The high-melting material reduces the conductivity, and the conductivity results can be rationalized in terms of the amount of such material that is present. Thus, for both drying regimes, the acetonitrile/methanol-cast samples have higher conductivities at high temperatures because of a proportionally smaller amount of high-melting material. Our apparatus does not allow us to measure conductivities above 120°C and so we cannot confirm the expected conductivity enhancement above 140°C.

For the samples dried at 50°C, the acetonitrile-cast material retains its higher conductivity even at low temperatures. For the annealed samples however, the reduction in amount of PEO-like material in the water-cast films has enhanced the low temperature conductivities to levels above those for acetonitrile-cast samples. The conductivity change in the region of 60°C (Fig. 5(b)) is also much larger for the acetonitrile-cast sample, reflecting the larger DSC peak.

'Melt-Casting': Series D_{180} and D_{180A}

High temperature thermal treatment was applied to melt all spherulites in water-cast and acetonitrile-cast films. DSC and VTPM were then used to identify any differences between them. Unfortunately the $PEO_8:ZnCl_2$ samples discoloured at 180°C. This was taken as a sign of degradation and a more dilute composition ($n = 12$) was studied instead.

The VTPM micrographs of $PEO_{12}:ZnCl_2$ series D_{180} from 20°C to complete melting at 66°C showed a Maltese Cross pattern typical of regular spherulites. The $PEO_{12}:ZnCl_2$ series D_{180A} cast from acetonitrile shows a different type of crystalline structure having a foreground phase which melts at 65°C leaving a few small spherulites which melt between 120°C and 130°C. The DSC and Arrhenius plots of both samples are presented in Fig. 6, where the small amount of high-melting material is undetectable by DSC. As expected the conductivities are essentially similar.

No discolouration was observed in the 8:1 bromide and iodide samples; the DSC and conductivity results are presented in Fig. 7. For the bromide, it can be seen from Fig. 7(a) that essentially the same behaviour is observed as in the chloride system.

The iodide case is somewhat more complicated. A melting event in the acetonitrile-cast films is seen at about 120°C, and this is presumably responsible for the lower conductivity found at high temperatures (Fig. 7(b)). The conductivity enhancement on melting can just be seen at the highest temperatures reported. The low temperature conductivities are similar for the samples from both the casting solvents because the total spherulite content is similar.

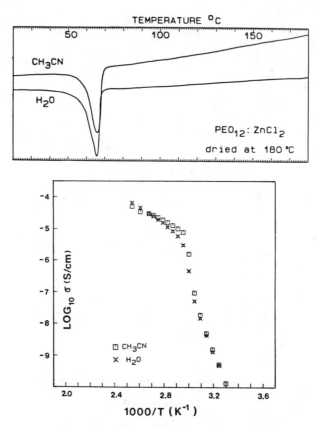

Figure 6. DSC (upper figure) and log (conductivity) versus $1/T$ plots (lower figure) for $PEO_{12}:ZnCl_2$ cast from different solvents and dried at 50°C for 7 days, then at 180°C for 2 h, under vacuum.

Determination of Water Content

Our water-cast films are necessarily wet, but the success of attempts to dry them at 50°C, 140°C and 180°C needs to be assessed. Karl–Fischer titration is claimed to give the lowest detection limits but it is difficult to prevent adventitious water reaching the film during the transfer process: consequently we used the more convenient technique of Fourier Transform Infrared Spectroscopy. We chose to use the

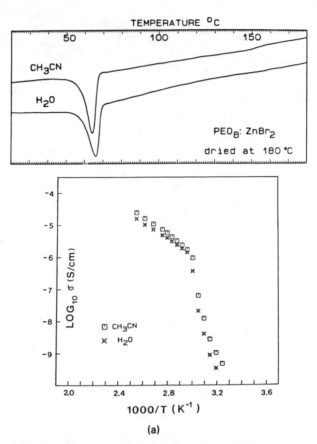

Figure 7. DSC traces (upper figures) and log (conductivity) versus $1/T$ plots (lower figures) for $PEO_8 : ZnX_2$ film cast and dried at 50°C for 7 days under vacuum, then at 180°C for 2 h. (a) X = Br; (b) X = I.

H—O—H bond at about $1650 \, cm^{-1}$ since this gives a clearer peak than the broad absorption around $3500 \, cm^{-1}$. Figure 8(a) shows the results of exposure of a 'dry' D_{50} sample of $PEO_8 : ZnCl_2$ to the atmosphere. Weighings were carried out for each spectrum and a calibration graph is shown in Fig. 8(b). This produces an absolute moisture content if one can assume that the baseline in the $1600–1740 \, cm^{-1}$ region is flat.

Extrapolation of Fig. 8(b) gives an approximate result of about 0·6% or less, which we regard as the minimum detection limit here.

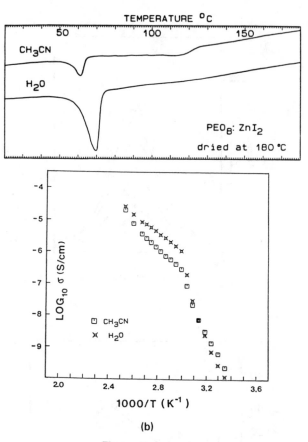

(b)

Figure 7—*contd.*

CONCLUSIONS

Water has been shown to be an acceptable casting solvent. It produces films which can be dried to be as free from water as those produced by other methods.

When water is used, it produces films which can be different from those cast from 75/25 acetonitrile/methanol. At room temperature water-cast films can be more (Fig. 5(b)) or less (Fig. 5(a)) conducting than those cast from other solvents.

The chloride films are the most hygroscopic, especially the concentrated ones, and here the presence of moisture has two effects.

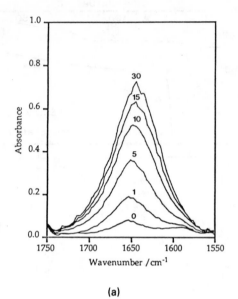

(a)

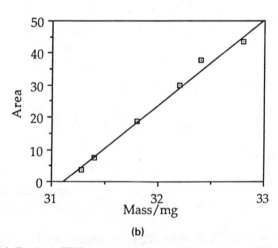

(b)

Figure 8. (a) Typical FTIR traces in the region of the 1650 cm^{-1} water peak for samples of $PEO_8:ZnCl_2$ cast from water and dried for 7 days at 50°C and subsequently exposed to laboratory air for the number of minutes specified on the figure. (b) Peak areas from (a) plotted as a function of total mass of sample.

Firstly, it depresses the melting point of the high-melting phase until it is no longer formed in films stored at room temperature. Secondly, moisture eliminates any reduction in conductivity in the concentrated films either by plasticization or because the crystallinity is lower.

Heating to 180°C clearly has the potential to eliminate the effect of the casting solvent (Fig. 6) whereas 140°C does not (Fig. 5). Even heating above the highest melting point of a polymer does not necessarily destroy all memory effects. The melted chains and the associated salt need time to diffuse somewhat, otherwise spherulites tend to reappear with the same form and in the same place. Only prolonged heating, or even higher temperatures, can eliminate such memory effects.

The memory of the solvent seems to have been destroyed for the bromide (Fig. 7(a)), but there is some evidence of memory in the iodide (Fig. 7(b)). The only other way to explain differences in Fig. 7(b) is to suppose that recrystallization during cooling from 180°C was affected not by memory but by different cooling conditions; however the samples here were cooled together.

The preceding results have implications for the construction of 'phase diagrams' for these and similar materials. There is no doubt that such diagrams can condense a great deal of relevant information into a convenient form, but they cannot be regarded as true equilibrium phase diagrams since they depend upon both the casting solvent and the thermal history.

We are extending these studies on water-cast films in order to complete the Zn and Ca systems and to compare with other appropriate divalent polymer electrolytes, e.g. Ni and Pb.

ACKNOWLEDGEMENTS

Walkiria S. Schlindwein thanks CNPq (Brazil) for the award of a scholarship. The UK National Advisory Board is thanked for financial support under its research initiatives.

REFERENCES

1. Farrington, G. C. & Linford, R. G., Divalent polymer electrolytes. In: *Polymer Electrolyte Reviews 2*, ed. C. A. Vincent. Elsevier Applied Science, London, 1989, Ch. 8, p. 254.

2. Cole, M., Sheldon, M. H., Glasse, M. D., Latham, R. J. & Linford, R. G., EXAFS and thermal studies on zinc polymeric electrolytes. *Appl. Phys. A.*, **49** (1989) 249–57.
3. Glasse, M. D. & Linford, R. G., Polymer structure and conductivity. In: *Electrochemical Science and Technology of Polymers 1*, ed. R. G. Linford. Elsevier Applied Science, London, 1987, pp. 23–43.
4. Sheldon, M. H., CNAA PhD Thesis, Leicester Polytechnic, 1989.
5. Neat, R. J., Hooper, A., Glasse, M. D. & Linford, R. G., A structural model for the interpretation of composition dependent conductivity in polymeric solid electrolytes. In: *Transport-Structure Relations in Fast Ion and Mixed Conductors*, ed. F. W. Poulsen, N. H. Anderson, K. Clausen, S. Skaarup & O. T. Sorensen. Riso National Laboratory, Roskilde, Denmark, 1985, pp. 341–6.
6. Sheldon, M. H., Glasse, M. D., Latham, R. J., Linford, R. G., Yang, H. & Farrington, G. C., Impedance and other studies of zinc polymeric electrolytes. In: *Extended Abstracts of the First International Symposium on Polymer Electrolytes, St. Andrews*, ed. G. C. Cameron, M. D. Ingram, J. R. MacCallum & C. A. Vincent. University of St. Andrews, 1987, pp. 30.1–30.4.
7. Yang, L.-L., Huq, R. & Farrington, G. C., Preparation and properties of PEO complexes of divalent cation salts. *Solid State Ionics*, **18, 19** (1986) 291–4.
8. Neat, R., Glasse, M., Linford, R. G. & Hooper, A., Thermal history and polymer electrolyte structure: implications for solid-state battery design. *Solid State Ionics*, **18, 19** (1986) 1088–92.

The Degree of Crystallinity in PEO-based Electrolytes

Åsa Wendsjö

Institute of Chemistry, University of Uppsala, Box 531, S-751 21 Uppsala, Sweden

&

Hong Yang

Department of Materials Science and Engineering, University of Pennsylvania, 3231 Walnut Street, Philadelphia, Pennsylvania 19104, USA

ABSTRACT

A wide range of salts can be dissolved into poly(ethylene oxide) (PEO) to form solid polymers whose anion and/or cation transport numbers fluctuate greatly with ion type and concentration. It is also a widely held view that ionic mobility is confined to amorphous regions. This suggests that a well-defined group of polymer electrolytes will provide valuable information concerning the relationship between structure and ionic conductivity in these materials. In this work, we study systematically the degree of crystallinity within the family of divalent cation halides: $M^{2+}X_2^-$ (M = Pb, Zn, Ca, etc., and X = Cl, Br, I), whose transport properties have already been studied extensively. The effects of using different solvents, different heat treatment, and exposure to water are also examined.

INTRODUCTION

It was earlier found that alkali metal salts can be dissolved in poly(ethylene oxide) (PEO) to produce polymer electrolytes with high ionic conductivities, e.g. $\sigma \sim 3 \times 10^{-2}$ $(\Omega\,cm)^{-1}$ in $LiCF_3SO_3$ at

225

25°C.[1] These have a clear potential in the construction of polymer batteries.[2-4]

A great variety of salts have since been dissolved into PEO to form solid electrolytes in which the anion and/or cation transport numbers vary greatly dependent on ion type and concentration. See, for example, Ref. 5. It is now known that thin films cast from these materials can often have a multiphase character consisting of: (i) polycrystalline PEO, (ii) one or more polycrystalline complexes, and (iii) amorphous components. Some phase diagrams have been established which indicate that more than one polycrystalline complex can coexist under certain salt concentration and temperature conditions.[6,7]

Since it is generally accepted that the high ionic conductivity occurs in the amorphous phase, the amount of crystalline material present will be an important factor in determining the bulk conductivity of the material. In this work, the degree of crystallinity within the system $MX_2(PEO)_n$ for $n = O/M$, $M = Pb^{2+}$, Zn^{2+}, Ni^{2+}, Mg^{2+}, Ca^{2+} and $X = Cl^-$, I^-, Br^- has been studied systematically at room and elevated temperatures using X-ray powder diffraction. Some aspects of hydration processes in these materials will also be illustrated using X-ray powder diffraction and infra-red spectroscopy.

EXPERIMENTAL

General

A two-solvent polymer film casting technique has been used in which PEO (Polyscience: MW = 5×10^6) was dissolved in acetonitrile (Aldrich 99·9%), and the metal salts in ethanol. An exception was the lead halides where dimethylsulphoxide (DMSO) was used as solvent.[8] After dissolving, the solutions were mixed and cast on Teflon plates, and the solvent allowed to evaporate at 25°C or at some higher temperature. This procedure yielded films typically 100 μm thick, which were then dried under vacuum for some days and stored in dry air.

The X-ray diffraction investigations were performed using a STOE position sensitive detector (PSD) system. All experiments used Ge monochromatised CuKα_1, radiation. Two different types of detector were used: a curved- ($r = 130$ mm) and a straight-wire detector covering angular ranges 45° and 7° in 2θ, respectively. The curved detector was used in a stationary mode for crude characterisation of the polymer films. The higher resolution straight-wire detector was used when more accurate data were required, typically for cell determination and refinement. Minimum attainable full-widths at half-height are ~0·15° for the curved and ~0·07° for the linear PSD. Exposure times are typically 180 s per scan range.

Room-temperature runs were performed by rotating the polymer film about the normal to its plane in an $\omega = \theta$ transmission mode. Measurements were made at elevated temperatures using the STOE high-temperature attachment equipped with a high-temperature controller (HTC). Strips of film were enclosed in 0·7 mm diameter glass capillaries, and the measurements made in a stationary transmission mode. Temperature was measured by a thermocouple set in the graphite heating block surrounding the capillary. Ramp-rates, heating ranges, equilibration and measuring times could all be defined in the HTC. Temperature fluctuations at a given temperature were less than 0·5°C. All measurements were made in air, with exposure times of 180 s. The hydration process in air was studied using X-ray diffraction, where the polymer film was mounted between two thin mylar films, and with infra-red spectroscopy using a DIGILAB FTS-45 FTIR spectrometer (resolution: 16 cm^{-1}, scan number: 8; measuring time: 10 s). The polymer film was exposed to air with a relative humidity of 45% at room temperature before observations were made.

RESULTS AND DISCUSSION

Various aspects of crystallinity were examined in this work. It was found that the degree of crystallinity was highly dependent on a number of factors.

(i) Preferred Orientation

The polymer film can be treated as a 'powder' specimen if the crystallites are randomly oriented. The presence of preferred orientation was thus tested for. With the X-ray beam perpendicular to the polymer film, diffractograms were taken in a stationary mode at 10° intervals about a normal to the film. The intensity variation for the different reflections over a complete 180° range was generally very

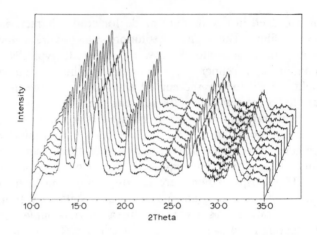

10·0   15·0   20·0   25·0   30·0   35·0
                        2Theta

Figure 1. Study of preferred orientation in $ZnCl_2(PEO)_4$ (see text).

small (<5%). This is illustrated by $ZnCl_2(PEO)_4$ (Fig. 1). On the other hand, significant preferred orientation was found for the case of $PbI_2(PEO)_{16}$, where intensity variations of up to 30% were observed. This suggests that preferred orientation may not be too serious a problem in general, but must be checked from case to case.

(ii) Cation and Anion Type

The degree of crystallinity of PEO, doped with different salts was studied in detail at room temperature, using the straight-wire PSD. The full range of crystallinity is exemplified by the following examples.

— An X-ray powder diffractogram for pure PEO is shown in Fig. 2(a). The peaks represent the reflections from the crystalline phase.

— Only these same peaks were then observed for $MgCl_2(PEO)_{16}$ (Fig. 2(b)), implying that only one crystalline phase is present, that of pure PEO, while the salt–PEO complex occurs in a totally amorphous phase.

— No crystalline component of PEO is found for $ZnCl_2(PEO)_4$; instead, a single phase due to the complex formed between $ZnCl_2$ and PEO (Fig. 2(c)).

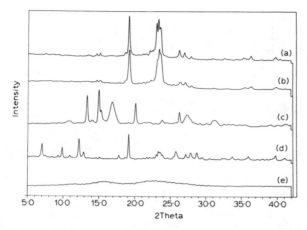

Figure 2. X-ray diffractograms illustrate the broad range of crystallinity in different PEO-based electrolytes at room temperature. (a) PEO; (b) $MgCl_2(PEO)_{16}$; (c) $ZnCl_2(PEO)_4$; (d) $PbI_2(PEO)_{16}$; (e) $CaI_2(PEO)_8$.

—The most complicated situation we have found is exemplified by $PbI_2(PEO)_{16}$. It contains crystalline PEO together with at least one (probably two) PbI_2–PEO complex phase which coexists with PEO (Fig. 2(d)).

—Finally, $CaI_2(PEO)_8$ was found to be completely amorphous, as evidenced by the total absence of Bragg peaks (Fig. 2(e)).

From these five examples, it is clear that the degree of crystallinity within the different electrolytes is very dependent on cation and anion type, and that several crystalline phases may coexist.

(iii) Salt Concentration

The degree of crystallinity was also studied as a function of salt concentration in PEO for a number of cases. Figure 3 shows $PbI_2(PEO)_n$, for $n = 8, 16, 24$ and 40. It is observed that the fraction of crystalline PEO decreases for increasing salt concentration. Inspection of the variation in the relative intensities of the reflections belonging to the complex suggests that there are, indeed, two coexisting complex phases, probably in different concentrations and probably with different compositions. The overall concentration (as denoted by n) determines the relative amount of these phases.

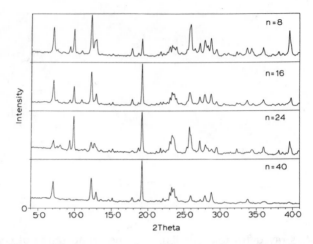

Figure 3. X-ray diffractograms for different salt concentrations (*n*) in $PbI_2(PEO)_n$ at room temperature.

(iv) Solvent

The choice of solvent for the salt may also influence the degree of crystallinity of the polymer film. This effect is clear for $PbI_2(PEO)_{16}$, when both dimethylformamide (DMF) and dimethylsulphoxide (DMSO) are used to dissolve PbI_2 (Fig. (4)). Different complex phases form in the two cases. Furthermore, the use of DMF inhibits the formation of crystalline PEO.

(v) Hydration

Water uptake in the films also varies greatly with ion type. With $ZnCl_2(PEO)_4$ mounted in the sample holder between two thin mylar sheets, the X-ray diffractograms after 5, 30 and 60 min exposure to air with a relative humidity of ~45% are shown in Fig. 5(a). Water uptake clearly destroys the crystalline phase, as evidenced by the disappearance of the Bragg peaks. A preliminary study of the hydration/dehydration process in $NiBr_2(PEO)_8$ is illustrated in Fig. 5(b). An 'as cast' film was heated to 135°C for 60 min. Peaks belonging to crystalline PEO disappear as the PEO melts. The crystalline

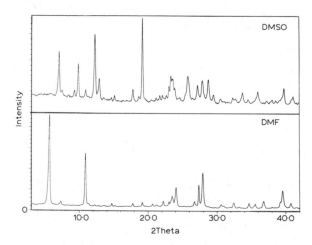

Figure 4. The effect on crystallinity in $PbI_2(PEO)_{16}$ at room temperature, using different solvents: (top) DMSO, (bottom) DMF.

complex phase remains more or less unchanged. The film was then returned to room temperature, and a diffractogram taken after 15 min. The peaks due to crystalline PEO had returned, while the intensities from the crystalline complex had decreased. After 4 h exposure to moisture, no crystalline complex could be detected. The sample was finally annealed for 16 h at 135°C, and a diffractogram taken at room temperature. Recrystallisation of both complex and PEO had occurred. To summarise: although water destroys the crystallinity in $NiBr_2PEO_8$, it subsequently returns on dehydration.

The water uptake in $PbX_2(PEO)_n$ for $X = I^-$, Br^-; $n = 8$, 16, 30 and $X = Cl^-$; $n = 16$ was studied using IR spectroscopy. Films were exposed to moisture at room temperature for 1 and 60 min. Spectra for two typical cases: $PbBr_2(PEO)_{30}$ and $PbI_2(PEO)_8$ are given in Fig. 5(c) and Fig. 5(d). Had water been present, bands due to symmetric stretching and bending of O—H bonds in the water molecules would have been observed at 3400 cm^{-1} and 1600 cm^{-1} respectively. No such bands were found, even after 1 h exposure to moisture. This adds to the earlier diffraction result that the crystallinity of films containing Pb^{2+} ions was not affected by water. We see here that, indeed, no water is taken up by any part of the film, crystalline or amorphous.

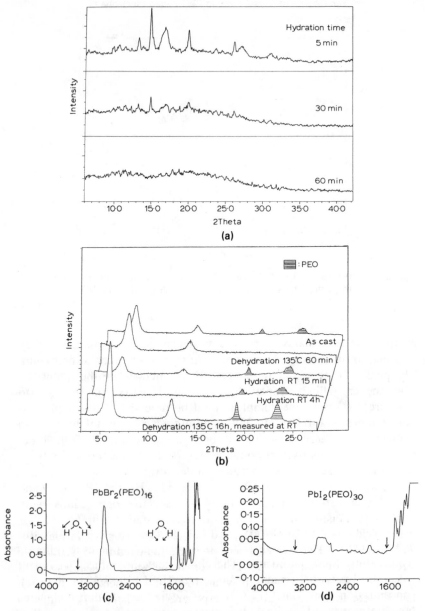

Figure 5. X-ray diffractograms showing the effect of hydration on crystallinity in (a) $ZnCl_2(PEO)_4$ (at room temperature), (b) $NiBr_2(PEO)_8$. IR spectra of (c) $PbBr_2(PEO)_{30}$ and (d) $PbI_2(PEO)_8$ after 60 min exposure to moisture at room temperature. The arrows mark the positions of expected OH-stretching and OH-bending bands from water.

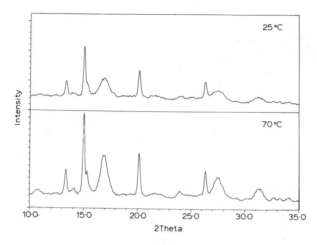

Figure 6. The effect on crystallinity in $ZnCl_2(PEO)_4$ using different annealing temperatures during film preparation.

(vi) Annealing Temperature

The annealing temperature during film preparation also influences the degree of crystallinity. Figure 6 shows $ZnCl_2(PEO)_4$ made using two different annealing temperatures: 25°C and 70°C. We see sharper and stronger peaks and lower background for the sample annealed at 70°C, implying that the relative amount of crystalline to amorphous material is higher. The higher annealing temperature appears to accelerate the crystallization process.

(vii) Temperature

The temperature dependence of the degree of crystallinity for the different electrolytes was also studied. The annealing times at different temperatures were all 30 min. The choice of temperatures was often based on differential scanning calorimetry (DSC) data.[9] The high salt concentration film $ZnCl_2(PEO)_4$ was studied at different elevated temperatures (see Fig. 7(a)). It is clearly seen that a single crystalline complex phase melts homogeneously into an amorphous phase. $PbBr_2(PEO)_{16}$ shows a quite different behaviour (Fig. 7(b)). Several crystalline phases coexist at room temperature; PEO and at least one

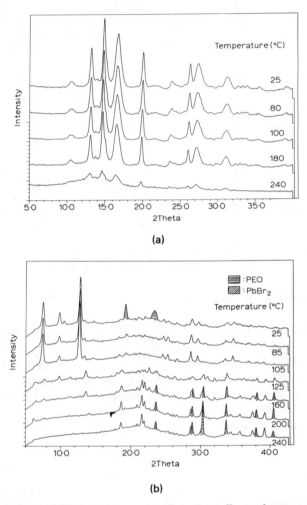

Figure 7. X-ray diffractograms showing the effect of temperature on crystallinity for (a) $ZnCl_2(PEO)_4$, (b) $PbBr_2(PEO)_{16}$, (c) $PbI_2(PEO)_{16}$, (d) $PbI_2(PEO)_{30}$.

complex between $PbBr_2$ and PEO. PEO melts between 25° and 85°C, and a new phase appears around 160°C where precipitation of $PbBr_2$ also begins. In $PbI_2(PEO)_{16}$ (Fig. 7(c)), crystalline PEO is seen to melt at 80–100°C, but the crystalline complex phase remains. A new phase appears at 200°C and another is seen to be developing at 240°C. None of these new reflections can be attributed to PbI_2 precipitation.

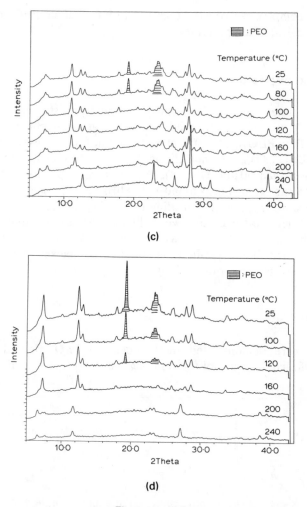

(c)

(d)

Figure 7—*contd.*

In $PbI_2(PEO)_{30}$, a more dilute sample, crystalline PEO is present up to 120°C (Fig. 7(d)). A new phase of the PbI_2–PEO complex is also observed between 160 and 200°C, which is stable up to at least 240°C. This same phase is found in $PbI_2(PEO)_{16}$ at about 200°C.

The degree of crystallinity is thus highly dependent on temperature, and often extremely complicated, varying significantly for different concentrations of the same salt type. We see that more than one salt–PEO complex can coexist, together with crystalline PEO, at both room and elevated temperatures.

(viii) Annealing Time

Annealing times used in the above temperature studies were all nominally 30 min. It was observed, however that a new phase began to appear in $PbI_2(PEO)_{16}$ at 240°C after 15 min at the expense of the old phase. $PbBr_2(PEO)_{16}$ was then chosen to probe the effect of annealing times longer than 30 min (Fig. 8). A new phase appeared at 120°C after 30–45 min. It was also found that $PbBr_2$ precipitated at a lower temperature using a longer annealing time. This result serves to stress the general need to ensure that we really have achieved equilibrium in these studies of crystallinity. This can, in practice, be an unrealistic ambition, since the crystallisation of the complexes can be an extremely slow process, often taking months or even years.

SUMMARY

We have here demonstrated the strength of PSD-oriented X-ray powder diffraction as a tool for studying the degree of crystallinity of

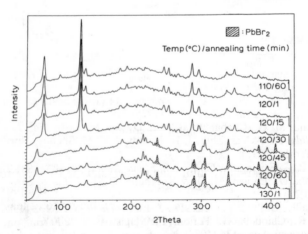

Figure 8. The effect of annealing time on crystallinity in $PbBr_2(PEO)_{16}$. After 30 min a new phase appears at 120°C.

polymers in general, and of PEO-based electrolytes in particular. Different processes, such as hydration and thermal behaviour, are readily observed. We also see that PEO–salts are exceptionally complicated systems, in which the degree of crystallinity is particularly dependent on cation and/or anion type, salt concentration, water content, temperature and thermal history.

The observations recorded here will form the *qualitative* basis for future work aimed at elucidating the subtle relationship between the crystalline and amorphous phases; the high ionic conductivity is believed to occur in the amorphous phase. This should lead us to a better understanding of conduction mechanisms. Efforts will also be made to solve the structures of a number of these crystalline PEO–salt complexes and, in so doing, bring our understanding onto an atomic level.

ACKNOWLEDGEMENTS

This work was supported in the United States by a grant from the Defence Advanced Project Agency through a contract monitored by the Office of Naval Research (ONR) under contract N00014-88-K-0732 and by NSF grant DMR88-19885, and in Sweden by the Swedish Natural Science Research Council (NFR). These are all hereby gratefully acknowledged.

REFERENCES

1. Ito, Y., Syakushiro, K., Hiratani, M., Miyauchi, K. & Kudo, T., *Solid State Ionics*, **18, 19** (1986) 277.
2. Parker, J. M. & Wright, P. V., *Polymer*, **14** (1973) 589.
3. Wright, P. V., *Brit. Polym. J.*, **7** (1975) 319.
4. Armand, M. B., Chabagno, J. M. & Duclot, M. J., In: *Fast Ion Transport in Solids*, ed. P. Vashishta, J. N. Mundy & V. M. Shenoy. North-Holland, New York, 1979, p. 131.
5. Huq, R. & Farrington, G. C., *Solid State Ionics*, **28–30** (1988) 990.
6. Fauteux, D., Lupien, M. D. & Robitaille, C. D., *J. Electrochem. Soc.*, **134** (1987) 2761.
7. Robitaille, C. D. & Fauteux, D., *J. Electrochem. Soc.*, **133** (1986) 315.
8. Huq, R., Chiodelli, G., Ferloni, P., Magistris, A. & Farrington, G. C., *J. Electrochem. Soc.*, **134** (1987) 364.
9. McGhie, A. R., Private communication.

Thermal Stability of Poly(Ethylene Oxide)(PEO) and PEO/Metal Salt Electrolytes

G. K. Jones,[a] G. C. Farrington[a] & A. R. McGhie[b]

[a] Department of Chemical Engineering, [b] Laboratory for Research on the Structure of Matter, University of Pennsylvania, 3231 Walnut Street, Philadelphia, Pennsylvania 19104, USA

ABSTRACT

The decomposition of PEO is influenced by atmosphere, catalyst residue, and the presence of multivalent cation salts. Both 'pure' PEO and PEO electrolytes containing multivalent cation salts are very sensitive to heat treatment in air, and significant decomposition reactions occur below 200°C. Most electrolyte compositions yield 1,4 dioxane as a major decomposition product, although the specific decomposition mechanisms are cation dependent. Lead and cadmium salts have little effect on the decomposition reactions of PEO, behavior which may correlate with the high cation transport observed in these electrolytes. In contrast, electrolytes containing Cu salts have rich decomposition chemistries, in which rich is a euphemism for complicated.

INTRODUCTION

Poly(ethylene oxide) (PEO) solutions of metal salts are unusual solid ionic conductors[1-3] with potential applications as battery electrolytes and electrolytes for electrochromic devices, among others. PEO electrolyte compositions formed by casting linear, high molecular weight PEO from solution often have multiphase structures. The multiphase components include amorphous phases, a pure PEO crystalline phase, and high temperature crystalline complex phases, all in proportions that depend on the initial salt concentration and the conditions of formation.

The specific composition of an electrolyte sample at any particular temperature is also influenced by its thermal history. For example, heating PEO electrolytes at elevated temperatures (up to 200°C) often results in an increased fraction of amorphous material,[4] improved electrolyte conductivity, and a decreased melting point of the nominally 'pure' PEO phase. Interestingly, these changes are similar to those observed when low molecular weight PEO is added to the same electrolyte composition.[5] The question logically arises as to whether divalent metal ions catalyse chain scission in high molecular weight PEO, producing low molecular weight PEO fractions by an in situ reaction.

A number of studies of the thermal stability of PEO have already been reported. Madorsky & Straus[6] observed PEO to decompose above 300°C into a variety of substances, among them low molecular weight PEO, ethylene oxide, ethanol, formaldehyde, carbon dioxide and water. Bortel & Lamot[7,8] reported that samples of PEO degraded until they reached 70% crystallinity, and concluded that amorphous PEO is unstable. Grassie & Perdomo-Mendoza[9,10] studied the degradation of PEO, both alone and complexed with ammonium polyphosphate, and proposed different mechanisms for the degradation reactions in each case. They analysed the degradation products and obtained results comparable to those of Madorsky and Straus for pure PEO, but found that water and 1,4 dioxane were the major degradation products of PEO complexed with ammonium polyphosphate. Taoda et al.[11] studied the effects of thermal cycling on different molecular weights of pure PEO and showed that moderate temperature cycling (e.g. 30–150°C daily in air) could almost completely volatilize high molecular weight PEO in a few weeks. These authors isolated degradation products similar to those reported by Madorsky and Grassie.

The present work represents a survey of the thermal decomposition behavior of PEO containing high concentrations of (mostly) divalent cation salts as well as of pure PEO of different molecular weights. These are unusual electrolytes with some properties characteristic of liquids and others of solids. The principal experimental technique used was combined thermogravimetry-mass spectrometry (TG-MS).

EXPERIMENTAL PROCEDURES

Samples of pure PEO and divalent metal salt electrolytes were prepared by solution casting.[2] Appropriate amounts of PEO and salt were dissolved in the necessary solvents (usually acetonitrile for the PEO and either ethanol, dimethylformamide, or dimethylsulfoxide for the salt), then mixed and stirred, forming solutions of 2–3 wt% in solids. The solutions were cast, typically onto PTFE or glass substrates, and allowed to 'dry'. Films, typically 50–150 μm thick, were further heated under vacuum above 100°C to remove the last traces of solvent. The resulting electrolyte films were stored in an inert atmosphere glovebox. In addition, pure PEO was examined in powder form 'as received' from two different suppliers: Aldrich (PEO-A) and Polysciences (PEO-B).

TG-MS was carried out using a DuPont 951 TGA, with the DuPont TA 2100 control system, and a VG Micromass 300 quadrupole mass spectrometer coupled through a silica capillary interface heated to about 150°C. Samples, 10–40 mg in weight, were heated in air or inert gas up to 700°C at 10°C/min. Mass spectra were collected every 2 or 5°C in the range 12–100 amu in the log histogram mode; a data conversion program allowed up to 16 individual masses to be selected and displayed as a function of time.

RESULTS

As-Received PEO

Thermogravimetry
Of all the materials studied, high molecular weight PEO (molecular weight, 5×10^6) was found to be the most stable in nitrogen and the least in air. Under nitrogen, high MW PEO did not begin to lose weight until nearly 350°C (Fig. 1) at which point rapid weight loss began and the polymer completely volatilized in a single step. Low molecular weight PEO samples (MW = 400 and 600) were also quite stable, showing little weight loss below 250°C.

The stability of pure PEO in air, however, depended greatly on the source and form of PEO studied. For example, Fig. 1 shows the differences in stability between samples of 5×10^6 MW PEO-A and

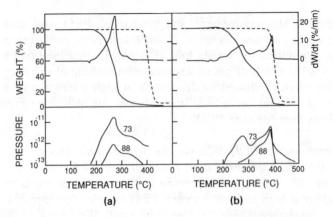

Figure 1. TG-MS of pure PEO powder from suppliers (a) A and (b) B in air. Dashed lines are TG of pure PEO.

PEO-B powders. In air, both PEO samples lost >5% weight by 200°C. PEO powder was also found to oxidize and gain up to 0·4% in weight at 175°C (Fig. 2). Cast PEO films did not show any weight gain at 175°C but had a very sharp onset of weight loss.

Another interesting observation was made with PEO below 100°C. All of the pure PEO samples and many of the PEO–salt electrolytes were found first to gain and then lose about 0·05% weight between room temperature and 100°C, as shown in Fig. 2. This occurred

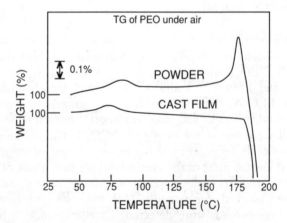

Figure 2. Weight gain and loss of PEO powder and cast film under air (both PEO-A).

regardless of the atmosphere and source of PEO and could be reproduced on the same samples. In some instances, evolution of CO_2 was detected during the weight loss, so it may be that the small weight gain/loss is the result of gas absorption upon melting of the pure PEO crystalline phase, followed by desorption at higher temperatures. The amount of weight gain/loss was always smaller for PEO–salt electrolytes than for pure PEO.

Mass Spectrometry (In the following discussion of mass spectrometry results, note that a species of mass ## amu is denoted as [##])
Significant differences were observed in the volatile products evolved when different high molecular weight PEO samples decomposed under different atmospheres (Fig. 1). Under nitrogen, the ratio [73]/[88] was about 20 for PEO-A and 10 for PEO-B. In air, the ratio was again 20 for PEO-A, but dropped below 2 for PEO-B (Fig. 1). The same ratio was observed for a sample of 3×10^5 MW PEO-B. The species [88] appears to be 1,4 dioxane, which is known to be generated by the action of certain cationic species on PEO.[10] Lower MW samples, while having stability similar to high MW PEO, yielded different spectra. For example, while PEO (MW 18 500) had [88] and [73] in equal amounts, there was more [88] than [73] in the spectra of 400 and 600 MW PEO.

Divalent Salt Electrolytes

The thermal stabilities of a large number of PEO electrolytes containing divalent metal salts were examined. All had the general formula MX_mPEO_n, where M = Cd, Co, Cu, Fe, Ni, Pb, Sn, Zn; X = Cl, Br and I; $m = 2$, 3 and $n = 8$, 16 and were prepared from 5×10^6 MW PEO-A. These electrolytes are known to have a wide range of conductivities and ion transport characteristics. Under nitrogen, all were less stable than pure PEO, but more stable in air. Figure 3 shows the temperature at the maximum rate of weight loss, in nitrogen and air, for most of the electrolytes studied. In all of the mass spectra of the products of decomposition, similar low molecular weight species were observed, for example, [44], [57] and [58]; however, no halogenated compounds were detected. Carbon dioxide, ethanol and ethylene oxide are all [44], and [57] and [58] could be derived from an unsaturated methylethylether [58], which has been detected among the

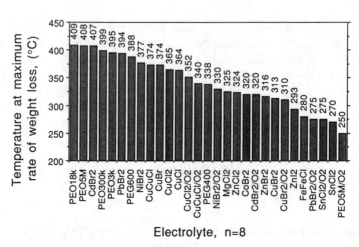

Figure 3. Temperature at maximum rate of weight loss for selected electrolytes. All electrolytes are composition $n = 8$.

decomposition products of pure PEO.[9] As mentioned, [88] appears to be 1, 4 dioxane. The species [73] is probably from methoxyacetaldehyde, [74]. As Grassie and Perdomo-Mendoza described the formation of 1, 4 dioxane as an acid catalysed reaction, the ratio [73]/[88] may be an indication of chemical conditions during decomposition. Following are brief summaries of the thermal characteristics of selected electrolytes.

1. $PbBr_2PEO_8$: Of all the electrolytes studied, the decomposition characteristics of this material were the most similar to that of pure PEO. Under inert atmosphere, $PbBr_2PEO_8$ was almost as stable as PEO, and its decomposition products gave mass spectra similar to those of PEO. These mass spectra show [73] as the principal fragment evolved (Fig. 4(left)), with a [73]/[88] ratio about 8. Under air, $PbBr_2PEO_8$ had a rapid onset of weight loss at around 165°C, behavior similar to that of PEO. Again, the mass spectra seen are similar to those of PEO and contain mostly [73].

2. $CdBr_2PEO_8$: This electrolyte behaved similarly to $PbBr_2PEO_8$ in certain respects. Under nitrogen, $CdBr_2PEO_8$ was the most stable of all the electrolytes, but produced mass spectra different from $PbBr_2PEO_8$ or PEO. The amount of [73] formed was less and the amount of [88] more (ratio about 0·4) than that formed in the previous two materials (Fig. 4(right)). In air, $CdBr_2PEO_8$ was much less stable,

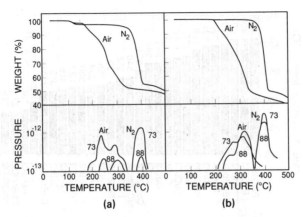

Figure 4. TG-MS of (a) $PbBr_2PEO_8$ and (b) $CdBr_2PEO_8$ under air and nitrogen.

and like $PbBr_2PEO_8$ and PEO its decomposition had a sharp onset. Here, [88] and [73] had comparable concentrations (ratio nearly 1).

3. CuX_nPEO_8: Systems based on Cu(I) and Cu(II) chlorides and bromides generally decomposed in three steps (Fig. 5). For $CuCl_2PEO_8$, the first step around 190°C resulted in 0·7–1·0% weight loss and produced only a trace of [88]. In the second step, around 200°C, about 7% was lost, producing a mass spectrum similar to those of many other PEO electrolytes (ratio [73]/[88] about 0·1). The first and second steps were the same regardless of whether the atmosphere was nitrogen or air, but they did vary with the Cu(I)/Cu(II) ratio and the halogen type. In the third and major step occurring at 370°C, the remainder of the PEO volatilized, again producing more [88] than [73] (ratio 0·3). In air, this process shifted to about 340°C. $CuBr_2PEO_8$ behaved similarly to $CuCl_2PEO_8$, but its decomposition products from the low temperature process produced more [73] than [88] (ratio 4).

4. $SnCl_2PEO_{8,16}$: These electrolytes were fairly unstable both in air and in nitrogen, and their decomposition produced the same mass spectra in both atmospheres (Fig. 6). The spectra were characterized by less [73] than [88] (ratio 0·2). However, for these particular materials, it is believed that the molecular weight of PEO in the electrolyte was much lower than the starting value of 5×10^6 because of solution degradation during the casting step, a process known to occur in solution in the presence of certain multivalent cations.[12] The degradation was manifest in a greatly reduced viscosity of the polymer–salt solutions. In addition, the two electrolyte compositions,

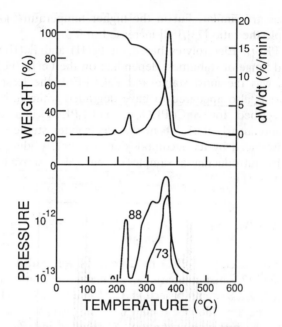

Figure 5. TG-MS of $CuCl_2PEO_8$ under nitrogen.

$n = 8$ and $n = 16$, showed intriguing differences in their behavior. The less concentrated electrolyte ($n = 16$) has low temperature weight losses at the same temperatures as the $n = 8$ system, but only half of the PEO (in the $n = 16$ electrolyte) was volatilized in these steps. The remainder was evolved in a separate process occurring around 360°C. The mass spectra produced by the products of both lower temperature

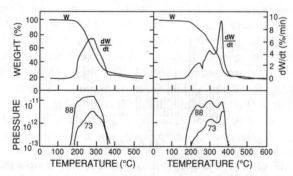

Figure 6. TG-MS of (a) $SnCl_2PEO_8$ and (b) $SnCl_2PEO_{16}$ in nitrogen.

weight losses are similar, but in the higher temperature loss for the $n = 16$ sample the ratio [73]/[88] increased to 0·5.

5. $FeCl_nPEO_8$: Electrolytes based on Fe(II) and Fe(III) chlorides had a broad range of stabilities, depending on the Fe(II)/Fe(III) ratio. $FeCl_2PEO_8$ was the most stable and $FeCl_3PEO_8$ the least. Some of these materials also appeared to have degraded while in solution, as already described for $SnCl_2PEO_{8,16}$. $FeCl_3PEO_8$ had the lowest [73]/[88] ratio, nearly 0·01, of all the electrolytes studied. The mass spectrum produced by its decomposition was nearly identical to that produced by introducing a sample of pure 1, 4 dioxane into the TG-MS system (Fig. 7).

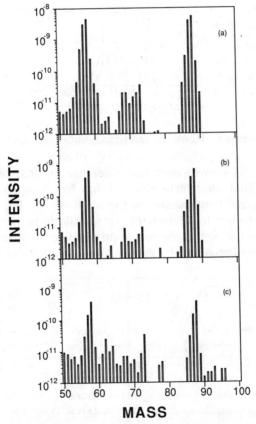

Figure 7. MS of (a) 1, 4 dioxane, (b) $FeCl_3PEO_8$ low temperature weight loss and (c) $FeCl_3PEO_8$ high temperature weight loss, under nitrogen.

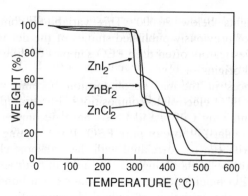

Figure 8. TG of $ZnCl_2$, $ZnBr_2$ and ZnI_2PEO_8 in nitrogen.

6. $ZnCl_2$, $ZnBr_2$, ZnI_2, $NiBr_2$ and $CoBr_2PEO_8$: These materials showed the simplest decomposition patterns of the electrolytes studied. In nitrogen and air, they degraded in single-step processes, with lower onset temperatures in air. Most of the compositions produced mass spectra containing more [88] than [73] (ratio of c. 0·2–0·5). It is interesting to note that the multiple decomposition steps observed with other electrolyte compositions often include one step similar to the single decomposition step observed with these electrolytes. One such example is the third loss process observed for $CuCl_nPEO_8$. An anion effect can be seen in the ZnX_2 series (Fig. 8). Stability decreased from the chloride to the iodide, although the mass spectra did not change significantly. This agrees with the CuX_nPEO_8 stability results for X = Cl, Br, although the mass spectra for the Cu electrolytes did show a halide dependence.

DISCUSSION AND CONCLUSIONS

PEO electrolytes have very different thermal decomposition characteristics depending on their specific composition. The differences begin at the level of the nominally 'pure' PEO used to prepare them. Samples of PEO obtained in various molecular weights from different suppliers were found to decompose differently, suggesting that the process of decomposition varies with catalyst type and its residual concentration. It is well known that high molecular weight PEO is prepared using several catalyst types (strontium and calcium amides and carbonates are common), and other catalysts are used for molecular weights below 20 000. This variability complicates the interpretation of previously published studies of the decomposition of PEO, since investigators often used PEO samples of different molecular weights and origins.

The differences in thermal decomposition become very obvious among specific PEO electrolyte compositions. For example, the PEO electrolytes containing $CdBr_2$ and $PbBr_2$ have decomposition characteristics quite similar to those of pure PEO. It is tempting to speculate that the similarity can be correlated with the transport characteristics of these electrolytes, whose high cation mobility suggests relatively weak cation/PEO binding. Since the carbon–oxygen bond is the most likely point for the initiation of radical decomposition processes,[10] weaker ion/polymer interaction may well influence polymer decomposition, though many other factors are surely quite important.

The second group of divalent electrolytes studied here, those containing Sn(II), Cu(I, II) and Fe(II, III), have thermal decomposition characteristics that are not easily correlated with conductivity or other properties. All degrade to some extent at relatively low temperatures and behave similarly in air and nitrogen.

The third group of electrolytes studied, the Zn(II), Ni(II) and Co(II) compositions, also have thermal decomposition characteristics which are not easily correlated with conductivity or other properties. These all degrade in one process at intermediate temperatures, and are slightly destabilized by air.

Certainly one of the more interesting observations from this study is that high concentrations of most of the metal salts investigated stabilize PEO towards oxidative degradation. To our knowledge, this has not been previously reported, though it is known that certain metal cations inhibit oxidative degradation by slowing the induction period for the reaction but then accelerate degradation[13] by forming radicals from hydroperoxides.

This study underscores the point that PEO samples from different origins can have quite different chemical characteristics and illustrates the considerable influence of the presence of multivalent cation salts on those properties.

ACKNOWLEDGEMENTS

This work was supported by a grant from the Defense Advanced Research Projects Agency through a contract monitored by the Office of Naval Research. Additional support from the NSF-MRL program under grant no. DMR88-19885 is gratefully acknowledged. Special thanks to H. S. Choe and W. A. Freeman, who synthesized the $CuCl_nPEO_8$, and to both C. R. Miao, who synthesized the $FeCl_nPEO_8$, and C. Gell, who prepared the $CuBr_nPEO_8$ samples as part of their senior research projects.

REFERENCES

1. Armand, M. B., Chabagno, J. M. & Duclot, M. In: *Fast Ion Transport in Solids,* ed. P. Vashishta, J. N. Mundy & G. K. Shenoy. North-Holland, New York, 1979, p. 131.
2. Huq, R. & Farrington, G. C., Ion transport in divalent cation complexes of poly(ethylene oxide). *Solid State Ionics,* **28–30** (1988) 990–3.
3. Vincent, C. A., *Prog. Solid State Chem.,* **17** (1987) 145.
4. Huq, R. & Farrington, G. C., Solid polymeric electrolytes formed by poly(ethylene oxide) and transition metal salt. *J. Electrochem. Soc.,* **135** (1988) 524–8.
5. Yang, H. & Farrington, G. C., unpublished results.
6. Madorsky, S. L. & Straus, S., Thermal degradation of polyethylene oxide and polypropylene oxide. *J. Polym. Sci.,* **36** (1959) 183–94.
7. Bortel, E. & Lamot, R., Untersuchung des abbaus hochmolekularer polyathylenoxide im festzustand. *Makromol. Chem.,* **178** (1977) 2617–28.
8. Bortel, E., Hodorowicz, S. & Lamot, R., Relation between crystallinity degree and stability in solid state of high molecular weight poly(ethylene oxide)s. *Makromol. Chem.,* **180** (1979) 2491–8.
9. Grassie, N. & Perdomo-Mendoza, G. A., Thermal degradation of polyetherurethanes: Part 1—Thermal degradation of poly(ethylene glycols) used in the preparation of polyurethanes. *Poly. Deg. Stab.,* **9** (1984) 155–65.
10. Grassie, N. & Perdomo-Mendoza, G. A., Thermal degradation of polyetherurethanes: Part 2—Influence of the fire retardant, ammonium polyphosphate, on the thermal degradation of poly(ethylene glycol). *Poly. Deg. Stab.,* **10** (1985) 43–54.
11. Taoda, H., Hayakawa, K., Kawase, K. & Kosaka, M., Thermal oxidative aging of poly(ethylene oxide) used as latent-heat-type thermal storage material. *Kobunshi Ronbunshu,* **43(6)** (1986) 353–9.
12. Bailey, F. E. Jr & Koleske, J. V., *Poly(ethylene oxide).* Academic Press, New York, 1976, p. 54.
13. Kelen, T., *Polymer Degradation.* Van Nostrand, New York, 1983, p. 182.

Polymer Electrolytes Based on Poly(Ethylene Oxide) and Zinc Chloride

V. C. Z. Bermudez, J. Morgado, T. M. A. Abrantes, & L. Alcácer

Dipartimento Química, Instituto Superior Técnico, Av. Rovisco Pais, P-1096, Lisboa Codex, Portugal

ABSTRACT

Differential scanning calorimetry, complex impedance spectroscopy and electrochemical studies have been carried out on polymer electrolyte films obtained by the addition of zinc chloride to poly(ethylene oxide). Detailed studies were performed in the $(PEO)_n-ZnCl_2$ system, in the range of concentrations $n = 4$–40, for the above general formula. From the differential scanning calorimetry results and X-ray data, a tentative pseudo-equilibrium phase diagram has been drawn. Two defined complexes appear to exist and two eutectics were detected for $n = 14$ and $n = 34$. Conductivities of the order of $10^{-7} \, S \, cm^{-1}$ at room temperature were found for $n = 8$ and $n = 20$.

Two different methods of preparation were used, in order to determine their influence on the properties of the electrolytes.

INTRODUCTION

The great potentialities of polymer electrolytes have been demonstrated by many investigations, since the initial reports by Fenton *et al.*[1] and Armand *et al.*[2] More recently, divalent cation polymer complexes have recieved considerable attention.[3–6]

The properties of these materials are strongly dependent on the phase diagram which determines the proportions of crystalline complexes and amorphous solid solutions. Ionic conduction only occurs in the amorphous phase and both cations and anions are mobile.[7]

In continuation of our previous work on multivalent cation polymer electrolytes we investigated the PEO–zinc chloride system in more detail.

In order to determine the influence of the method of preparation on the properties of these materials, we used two different methods, namely the usual solvent casting technique and a diffusion method previously referred.[8]

The main purpose of the use of the diffusion method has been to determine how the ionic conductivity changes with the salt concentration in the $PEO + ZnCl_2$ system, at temperatures below the melting point of the pure PEO (63°C).

The solvent casting method of preparation of these polymeric electrolytes consists in the mixing of the polymer(s) and the salt(s) in the same or different solvents, followed by their evaporation. This method gives a material which has a heterogeneous salt distribution. Generally, one obtains three phases:[8]

(i) a crystalline phase due to the PEO crystallites;
(ii) one or more crystalline phases, due to the crystallites of the polymer–salt complexes;
(iii) an amorphous phase with some 'free' salt.

At temperatures not high enough to dissolve the crystalline phase containing the salt, the amorphous phase will determine the ionic conductivity of the sytem.

By looking at the pseudo-equilibrium phase diagram for the $PEO + ZnCl_2$ system which we report in this communication, we can see that only for temperatures above approximately 130°C, there will be no more crystalline phases. Therefore, only for temperatures above this value, can a direct correspondence between conductivity and salt concentration be established. For lower temperatures one must have the phase diagram to know how much salt really contributes to the ionic conductivity.

Thus, if we want to know how the conductivity changes with the salt concentration near room temperature, we must avoid the formation of the crystalline complexes. This would allow for all the salt to participate in the conduction process. In trying to prevent their formation, we used a preparation procedure based on the diffusion method. We expected that by introducing the salt in a predefined polymer microstructure it would diffuse only into the amorphous phase, preventing the formation of crystalline complexes. A crystallisation process may occur for the more concentrated samples.

EXPERIMENTAL

Preparation of the Polymer Electrolyte Films

Two different methods were used to prepare the films. The conventional solvent casting technique was used to prepare a set of films on which differential scanning calorimetry (DSC) measurements and X-ray diffraction, as well as impedance spectroscopy and electrochemical studies were performed. Another set of films was prepared by introducing the zinc salt by diffusion in a non-charged poly(ethylene oxide) film immersed in an ethyl acetate solution containing the salt. On those, impedance spectroscopy was performed and the diffusion coefficient of $ZnCl_2$ in PEO was estimated.

(a) Solvent Cast Films

$(PEO)_n ZnCl_2$ films, in the range $n = 4$–40 were prepared by casting acetonitrile solutions containing the appropriate amounts of PEO ($MW = 5 \times 10^6$, Aldrich) into glass formers on Teflon sheets. Residual solvent was removed in vacuum for 48 h and the films were then put in a dry box where they were kept for approximately 4 weeks until used. All subsequent handling and studies were performed in a dry atmosphere.

(b) Diffused films

Using commercial PEO ($MW = 5 \times 10^6$, Aldrich) and acetonitrile as solvent, a large film of pure PEO was prepared by the solvent casting method. Then, nine circular films of 19 mm diameter were cut and maintained in vacuum for 48 h. Subsequently the films were weighed in a glove bag under argon, this procedure being repeated until constant weight was attained. The films were then immersed in a solution of zinc chloride in ethyl acetate (pure PEO is insoluble in this solvent), allowing diffusion through both sides. The amount of salt incorporated in each film was controlled by the immersion time. It is assumed that the salt diffuses only into the amorphous phase, which we considered as 20% of the initial weight (Table 1). The films

Table 1. Set of Films Obtained by the Diffusion Method

Film	Diffusion time (min)	Weight increase (%)	O/Zn in 20%	Concentration in 20% (mol dm^{-3})	Bulk concentration in 100% (mol dm^{-3})
1	5	5·21	11·88	1·754	0·376
2	10	9·39	6·59	2·966	0·667
3	20	12·71	4·87	3·827	0·894
4	60	25·12	2·46	6·437	1·697
5	90	26·48	2·34	6·677	1·781
6	180	39·06	1·58	8·575	2·527
7	290	49·51	1·25	9·815	3·105
8	420	49·87	1·24	9·853	3·124
9	6139	53·77	1·15	10·254	3·330

removed from the solution were kept under vacuum for several days before they were weighed, until constant weight, in the glove bag. During handling, the films were never heated, to prevent destruction of the microstructure.

Differential Scanning Calorimetry and X-Ray Diffraction

DSC curves were obtained in a Mettler TA3000 System, calibrated with an iridium standard. The samples were annealed at 210°C and then left at ambient temperature for 6 days. After this pre-conditioning treatment, they were quenched from room temperature to −115°C and then thermally analysed at a heating rate of 20°C min^{-1}.

Powder X-ray diffraction patterns were obtained in a Philips PM 8000 diffractometer.

Temperature Dependence of Electrical Conductivity

Electrical conductivity data were obtained by the usual complex impedance analysis (Hewlett-Packard HP 4192A Impedance Analyzer controlled by a HP85B microcomputer) over the frequency range of 5 Hz to 13 MHz. The films were cut into 1·3 cm^2 discs, pressed between two platinum electrodes and mounted in a home-made conductivity co-axial cell provided with a temperature controller.

(a) Solvent Cast Films

Impedance spectra were taken at temperatures in the range 20–100°C, at 5°C intervals along three heating/cooling cycles. At each temperature, the system was left for at least 15–20 min to reach thermal equilibrium. It was found that the data from the first heating are in some cases irregular at the lower temperatures. This was probably due to contact problems in the electrode/electrolyte interface, derived from the spherulitic microstructure of the films. Subsequent cycles gave, in general, reproducible results. The conductivity values reported in this communication correspond to the 2nd thermal cycle (heating branch), this meaning that data were obtained from samples with previous thermal treatment.

(b) Diffused Films

The conductivity cell and the films were handled in a glove bag under argon. Conductivity measurements were made for samples 1, 3, 4, 5, 6, 8 and 9. They were submitted to a first heating/cooling cycle up to a maximum temperature of 60°C in order to minimise contact problems between the films and the platinum electrodes. This also allows the salt concentration along the film length to become more uniform, destroying the possible parabolic profile resulting from the diffusion process through both sides of the film. In this first cycle, the temperature was never higher than 60°C to prevent melting of the pure PEO crystallites present in the films and thus not destroying the microstructure.

Usually, the first heating gives conductivity values much lower than those obtained in the first cooling. Data from the first cooling are similar to those of the second heating.

During the measurements, the time needed for stabilising the impedance curve was about 2–3 h.

RESULTS AND DISCUSSION

Pseudo-equilibrium Phase Diagram

A pseudo-equilibrium phase diagram of the PEO rich region based on DSC and X-ray diffraction data is presented in Fig. 1, where component 1 is the monomeric unit of PEO and the compositions are expressed as the molar fraction of zinc chloride. Only DSC heating traces taken after the preliminary treatment described in the ex-

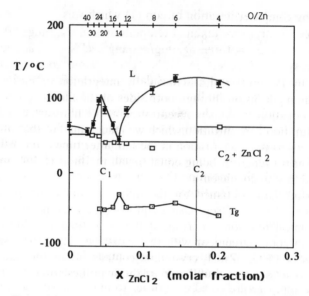

Figure 1. The PEO–ZnCl$_2$ pseudo-equilibrium phase diagram. Note the presence of two complexes and two eutectics for $n = $ O/Zn > 4.

perimental section were taken into account. The error bars shown in the figure should be taken as the estimated error in the determination of the melting peaks under the conditions described above and not in the position of those peaks in a real equilibrium phase diagram. The compositions of the two complexes C_1 and C_2 correspond to values of $x = 0.043$ ($n = 22$) and x between $x = 0.143$ and $x = 0.2$ (n between 4 and 6). The two eutectic compositions were evaluated at $x = 0.0286$ or $n = 34$ and $x = 0.066$ or $n = 14$, respectively. The eutectic temperatures are approximately 50°C and 42°C respectively. Glass transition temperatures are also shown for some compositions. X-ray diffraction data are in agreement with this diagram and show that for the composition of $x = 0.2$ ($n = 4$) there is already evidence of crystalline zinc chloride.

Diffusion Coefficient of ZnCl$_2$ in PEO

Using the values of the weight changes and the diffusion times, we have made an estimate of the diffusion coefficient of ZnCl$_2$ in the polymer by computer iteration of eqn (1) which was

$$\frac{\langle c \rangle}{c_s} = 1 - \frac{8}{\pi^2} \sum_{v=0}^{\infty} \frac{1}{(2v-1)} e^{-(2v-1)^2 \pi^2 \frac{Dt}{a^2}} \qquad (1)$$

obtained by Perlmutter and Scrosati[9] by integration of Fick's second law, assuming diffusion through both sides of the films.

In this equation, c_s is the concentration at the film surface (assumed as the saturation concentration which we made equal to that of film 9), $\langle c \rangle$ is the average concentration in the film after time t, a is half of the film thickness ('length'), being equal for all the films (0·18/2 mm) and D is the diffusion coefficient.

The mean value obtained for the diffusion coefficient of $ZnCl_2$ in PEO was 8×10^{-10} cm^2/seg. Using eqn (1) we are assuming that there is no chemical reaction occurring at the same time as diffusion. The data are in good agreement with that equation. Therefore, since the formation of PEO + $ZnCl_2$ crystalline complexes has the nature of a chemical reaction, this could support the hypothesis that their formation does not occur.

Temperature Dependence of Electrical Conductivity

(a) Solvent Cast Films

Representative a.c. conductivity results are summarised in Fig. 2. The conductivities of compositons corresponding to $n = 14$, 16 and 20 show three Arrhenius type regions with different activation energies.

From the analysis of the pseudo-equilibrium phase diagram shown in Fig. 1 we would indeed expect three different conductivity patterns for compositions $n = 16$ and 20, when going from room temperature to 100°C. They should correspond to: (i) a solid phase at low temperatures in which the conductivity would be restricted to the usually low percentage (20%) amorphous PEO–$ZnCl_2$ solution; (ii) a two-phase region with steeply increasing conductivity due to the growing amount of liquid phase added to the effect of its characteristic activation energy; (iii) a completely liquid phase with its own activation energy.

For $n = 14$ (a eutectic composition) we would expect two separate regions ($T > 42$°C and $T < 42$°C) with different activation energies or a true VTF behaviour above 42°C. One of the samples has indeed shown a VTF type behaviour above 40°C and very high resistances below this temperature, however the data were scattered and therefore dis-

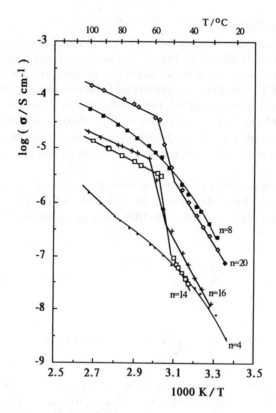

Figure 2. Arrhenius plots of low conductivity versus $1/T$ for $(PEO)_nZnCl_2$

carded. The data reported in Fig. 2 have similar behaviour to those found for $n = 16$ and $n = 20$. Further data are presently being collected. For the $n = 4$ and $n = 8$ compositions we would expect two region plots in the temperature range studied—an Arrhenius behaviour in the solid phase and a different pattern in the two-phase region. In the two-phase region, the conductivity should increase due to both the increase of the quantity of liquid phase and to the liquid phase characteristic activation energy. This could eventually lead to a VTF-like curvature due to the accentuated curvature of the liquidus line. This hypothesis needs, however, more experimental evidence and more detailed analysis which we are presently pursuing.

(b) Diffused Films

The temperature dependence of the conductivity for the second heating cycle is shown in Fig. 3. As in the second heating cycle the temperature was raised well above the melting temperature of pure PEO (63°C), it would be expected that the conductivity values for the subsequent cycles would be quite different as a consequence of the diffusion of the salt into the molten PEO initial crystallites. But, the obtained results are not in agreement with these expectations. Usually, the conductivity curves for the second and third heating cycles are very similar.

It is possible that salt diffusion into the molten PEO in the second heating may be insignificant, in spite of the 2–3 h wait at each temperature. It is also possible that the redistribution of the salt will

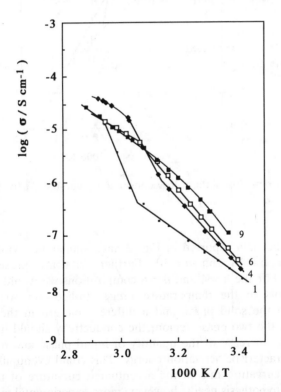

Figure 3. Arrhenius plots of log conductivity versus $1/T$ for the PEO–ZnCl$_2$ diffused films (samples 1, 4, 6 and 9).

lead to a situation in which the number of mobile carriers will remain the same.

Since only the amorphous phase conducts and if the salt only diffuses into this phase as we expect, the conductivity plots should all be of the same type, probably a true VTF. As is shown in Fig. 3, only film 9 (the saturated one) has this behaviour.

Concentration Dependence of Conductivity

(a) Solvent Cast Films

The evolution of the conductivity with n in $(PEO)_n ZnCl_2$ is shown in Fig. 4. The highest conductivities are observed for $n = 8$ at low temperatures, for $n = 20$ between 55 and 80°C and for $n = 14$ above 80°C. The lowest conductivity is found for $n = 4$.

(b) Diffused Films

As it has already been pointed out, we assumed that the salt diffuses only into the amorphous phase of the films. Thus, to calculate the real

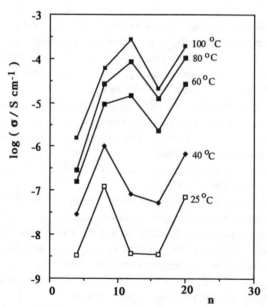

Figure 4. Conductivity as a function of n in $(PEO)_n ZnCl_2$ (solvent cast films).

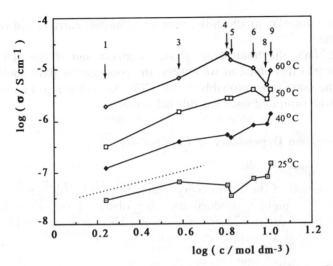

Figure 5. Conductivity as a function of concentration (mol dm^{-3}). The numbers pointing at the experimental points indicate the sample number. The dotted line (slope 1) corresponds to complete dissociation.

salt concentration we should know their degree of crystallinity. Since this has not been accurately determined we assumed that in all the films (because they are, all of them, parts of a larger single film) we have approximately 80% crystallinity. Thus, only 20% of the initial weight of each film is accessible to the salt.

We note that, if the real crystallinity is different (as it probably is), then the concentrations to be determined will all be affected by a nearly constant factor. In Table 1, the estimated concentration values are shown for the studied samples.

The isothermal variations of the conductivity with salt concentration (in mol dm^{-3}) are shown in Fig. 5.

Similar studies about the concentration dependence of the conductivity, have been done by Le Nest,[8] Cheradame & Le Nest[10] and Killis et al.[11] for macromolecular clusters based on PEO and LiClO$_4$. The method of preparation was the solvent casting one, and the study was done for constant free volumes (constant values of $T - T_g$, from $T - T_g = 50°C$ to $T - T_g = 125°C$). The curves that these authors obtained are very similar to those reported here.

Without the formation of crystalline complexes, all the salt would be free to participate in the conduction process. Thus, the dependence of the conductivity on the salt concentration should be similar to that observed in liquid electrolytes.

For PEO networks, the conductivity usually goes through a maximum with increasing salt concentration. This is due to the fact that, at low concentrations, the conductivity is proportional to the number of charge carriers, while at high concentrations the mobility of the charges decreases with increasing ionic interactions.[10] For low salt concentrations there should be a linear variation, with a slope equal to unity corresponding to total dissociation of the salt.[8,10–12]

In our study of the PEO–$ZnCl_2$ system (Fig. 5), we observe that, near room temperature, the conductivity is proportional to the salt concentration up to concentrations of the order of $4 \, mol \, dm^{-3}$, which is a rather unexpected high value. Apparently the elementary units which generate the charge carriers ($ZnCl_2$ molecules or higher aggregates) are completely dissociated. We also find that for higher temperatures the conductivity increases with salt concentration raised to a power higher than one. Similar behaviour is described for the PPO–sodium tetraphenylborate[10] system and is accounted for by the assumption that the salt is present in the form of multiplets in equilibrium with ions and ion pairs. A slope greater than unity is explained by the combination of these equilibria.

The concentration for which there is a maximum conductivity varies with temperature, this meaning that the influence of ionic interactions in the mobility of the charges also varies with temperature. The possible reason for the more irregular evolution of the curves for samples 8 and 9 could be excess of salt. Presently, we do not know if in this concentration range there can be formation of portions with definite stoichiometries. They would not be crystalline as no evidence of melting is obtained from the conductivity curves which, for these samples, show a VTF behaviour in the whole temperature range. The conductivity versus temperature curves, as we can see in Fig. 3, show a transition from an Arrhenius to a VTF behaviour, at a concentration corresponding to that of the sample 6.

As was already pointed out, the conductivity versus salt concentration studies in the samples prepared by the usual solvent casting process should be done at temperatures above those needed to dissolve the rich salt crystallites present. Therefore, a direct comparison of the results obtained for these samples with those obtained for the samples prepared by the solvent casting process and described in the previous section is not possible.

There is also another point which has not yet been referred. The commercial polymers usually have impurities such as catalytic residues or inhibitors. This may be a very important factor in this kind of study as has been shown by D. Fauteux *et al.*[13] for the systems PEO–LiX $(X = (ClO_4)^-, (CF_3SO_3)^-)$.

CONCLUSIONS

The phase diagram of the (PEO)–$ZnCl_2$ system proves the existence of two complexes ($n = 22$ and $n = 4$–6) and two eutectics ($n = 14$ and $n = 34$). The highest ionic conductivities at room temperature are found in solvent cast films for $n = 8$ and for the saturated diffused film. The conductivity values are of the same order of magnitude and the temperature dependence is very similar. It is shown that in the films obtained by diffusion the conductivity near room temperature is approximately proportional to the salt concentration in the low concentration range. This indicates that at room temperature the salt is completely dissociated in the polymer for the lower concentrations, unless some complex equilibria take place between multiplets, ions and ion pairs.

The diffusion technique is of some interest, providing that several other studies, such as DSC or X-ray diffraction can be done to clarify the possibility of formation of crystalline complexes.

A study in the range of lower concentrations should, and will be undertaken in the near future. Special attention will be paid to the kinetics of the salt dissociation and DSC characterisation.

ACKNOWLEDGEMENT

This work was supported by JNICT contract 713-85-51.

REFERENCES

1. Fenton, B. E., Parker, J. M. & Wright, P. V., *Polymer,* **14** (1973) 589.
2. Armand, M., Chabagno, J. M. & Duclot, M., In: *Fast Ion Transport in Solids,* ed. P. Vashishta, J. N. Mundy & G. K. Shenoy. North Holland, New York, 1979, p. 131.

3. Armand, M. B., *Solid State Ionics, Lake Tahoe,* 1985, Extended Abstract, SSI 85.
4. Abrantes, T. M. A., Alcacer, L. J. & Sequeira, C. A., *Solid State Ionics* **18, 19** (1986) 315–20.
5. Yong, L. L., Huq, R., Farrington, G. C., Chiadelli, G. & McGhie, A. R., *Solid State Ionics,* **18, 19** (1986) 291.
6. Patrick, A. J., Glasse, M. D., Latham, R. J. & Linford, R. G., *Solid State Ionics,* **18, 19** (1986) 1063–7.
7. Armand, M. B., In: *Polymer Electrolyte Reviews,* ed. J. R. MacCallum & C. Vincent. Elsevier Applied Science, London, 1987, pp. 1–22.
8. Le Nest, J. F., These d'Etat Es-Sciences Physiques, Grenoble, 1985.
9. Perlmutter, D. & Scrosati, B., *Solid State Ionics,* **27** (1988) 115–23.
10. Cheradame, H. & Le Nest, J. F., In: *Polymer Electrolyte Reviews,* ed. J. R. MacCallum & C. Vincent. Elsevier Applied Science, London, 1987, pp. 103–38.
11. Killis, A., LeNest, J. F., Gandini, A. & Cheradame, H., *Macromolecules,* **17** (1984) 63.
12. Ratner, M. A., In: *Polymer Electrolyte Reviews,* ed. J. R. MacCallum & C. Vincent. Elsevier Applied Science, London, 1987, pp. 173–236.
13. Fauteux, D., Prud'Homme, J. & Harvey, P. E., *Solid State Ionics,* **28–30** (1988) 923–8.

Effect of a Low Molecular Weight Plasticizer on the Electrical and Transport Properties of Zinc Polymer Electrolytes

H. Yang & G. C. Farrington

*Department of Materials Science and Engineering,
University of Pennsylvania, 3231 Walnut Street,
Philadelphia, Pennsylvania 19104, USA*

ABSTRACT

PEO-based electrolytes containing $ZnBr_2$ and poly(ethylene) glycol dimethyl ether (PEGM) as a plasticizer were prepared by solution casting. The presence of PEGM results in a significant increase in electrolyte conductivity, particularly at room temperature. The transport number of Zn(II), obtained by DC polarization measurements, was about 0·4 at 53°C for a sample of composition, $ZnBr_2(0\cdot5PEGM + 0\cdot5PEO)_{16}$.

INTRODUCTION

Polymer electrolytes are a new class of ionic conducting materials formed by incorporating inorganic salts into ion-coordinating polymers. High molecular weight, typically 5×10^6, poly(ethylene oxide) (PEO) is the most popular polymer host for such electrolytes because of its exceptional ability to solvate ionic salts.[1] Most PEO-based electrolytes have good mechanical strength and achieve high conductivity at about 100°C. However, they suffer from low conductivity below 65°C because of high crystallinity. The incorporation of a plasticizer into the electrolyte composition may reduce crystallinity at lower temperatures and therefore improve the conductivity of PEO-based electrolytes, particularly at room temperature.

Ito et al.[2] have studied the effect of the addition of poly(ethylene) glycol (PEG) on the conductivity of PEO-based $LiCF_3SO_3$ electrolytes. They found that the incorporation of PEG with a molecular weight 600 or less could significantly improve the total conductivity of the electrolytes at room temperature. However, as Kelly et al.[3] have pointed out, the hydroxyl end groups of PEG react with lithium metal, making such electrolyte compositions unsuitable for battery applications.

This paper describes the use of low molecular weight poly(ethylene oxide) glycol dimethyl ether (PEGM) as a plasticizer for PEO-based electrolytes. PEGM has essentially the same repeat unit as high molecular weight $PEO(-(-CH_2-CH_2-O-)-)$ and can therefore be expected to be quite compatible with it. In addition, the ether end groups of PEGM should be more stable electrochemically than the hydroxyl end groups of PEG. The specific compositions studied were PEO-based $ZnBr_2$ electrolytes in which PEGM (MW = 400) comprised 0, 25%, 50% or 75% (wt) of the polymer fraction. The effect of PEGM on the electrical and transport properties of these plasticized polymer films has been studied.

EXPERIMENTAL PROCEDURES

Sample Preparation

$ZnBr_2((1 - x)PEGM + xPEO)_{16}$ films with $x = 1·0$, $0·75$, $0·50$ and $0·25$ were prepared by a two-solvent solution casting technique. PEO with an average molecular weight of 5×10^6, obtained from Polyscience Inc., was used as-received. PEGM with a molecular weight of 400 from Polyscience Inc. was annealed above 100°C under vacuum for more than 24 h before use. Stoichiometric amounts of $ZnBr_2$ (ultra-pure, Aldrich), PEO and PEGM were dissolved in a 1:5 mixture of anhydrous ethanol and acetonitrile (Aldrich) and stirred for about 20 h to form a homogeneous solution. The solution was cast onto a Teflon plate and the solvents were slowly evaporated. The resulting coherent films, typically 80–140 μm, were dried at about 90°C under vacuum for more than 24 h and then stored in an argon dry-box.

Thermal Analysis

The thermal stability of the Zn(II) polymer films was studied by thermogravimetry (TGA) using a DuPont 951 thermogravimetric analysis unit (TGA) at a heating rate of 5°C/min. Differential scanning calorimetry (DSC) was also carried out using a DuPont 910 DSC with a heating rate of 10°C/min. Samples were typically 6–8 mg.

AC Impedance

Conductivity measurements were performed using complex AC impedance analysis over the temperature range 30–130°C and the frequency range 10^2–10^5 Hz, using a Solartron 1174 frequency response analyser under computer control.

DC Polarization

DC polarization experiments using a symmetric cell with two non-blocking zinc electrodes were used to estimate the transport number of zinc cations in the sample electrolytes at 53°C. A 1286 Electrochemical Interface coupled with a HP computer was used to monitor the current change with time under a constant applied potential. The length of each experiment was about 80 min. Similar DC experiments were conducted at 53°C on a cell with two platinum blocking electrodes.

RESULTS AND DISCUSSION

Thermal Stability

Typical TGA curves for both $ZnBr_2(PEO)_{16}$ and $ZnBr_2(0.5PEGM + 0.5PEO)_{16}$ are shown in Fig. 1. The initial 2·5% weight loss is mainly the result of moisture absorbed during the process of sample loading. A pure PEO-based sample was found to be stable up to a final one-step decomposition around 345°C. However, a sample with 50% PEGM started to lose weight continuously around 150°C culminating with a final decomposition around 335°C. The weight loss at lower temperatures is most certainly the result of the evaporation of PEGM.

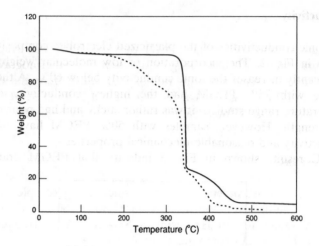

Figure 1. TGA curves for $ZnBr_2(PEO)_{16}$ (solid line) and $ZnBr_2(0.5PEGM + 0.5PEO)_{16}$ (dashed line).

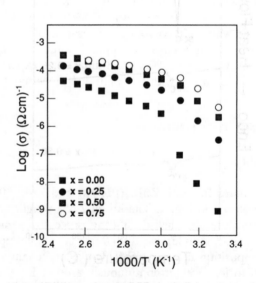

Figure 2. Conductivity of $ZnBr_2(xPEGM + (1 - x)PEO)_{16}$ with $x = 0.00$, 0.25, 0.50 and 0.75.

Conductivity

The ionic conductivities of the plasticized electrolyte compositions are shown in Fig. 2. The incorporation of low molecular weight PEGM significantly increases the ionic conductivity below 60°C. Although the sample with 75% PEGM had the highest conductivity over the temperature range studied, it was rather sticky and had poor mechanical strength. However, samples with 50% PEGM have both high conductivity and reasonable mechanical properties.

DSC results shown in Fig. 3 indicate that PEGM dramatically

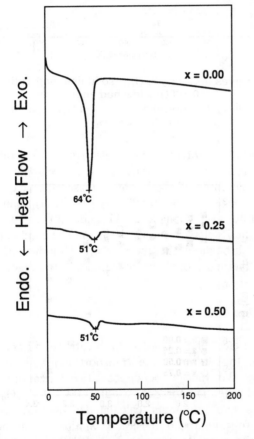

Figure 3. DSC curves of $ZnBr_2(xPEGM + (1 - x)PEO)_{16}$ with $x = 0.00$, 0.25 and 0.50.

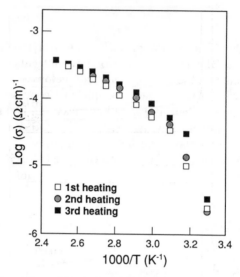

Figure 4. Conductivity of $ZnBr_2(0.5PEGM + 0.5PEO)_{16}$ during three temperature cycles.

reduces the crystallinity of PEO-based electrolytes at room temperature, even for PEGM concentrations as low as 25%. This appears to lead directly to the observed significant increase of ionic conductivity.

The conductivity of a $ZnBr_2(0.5PEGM + 0.5PEO)_{16}$ sample during three temperature cycles between 30 and 130°C has also been measured, and the results are shown in Fig. 4. The conductivity of this sample is high throughout the thermal cycles, indicating that such electrolytes are thermally stable in this temperature range.

Transport Number

Most PEO-based electrolytes conduct both cations and anions, so it is quite important to determine the transport numbers of each ion. Figure 5 shows the time dependence of polarization current at different applied potentials for $ZnBr_2(0.5PEGM + 0.5PEO)_{16}$ at 53°C. The Zn(II) transport numbers calculated from DC polarization measurements are summarized in Table 1. Similar measurements were performed on the pure PEO-based $ZnBr_2(PEO)_{16}$ sample, and the resulting transport number is about 0.18 at 82°C. Note that the reason

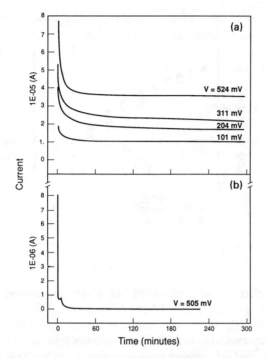

Figure 5. Current change with time at constant applied potentials for
(a) Zn/ZnBr$_2$(0·5PEGM + 0·5PEO)$_{16}$/Zn at 53°C; (b) Pt/ZnBr$_2$(0·5PEGM +
0·5PEO)$_{16}$/Pt at 53°C.

Table 1. Transport Number of Zn(II)

$V_{applied}$ (mV)	$^a I_{initial}$ (μA)	$^b I_{stable}$ (μA)	t_+
101	25·25	10·34	0·41
204	40·48	16·51	0·41
311	54·85	21·00	0·39
524	89·27	36·00	0·40

^{a} Calculated from the equation: $I_{initial} = V_{applied}/R_b + R_i$
^{b} Obtained from DC polarization measurements.

these transport numbers are given to two significant figures is to demonstrate the linearity of the measurement with applied voltage, not to suggest that the data truly justify such an absolute statement about the true values of the transport numbers.

Electronic Conductivity

Figure 5 shows that when a constant potential is applied to a sample of $ZnBr_2(0.5PEGM + 0.5PEO)_{16}$ sandwiched between two platinum blocking electrodes, the current rapidly drops to zero. This result indicates that there is no evidence of electronic conductivity or extraneous electrochemical reactions occurring in the plasticized zinc film.

CONCLUSIONS

The use of PEGM as a plasticizer is a very successful method of producing PEO-based electrolytes with relatively high ionic conductivities for Zn(II). It is reasonable to expect that the addition of PEGM might also significantly improve the conductivities of other PEO electrolytes containing dissolved salts of divalent cations.

ACKNOWLEDGEMENTS

This work was supported by a grant from the Defense Advanced Research Projects Agency through a contract monitored by the Office of Naval Research. Additional support from the NSF-MRL program under grant no. DMR88-19885 is gratefully acknowledged.

REFERENCES

1. Vincent, C. A., *Prog. Solid State Chem.*, **17** (1987) 145.
2. Ito, Y., Kanehori, K., Miyauchi, K. & Kudo, T., Ionic conductivity of electrolytes formed from PEO–LiCF₃SO₃ complex with low molecular weight poly(ethylene glycol). *J. Mater. Sci.*, **22** (1987) 1845.
3. Kelly, K. E., Owen, J. R. & Steele, B. H., Poly(ethylene oxide) electrolytes for operation at near room temperature. *J. Power Sources*, **14** (1985) 13.

PEO Electrolytes Containing Eu(III)

Rokeya Huq & Gregory C. Farrington

Department of Materials Science and Engineering, University of Pennsylvania, 3231 Walnut Street, Philadelphia, Pennsylvania 19104, USA

ABSTRACT

Trivalent and divalent europium salts dissolve easily in poly(ethylene oxide) (PEO) to form solid polymer electrolytes. The specific compositions studied in this work were $PEO_{8-20}EuX_3$, where $X = Cl^-$ and Br^-. $PEO_{16}EuCl_3$ and $PEO_{16}EuBr_3$ were found to have essentially identical conductivities, the highest of the compositions measured are comparable to those previously measured for PEO electrolytes containing Eu(II). The Eu(III) electrolytes are strongly fluorescent over the temperature range of -80 to $100°C$ and have well-defined emission spectra at room temperature and above. Initial results indicate that conventional fluorescence and excitation spectroscopy may yield considerable information about the coordination and local environment of Eu(III) ions in PEO and perhaps of other fluorescent ions as well.

INTRODUCTION

The variety of polymer electrolytes that PEO can form is large and diverse. A few years ago researchers[1-3] began to study electrolytes formed by dissolving salts of divalent cations in PEO and related polymer solvents. Since then investigations of these materials have expanded considerably.[4-6]

We recently became interested in the possibility that salts of divalent and trivalent europium ions might dissolve in PEO and were curious whether optical spectroscopy might be used to provide insight into the nature of cation/polymer bonding and cation/anion association within such materials. We were pleased to find that PEO electrolytes

273

containing $EuCl_3$ and $EuBr_3$ are strongly fluorescent over the temperature range of -80 to $110°C$ and have well-defined emission spectra at room temperature and above.

This paper summarizes the results of a study of the preparation, characterization, optical properties and conductivities of PEO electrolytes formed with Eu(III) halides. The specific compositions studied were $PEO_{8-20}EuX_3$ where $X = Cl^-$ and Br^-.

EXPERIMENTAL PROCEDURES

Electrolyte Preparation

Electrolytes were prepared by solution casting using a two-solvent technique. $EuCl_3$ and $EuBr_3$ (Cerac, Inc. 99·9%), PEO (Polysciences, $MW = 5 \times 10^6$), acetonitrile (Aldrich 99·9%) and anhydrous ethanol (Gold Label, Aldrich) were used as-received. Anhydrous $EuCl_3$ dissolves readily in alcohol to give a pale yellow solution and anhydrous $EuBr_3$ gives a yellowish-green solution. The solutions were filtered and then mixed with PEO dissolved in acetonitrile. Films were cast on a Teflon plate, and the solvent evaporated in the usual manner.[7] The resulting electrolytes were colorless to faint yellow in the case of $EuCl_3$ and yellow in case of $EuBr_3$. The dried samples were stored in an inert atmosphere dry-box, where all subsequent handling was done.

Sample Characterization

Conductivities were measured by sandwiching electrolyte films between blocking Pt electrodes and recording the complex impedance/admittance spectrum of the cell from 10^2 to 10^6 Hz or $0·01$ Hz to 10^6 Hz. Conductivity samples were 1 cm in diameter and about $0·01$ cm thick (measured at the start of the experiment). The electrodes were spring-loaded to maintain good contact with the polymer films. The entire sample holder was contained in a glass cell, and measurements were carried out in a stream of dried argon.

Other physical characterizations included differential scanning calorimetry (DSC) and thermogravimetric (TGA) analyses carried out using a DuPont 990 system. In addition, the microstructures of the samples were studied using a Reichert optical microscope equipped with a Mettler FP5/FP52 hot stage and camera.

A Perkin–Elmer Lambda 4C UV/visible spectrophotometer with a maximum resolution of about 1 nm ($40 \, cm^{-1}$ at 500 nm) was used to obtain the absorption spectra. The fluorescence and excitation data were obtained on a Perkin–Elmer MPF-66 fluorescence spectrophotometer with a maximum resolution of about 0·5 nm. Both instruments had an effective range of 200–850 nm. The data were collected and processed with a Perkin–Elmer computer and software supplied with the instruments. Temperature-dependent spectra were obtained with the aid of an Oxford Instrument DN-1704 liquid nitrogen cryostat.

RESULTS AND DISCUSSION

Thermogravimetric Analysis

Upon initial heating from room temperature to 200°C, freshly prepared samples of $EuCl_3$–PEO and $EuBr_3$–PEO showed a gradual weight loss of 3–7% below 90°C. This loss process did not recur during a second heating and therefore was considered the result of the loss of solvent and water from the electrolytes. No further weight loss occurred until the onset of decomposition around 260°C for $EuCl_3$ electrolytes and 240°C for those containing $EuBr_3$. Qualitative observations indicated that both compositions are quite hygroscopic, the $EuBr_3$ compositions in particular.

Differential Scanning Calorimetry

The DSC for dry $PEO_{16}EuBr_3$ recorded upon cooling from room temperature to −80°C, shows an exotherm around 0°C. Upon subsequent heating, an endotherm was observed around 58°C due to the melting of crystalline PEO, but no endotherm corresponding to the melting of a crystalline salt–polymer complex was observed up to 180°C (Fig. 1(a)). When the sample was cooled to room temperature and reheated immediately, only a small eutectic endotherm around 42°C was observed, which indicates that the crystalliztion of pure PEO is slow in these electrolytes (Fig. 1(b)). The DSC of $PEO_{16}EuCl_3$ is very similar except that the exotherm upon cooling occurs at −10°C and on reheating the endotherm appears around 40°C.

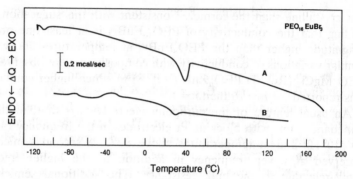

Figure 1. DSC scan for $PEO_{16}EuBr_3$ at 10°C/min, first heating (A), second heating (B).

Conductivity Measurements

EuBr₃ Electrolytes

Two compositions were prepared: PEO_8EuBr_3 and $PEO_{16}EuBr_3$. The conductivities of both during heating and cooling are shown in Fig. 2. A greater degree of hysteresis was observed with PEO_8EuBr_3 than with $PEO_{16}EuBr_3$, which may be explained by DSC and optical microscopy studies which found that the latter composition is much

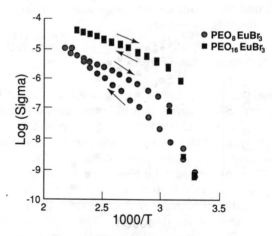

Figure 2. Conductivity of PEO_nEuBr_3 heating and cooling, $n = 16$ (■) and $n = 8$ (⊗).

less crystalline than the former. Consistent with this suggestion is the finding that the conductivity of $PEO_{16}EuBr_3$ is at least one order of magnitude higher than the PEO_8EuBr_3 at temperatures above 50°C. Similar variations of conductivity with composition were observed with PEO_nMgCl_2 (Ref. 7) PEO_nZnI_2 (Ref. 8) in which higher conductivity was reported for $n = 16$ than $n = 8$.

An usual feature of the europium electrolytes is that, in the a.c. impedance plots with blocking Pt electrodes in the frequency range of 10^2–10^6 Hz and at temperatures above 80°C, a second semicircle is observed at lower frequency in addition to the higher frequency semicircle due to the bulk resistance. The additional semicircle is possibly the result of redox reactions involving the Eu^{2+}/Eu^{3+} couple.

EuCl₃ Electrolytes

Compositions of $(PEO)_{8-16}EuCl_3$ have conductivities comparable to those of the corresponding bromide compositions. Figure 3 summarizes the conductivity results for the two compositions during heating and cooling. As with the bromide compositions, the conductivity of $(PEO)_{16}EuCl_3$ is much higher than that of $(PEO)_8EuCl_3$. Figure 4 compares the conductivities of chloride and bromide electrolytes for the same composition, $n = 16$. They are almost identical, indicating that there is little or no anion effect between the

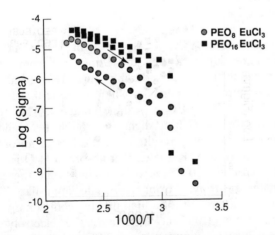

Figure 3. Conductivity of PEO_nEuCl_3 heating and cooling, $n = 16$ (■) and $n = 8$ (◉).

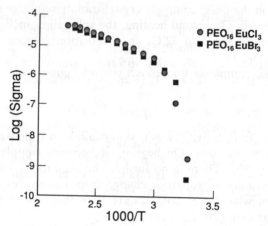

Figure 4. Comparison of conductivity of $PEO_{16}EuCl_3$ (⊙) and $PEO_{16}EuBr_3$ (■).

chloride and bromide complexes so far as the conductivity is concerned. Also, as mentioned earlier, there is very little difference in their thermal properties. These are the first cation complexes yet reported that show little or no anion effect.

Polarizing Microscopy

$PEO_{16}EuBr_3$

At room temperature, $PEO_{16}EuBr_3$ is light yellow in color and partially crystalline. Upon heating at 3°C/min, it becomes completely amorphous around 40°C, a considerably lower temperature than the melting point of pure PEO, behavior which suggests that PEO forms a eutectic with the $EuBr_3$–PEO crystalline complex. The only further color change up to 180°C is a slight shift to darker yellow around 90–100°C which presumably results from the loss of absorbed water. From TGA experiments it is known that absorbed H_2O is lost by that temperature. Around 210°C there is a change in the mechanical properties of the electrolyte, from rubbery to liquid-like. There is no indication of the 'salting-out' of a high melting complex up to 220°C.

On cooling from 210°C at 10°C/min, the electrolyte remains amorphous to 30°C. If maintained at 30°C for 5–10 min, it slowly crystallizes, beginning from the edges. After about 30 min, spherulites start to grow in the bulk; complete crystallization into large spherulites takes about 3–4 h. On second heating, the spherulites melt at an even lower temperature, around 32°C, an observation consistent with the DSC results (Fig. 1).

The bromide complexes fluoresce a yellowish-green color under UV illumination.

$PEO_{16}EuCl_3$

As with $PEO_{16}EuBr_3$, $PEO_{16}EuCl_3$ is crystalline at room temperature and faint yellow in color. On heating, it becomes completely amorphous around 38°C. There is no further change until about 185°C, at which the mechanical properties change from rubbery to liquid-like. This is a somewhat lower temperature than that observed with the bromide electrolyte (210°C).

On cooling, crystallization starts around 28°C, again from the edges, and takes about 2·5–3 h to crystallize completely. The crystals in this case are in the form of very small spherulites in contrast to the large, well-defined spherulites observed for $PEO_{16}EuBr_3$. As with the chloride complexes, upon re-heating the spherulites melt at a slightly lower temperature, around 35°C.

Under UV illumination the chloride complexes fluoresce in an orange-yellow color.

In summary, both $PEO_{16}EuBr_3$ and $PEO_{16}EuCl_3$ appear to form eutectics with PEO which melt around 38–40°C, much lower than the melting point of pure crystalline PEO, which is about 65°C. Both of the electrolytes are completely amorphous above 40°C, and there is a phase change from rubbery to liquid-like around 185°C for $PEO_{16}EuCl_3$ and 210°C for $PEO_{16}EuBr_3$. The recrystallization kinetics are slower than observed with most of the divalent cation complexes studied so far. The effect of different anions is seen only in the nature of crystal formation: $PEO_{16}EuBr_3$ forms big, well-defined spherulites, while the spherulites in $PEO_{16}EuCl_3$ are not so well-defined and very small.

Optical Spectroscopy

Since fluorescence and excitation spectroscopy experiments have yielded considerable information about the coordination and symmetry of Eu(III) and Eu(II) ions in Na–Eu(III)/Eu(II)-β''-Alumina,[9]

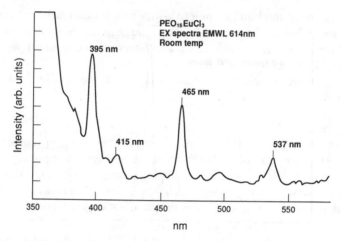

Figure 5. Excitation spectra of PEO$_{16}$EuCl$_3$ at room temperature, emission wavelength 614 nm.

it was interesting to explore whether they might be useful in obtaining information about local structure in polymer-based electrolytes containing Eu(II) and Eu(III). In fact, initial results are very encouraging. PEO–Eu(III) electrolytes yield sharp and reproducible spectra at room temperature and above, unlike Eu(III)-β''–alumina which must be cooled to 77 K to obtain sharp spectra.[9]

Figure 5 shows a typical excitation spectrum of PEO$_{16}$EuCl$_3$ at room temperature at an emission wavelength of 614 nm. In this composition, there seems to be only one type of excitation spectrum regardless of the emission wavelength and temperature. Figure 6 shows a typical emission spectrum at room temperature, the excitation wavelength being 395 nm. Again there is only one type of emission spectrum regardless of the peak excitation wavelength chosen and the temperature, which suggests that there is only one site which is occupied by the Eu(III), whether the material is in a multiphase crystalline or completely amorphous state. Hydration seems to have very little or no effect on the spectra, in contrast to what has been reported for Ni(II) and Co(II) electrolytes.[10] The group of peaks observed in the emission spectrum (Fig. 6) can be assigned to the $5D_0 \rightarrow 7F_2$ hypersensitive transition of Eu(III).[11,12] Additional spectroscopic studies of these electrolytes are in progress.

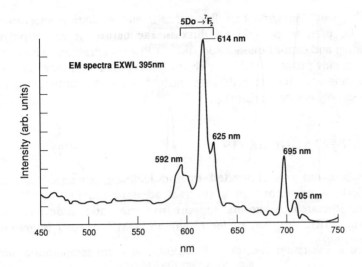

Figure 6. Emission spectra of $PEO_{16}EuCl_3$, at room temperature, excitation wavelength 395 nm.

CONCLUSIONS

PEO electrolytes containing Eu(III) salts form quite readily and are reasonably good conductors, presumably of anions. Whether the anion is chloride or bromide makes little difference to the conductivity. The main difference between the anions is in the nature of crystal formation. $PEO_{16}EuBr_3$ forms large, well-defined spherulites, while $PEO_{16}EuCl_3$ forms small spherulites that are not so well-defined. The electrolytes are thermally stable in nitrogen up to 240–260°C, behavior similar to that of some of the divalent cation electrolytes, e.g. Ni(II), Co(II)[4] and Zn(II).[8]

The most interesting aspects of these electrolytes are their optical properties. Those containing $EuCl_3$ fluoresce yellowish-orange from −80°C to 100°C with little or no apparent loss in intensity. Electrolytes containing $EuBr_3$ fluoresce yellowish-green. Preliminary spectroscopic studies suggest that Eu(III) is in the same coordination environment regardless of whether the material is in a multiphase crystalline or completely amorphous state. Hydration does not appear to alter the cation coordination as it does with those formed with Ni(II) and Co(II).[10]

This study suggests that fluorescence and excitation spectroscopy can be used to provide insight into the nature of cation/polymer bonding and cation/anion association in PEO electrolytes. The Eu(III) system may prove a model for electrolytes containing other lanthanide ions and provide considerable insight into the nature of local cation/polymer chain association.

ACKNOWLEDGEMENTS

This research was supported by the Defence Advanced Research Projects Agency through a contract administered by the Office of Naval Research. Additional support from the National Science Foundation, MRL Program under Grant no. DMR-8519059 is gratefully acknowledged.

REFERENCES

1. Yang, L. L., Huq, R., Farrington, G. C. & Chiodelli, G., *Solid State Ionics*, **18, 19** (1986) 291.
2. Patrick, A., Glasse, M., Latham, R. & Linford, R., *Solid State Ionics*, **18, 19** (1986) 1063.
3. Abrantes, T. M. A., Alcacer, L. J. & Sequeira, C. A. C., *Solid State Ionics*, **18, 19** (1986) 315.
4. Huq, R. & Farrington, G. C., *J. Electrochem. Soc.*, **135** (1988) 524.
5. Bonino, F., Pantaloni, S., Passerini, S. & Scrosati, B., *J. Electrochem. Soc.*, **135** (1988) 1961.
6. Bruce, P. G., Krok, F., Evans, J. & Vincent, C. A., *Solid State Ionics*, **27** (1988) 81.
7. Yang, L. L., McGhie, A. R. & Farrington, G. C., *J. Electrochem. Soc.* **133** (1986) 1380.
8. Yang, H. & Farrington, G. C., University of Pennsylvania, unpublished results.
9. Saltzberg, M., Ph.D. thesis, Department of Materials Science, University of Pennsylvania, 1988.
10. Huq, R. & Farrington, G. C., *J. Electrochem. Soc.*, **136** (1989) 1260.
11. Jorgensen, C. K. & Judd, B. R., *Mol. Phys.*, **8** (1964) 281.
12. Henrie, D. E., Fellows, R. L. & Choppin, G. R., *Coord. Chem. Rev.*, **18** (1976) 199.

Ionic Conductivity of Poly(Ethylene Oxide)-Nd(CF$_3$SO$_3$)$_3$

A. S. Reis Machado & L. Alcácer

Dipartimento Química, Instituto Superior Técnico, Av. Rovisco Pais, P-1096 Lisboa Codex, Portugal

ABSTRACT

Electrical conductivity, differential scanning calorimetry and electrochemical studies have been carried out on polymer electrolyte films obtained by addition of neodymium triflate to poly(ethylene oxide) in the concentration range of $n = 12$–40 in the general formula (PEO)$_n$–Nd(CF$_3$SO$_3$)$_3$. The films were prepared by the usual solvent casting process in anhydrous conditions. From differential scanning calorimetry data we can infer that an eutectic probably exists for $n = 23$ with a melting point in the range 35–45°C. The highest conductivity, $\sigma = 2 \times 10^{-7}\,\mathrm{S\,cm^{-1}}$ at 21°C, was found for this concentration, in agreement with the eutectic hypothesis. The temperature dependence of the conductivity follows the Vogel–Tamman–Fulcher (VTF) equation above 40°C, for most compositions.

INTRODUCTION

Polymers which dissolve salts and become ionic conductors have received considerable attention for potential applications in solid state devices, since the initial reports by Fenton et al.[1] and Armand et al.[2] More recently, the interest has extended to divalent cation salts.[3-6]

The properties of these materials are strongly dependent on the phase diagram which determines the proportions of crystalline complexes and amorphous solid solutions. Ionic conduction only occurs in the amorphous phase and both cations and anions are mobile.[7]

Attracted by the prospective potentialities of trivalent rare earth cation conductors, we decided to investigate the properties of complex compositions of neodymium triflate.

EXPERIMENTAL

Neodymium triflate, obtained by the reaction of neodymium carbonate with trifluoromethanosulphonic avid, was dissolved in a minimum volume of ethanol and added to solutions containing the appropriate amounts of PEO (MW $= 5 \times 10^6$, Aldrich) in propanol and acetonitrile.

The $(PEO)_n Nd(CF_3SO_3)_3$ films, for n in the range 12–40, were prepared by casting the above mentioned solutions into glass formers on Teflon sheets. Residual solvent was removed in vacuum for at least 3 days and the films were then put in a dry-box where they were kept for a few days until used. All subsequent handling and studies were performed in dry conditions.

DSC curves were obtained in a Mettler TA3000 System, calibrated with an iridium standard. The samples were transferred from the dry atmosphere to tight aluminium capsules in the air as fast as possible to avoid hydration and were thermally analysed at a heating rate of $20°C \, min^{-1}$.

Electrical conductivity data were obtained by the usual complex impedance analysis (Hewlett–Packard HP 4192A Impedance Analyzer controlled by a HP85B microcomputer) over the frequency range of 5 Hz to 13 MHz. The films were cut into $1·3 \, cm^2$ discs, $0·1$ mm thick, and then pressed between two platinum electrodes and mounted in a home-made conductivity co-axial cell which was provided with a temperature controller. Impedance spectra were taken at temperatures in the range of 20–80°C, at 5°C intervals along three heating/cooling cycles. The system was left for 2–3 h at each temperature to reach thermal equilibrium.

Cyclic voltammetry was used to characterize the electrochemical stability of the materials. The working and auxiliary electrodes were platinum and the reference was Ag/Ag^+.

RESULTS AND DISCUSSION

Differential Scanning Calorimetry

The DSC curves are presented in Fig. 1. Unfortunately the compositions studied were not enough to determine a phase diagram for this system. From the DSC curves we can, however, establish a few

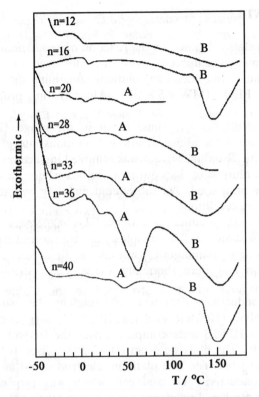

Figure 1. DSC curves for $(PEO)_n Nd(CF_3SO_3)_3$.

relevant facts. The degree of crystallinity was estimated for the different materials by comparing the intensity of the glass transition event with those found in completely amorphous samples. The values go from 27% for $n = 28$ to 63% for $n = 4$ going through a maximum value of 78% for the $n = 36$ composition. One of the main features of the DSC curves is the evidence of at least one PEO–Nd(CF₃SO₃)₃ complex, probably for $n = 36$ indicated in Fig. 1, associated with the A melting peak occurring at approximately 34°C. A second peak, indicated by B in the figure was associated to the possible dehydration of the material. In fact, Roberts & Bykowski[8] have studied the thermal properties of rare earth triflates and found that they are very hygroscopic, tending to form stable hydrides of formula RE(CF₃SO₃)₃.9H₂O (RE = rare earth). The dehydration of the neodymium salt starts at approximately 90°C and was observed by us as a peak similar to event B represented in Fig. 1. The hydration of our samples could have occurred during encapsulation which had to be done in air.

Temperature and Concentration Dependence of Electrical Conductivity

The conductivities of the films with compositions corresponding to $n = 33$, 36 and 40 show two different regions. A low temperature region with Arrhenius type behaviour up to 40°C, followed by a VTF type curve.[9] This transition between the two regions may correspond to the melting of a eutectic or of the complex mentioned above. For compositions $n = 20$, 23 and 28, the behaviour of the conductivity is VTF. This behaviour may correspond to the eutectic composition with a low degree of crystallinity. The $n = 16$ composition has a characteristic VTF behaviour which is attributed to the highly amorphous

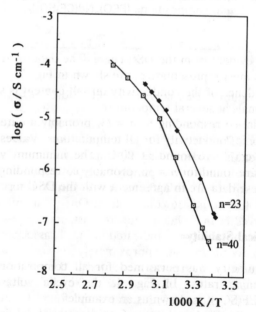

Figure 2. Arrhenius plots of log conductivity versus $1/T$ for $(PEO)_n Nd(CF_3SO_3)_3$.

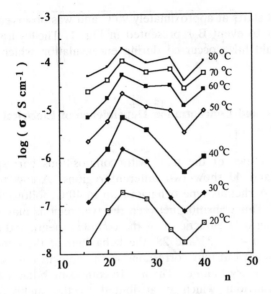

Figure 3. Ionic conductivity as a function of n for polymer electrolytes with compositions $(PEO)_n Nd(CF_3SO_3)_3$.

material also evident from the DSC curve. The data for the $n = 23$ and 40, being the most representative, are shown in Fig. 2.

The dependence of the conductivity on salt content is shown in Fig. 3 for isothermals at several temperatures.

The material corresponding to $n = 23$, probably a eutectic, exhibits the highest ionic conductivity for all temperatures. Values of the order of $2 \times 10^{-7}\,S\,cm^{-1}$ are found at 20°C. The minimum values of the conductivity are found for $n = 36$, probably corresponding to a defined complex. These data are in agreement with the DSC results.

Electrochemical Stability

Cyclic voltammetry was performed for all compositions studied at various scanning rates. In Fig. 4, the cyclic voltammogram of $(PEO)_{36}Nd(CF_3SO_3)_3$ is shown as an example.

The electrochemical stability of this material is limited at the cathodic side by the Nd^{3+} reduction which is observed at $-2.0\,V$

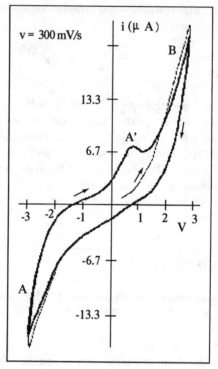

Figure 4. Cyclic voltammogram for $(PEO)_{36}Nd(CF_3SO_3)_3$ at ambient temperature.

versus Ag/Ag^+. The E_0 value reported in the literature is -2.32 V versus SCE.[10,11]

At the anodic side the oxidation of the anion occurs at a potential of $+2.0$ V versus Ag/Ag^+.

A peak referred as A' in Fig. 4 is attributed to the redissolution of Nd (0.6 V versus Ag/Ag^+) associated with the oxidation of the metal.

Table 1. Cathodic, E_c(V) and anodic, E_a(V) limits for various compositions of $(PEO)_n Nd(CF_3SO_3)_3$

n	40	36	20	16
E_a	1·8	1·5	2·2	1·0
E_c	−1·7	−1·7	−1·6	−1·9
$E_{A'}$	0·8	0·6	0·8	0·3

The values of the cathodic and anodic limits for some of the compositions are shown in Table 1.

CONCLUSIONS

This preliminary investigation on PEO–Nd^{3+} polymer electrolytes clearly demonstrates the existence of at least one well defined polymer–salt complex of approximate composition (PEO)$_{36}$Nd(CF$_3$SO$_3$)$_3$ and a eutectic with composition near (PEO)$_{23}$Nd(CF$_3$SO$_3$)$_3$. The highest conductivity exhibited by the studied compositions is of the order of $2 \times 10^{-7}\,\text{S cm}^{-1}$ at room temperature. A large stability window of approximately 4 V was demonstrated as well.

ACKNOWLEDGEMENT

This work was supported by JNICT contract 713-85-51.

REFERENCES

1. Fenton, B. E., Parker, J. M. & Wright, P. V., *Polymer,* **14** (1973) 589.
2. Armand, M., Chabagno, J. M. & Duclot, M. In: *Fast Ion Transport in Solids,* ed. P. Vashishta, J. N. Mundy & G. K. Shenoy. North-Holland, New York, 1979, p. 131.
3. Armand, M. B., *Solid State Ionics, Lake Tahoe,* 1985, Extended Abstract, SSI 85.
4. Abrantes, T. M. A., Alcácer, L. J. & Sequeira, C. A., *Solid State Ionics,* **18, 19** (1986) 315–20.
5. Yong, L. L., Huq, R., Farrington, G. C., Chiadelli, G. & McGhie, A. R., *Solid State Ionics,* **18, 19** (1986) 291.
6. Patrick, A. J., Glasse, M. D., Latham, R. J. & Linford, R. G., *Solid State Ionics,* **18, 19** (1986) 1063–7.
7. Armand, M. B. In: *Polymer Electrolyte Reviews,* ed. J. R. MacCallum & C. Vincent, Elsevier Applied Science, London, 1987, pp. 1–22.
8. Roberts, J. E. & Bykowski, J. S., *Thermochim. Acta,* **25** (1978) 233.
9. See for example Ratner, M. A. In: *Polymer Electrolyte Reviews,* ed. J. R. MacCallum, & C. Vincent, Elsevier Applied Science, London, 1987, pp. 173–236.
10. Johnson, M. D. A., *Dalton JCS* (1974) 1671.
11. Nugent, L. J., *J. Inorg. Nucl. Chem.,* **37** (1975) 1767.

Preliminary Results on PEO– Cu(ClO$_4$)$_2$ Polymer Electrolytes

K. Singh,* G. Chiodelli, A. Magistris & P. Ferloni

Centro di Studio per la Termodinamica ed Elettrochimica dei sistemi salini fusi e solidi del CNR and Dipartimento di Chimica Fisica della Universita', Viale Taramelli, 16, 27100 Pavia, Italy

ABSTRACT

Four compositions of the solid polymer electrolyte, poly(ethylene oxide)$_n$– Cu(ClO$_4$)$_2$, have been prepared, with the ratio of ethylene oxide units to cupric ions ranging from 49 to 8. The variation of conductivity with temperature does not follow a simple Arrhenius law. The films of composition 8:1, 12:1 and 16:1 are completely amorphous, but at room temperature they become crystalline with slow kinetics.

INTRODUCTION

In recent years solid polymer electrolytes containing salts with divalent metals have attracted growing attention.[1-5] These materials might conduct both cations and anions with a predominant possibility of anion transport.[6] In a previous work, the system formed by poly(ethylene oxide) (PEO) with lead perchlorate was studied.[5] In this paper, the properties of polymer electrolyte films containing PEO with Cu(ClO$_4$)$_2$ have been investigated.

Four electrolyte compositions, with the ratio of ethylene oxide units to cupric ions, *n*, being 49, 16, 12 and 8, have been prepared and are

---

* Present address: Department of Physics, Nagpur University, 440010 Nagpur, India

discussed in this communication. The results for additional compositions will be presented elsewhere.[7]

EXPERIMENTAL

Film Preparation

A solution of $Cu(ClO_4)_2.6H_2O$ (Aldrich, 98%) in acetonitrile was
added to a solution of PEO (BDH, MW = 4M) in the same solvent.
After casting on Teflon plates, the solvent was allowed to evaporate
slowly at room temperature. The films were dried under vacuum at
room temperature for 48 h, then heated stepwise up to 50°C for 24 h.
In the following, these samples will be called 'as cast and dried'.

Nearly transparent, pale blue films, typically 100–200 μm thick,
were obtained by this procedure. These films being hygroscopic, it
seemed worth checking the presence of water in them. Measurements
of electrical resistance versus time were taken on a 8 : 1 film held under
dynamic vacuum at room temperature for 4 days. Resistance increased
with time by three orders of magnitude, until it attained a saturation
value after about 2 days. The same experiment at 60°C gave the
corresponding saturation value after about 1 h. Thus, before the
electrical measurements, the 'as cast and dried' samples were kept
under vacuum for 1 h at 80°C, then slowly cooled to room
temperature.

Techniques

Thermo-optical analysis was performed by means of a Leitz hot-stage
polarizing microscope, coupled with a Thermal Analysis microscopy
cell Mettler FP 84, and connected to a XY recorder through a central
processor Mettler FP 80.

X-ray diffraction patterns, at room temperature, were recorded with
a Philips PW1710 diffractometer (CuK_α radiation), operated at 40 kV
and 20 mA.

AC conductivity measurements were carried out in the temperature
range 20–120°C, on samples (10 mm in diameter) held between gold
plates under vacuum. On heating, experiments were performed in the

frequency range covering from 10^{-2} to 10^6 Hz using a Solartron 1255 frequency response analyser equipped with a home made high impedance adaptor (10^{12} Ω, 3 pF) and controlled by a CompaQ 286 computer. On cooling, conductivities were determined at a fixed frequency of 1500 Hz using a Wayne–Kerr Universal Bridge RF B602.

RESULTS AND DISCUSSION

Phase Identification

Thermomicroscopic observation of the 'as cast and dried' samples in air gave evidence of the crystalline and amorphous phases as a result of the thermal history of the sample. The film of composition 49:1 showed a crystalline phase undergoing reproducible melting and crystallization phenomena. The remaining films after the first heating run and melting were found to be entirely amorphous. As an example, DTA curves for $n = 49$ and 16 are presented in Fig. 1. However, spherulite growth is observed in an amorphous sample at room temperature after a long time, e.g. some days.

Powder X-rays diffraction patterns of the 16:1 electrolyte, presented in Fig. 2, indicated that films of this composition before and after heat treatment show partially crystalline and fully amorphous nature, respectively.

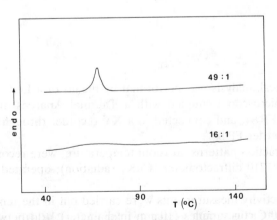

Figure 1. DTA curves for PEO$_n$–Cu(ClO$_4$)$_2$ electrolytes.

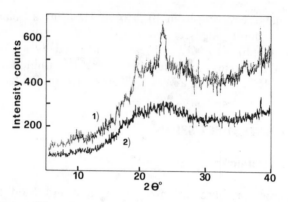

Figure 2. X-ray difftraction patterns on a $n = 16$ film: (1) 'as cast' sample; (2) 'as cast and dried' sample.

Electrical Characterization

The conductivity values recorded during the first heating were reproducible in further cycles. A remarkable difference is observed between the $n = 49$ composition and the rest of the films (Fig. 3). For the 49:1 electrolyte, starting from a low conductivity value at room temperature ($\sim 10^{-9}\,\Omega^{-1}\,cm^{-1}$), a linear increase is recorded up to a discontinuity at about 60°C, which corresponds to the melting of the

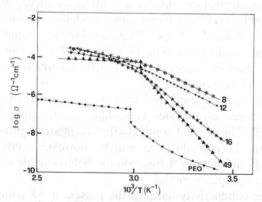

Figure 3. Conductivity data versus 1/T for PEO_n–$Cu(ClO_4)_2$ electrolytes (the numbers indicate n values). Data for pure PEO are reported for comparison.

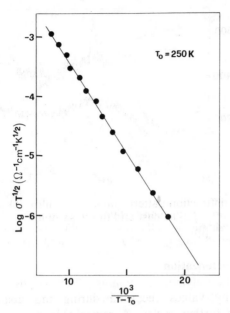

Figure 4. Vogel–Tamman–Fulcher plot for $PEO_8–Cu(ClO_4)_2$.

crystalline PEO. Beyond this temperature, up to 120°C a nearly constant value of about $10^{-4}\,\Omega^{-1}\,cm^{-1}$ is observed. On the other hand, the amorphous electrolytes display a continuous curve, which can be interpreted on the basis of a free-volume model of conductivity. As an example, for $n = 8$, the Vogel–Tamman–Fulcher equation gives a good fit of the conductivity data (Fig. 4), with a T_o value of 250 K, slightly lower than the experimental T_g (275 K) which is obtained by DSC in a second heating run.

These results can be compared with those reported by Abrantes *et al.*[1] on PEO-based electrolytes containing copper perchlorate, with $n = 4$, 8 and 12. The present conductivity values are slightly lower than those of Ref. 1, according to which, however, only the $n = 8$ composition should display a free-volume behaviour described by the Vogel–Tamman–Fulcher model.

In Fig. 5 the conductivity data for the present $n = 8$ composition are illustrated in comparison with those for other PEO-based films containing perchlorates.[4,5,7,8] In the higher temperature region, the conductivity of the copper salt doped film does not differ too much

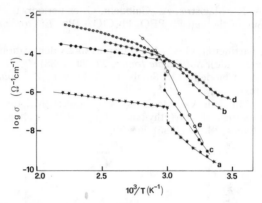

Figure 5. Comparison of conductivity data for PEO-based polymer electrolytes containing perchlorates with different cations (a) PEO; (b) $LiClO_4$, $n = 8$; (c) $Pb(ClO_4)_2$, $n = 20$; (d) $Cu(ClO_4)_2$, $n = 8$; (e) $Hg(ClO_4)_2$, $n = 20$.

from those of the polymers containing other cations, whereas at low temperature the $PEO_8.Cu(ClO_4)_2$ film shows the highest conductivity.

ACKNOWLEDGEMENTS

One of the authors, K. S., acknowledges the support of the 'ICTP Programme for Training and Research in Italian Laboratories', Trieste, Italy. This work has been supported by P. F. Energetica II, CNR-ENEA and by Ministero P.I., Rome, Italy (40% funds).

REFERENCES

1. Abrantes, T. M. A., Alcácer, L. J. & Sequeira, C. A. C., Thin film solid state polymer electrolytes containing silver, copper and zinc ions as charge carriers. *Solid State Ionics*, **18, 19** (1986) 315–20.
2. Patrick, A., Glasse, M., Latham, R. & Linford, R., Novel solid state polymeric batteries. *Solid State Ionics*, **18, 19** (1986) 1063–7.
3. Huq, R., Chiodelli, G., Ferloni, P., Magistris, A. & Farrington, G. C., Poly(ethylene oxide) complexes of lead halides. *J. Electrochem. Soc.*, **134** (1987) 364–9.
4. Bruce, P. G., Krok, F. & Vincent, C. A., Preparation and characterization of PEO–$Hg(ClO_4)_2$ complexes and some thoughts on ion transport in polymer electrolytes. *Solid State Ionics*, **27** (1988) 81–8.

5. Saraswat, A. K., Magistris, A., Chiodelli, G. & Ferloni, P., Electrochemical behaviour of the cell Pb/PEO$_{40}$Pb(ClO$_4$)$_2$/Pb. *Electrochimica Acta*, **34** (1989) 1745–7.
6. Huq, R. & Farrington, G. C., Ion transport in divalent cation complexes of poly(ethylene oxide). *Solid State Ionics*, **27, 30** (1988) 990–3.
7. Magistris, A., Chiodelli, G., Singh, K. & Ferloni, P., *Solid State Ionics* (1990) in press.
8. Ferloni, P., Chiodelli, G., Magistris, A. & Sanesi, M., Ion transport and thermal properties of poly(ethylene oxide)–LiClO$_4$ polymer electrolytes. *Solid State Ionics*, **18, 19** (1986) 265–70.

Multivalent Cation-containing Polymer Electrolytes: Morphology, Conductivity and Ion Mobility

F. M. Gray, C. A. Vincent

Polymer Electrolytes Group, Scotland, Department of Chemistry, University of St Andrews, St Andrews, Fife KY16 9ST, UK

P. G. Bruce, J. Nowinski

Polymer Electrolytes Group, Scotland, Department of Chemistry, Heriot-Watt University, Riccarton, Edinburgh EH14 4AS, UK

ABSTRACT

INTRODUCTION

Multivalent Cation-containing Polymer Electrolytes: Morphology, Conductivity and Ion Mobility

F. M. Gray, C. A. Vincent

Polymer Electrolytes Group Scotland, Department of Chemistry, University of St Andrews, St Andrews, Fife KY16 9ST, UK

&

P. G. Bruce, J. Nowinski

Polymer Electrolytes Group Scotland, Department of Chemistry, Heriot-Watt University, Riccarton, Edinburgh EH14 4AS, UK

ABSTRACT

Polymer electrolytes based on poly(ethylene oxide) and containing salts of multivalent cations have been investigated. Solutions of anhydrous perchlorate salts may be easily prepared but care must be taken to avoid cation-catalysed side reactions. PEO–Ca(ClO$_4$)$_2$ systems were found to form two high melting crystalline complexes, while only one was detected in PEO–La(ClO$_4$)$_3$ materials. Both systems have a cationic transference number of zero. The electrical behaviour of PEO–Cu(CF$_3$SO$_3$)$_2$ electrolytes was complex: reaction of the electrolyte with steel and copper make it difficult to extract data from a.c. impedance spectra which relate to the bulk properties only.

INTRODUCTION

The possibility of fabricating thin film all-solid-state electrochemical devices has resulted in a rapid expansion of research and development in the field of polymer electrolytes. In particular, polymer electrolytes offer specific advantages such as flexibility and ease of preparation of thin films. Research has, until recently, focused largely on the development of rechargeable lithium batteries and consequently, polymers incorporating lithium salts have received greatest attention.

There is now considerable diversification in the field of ion conducting polymers. One area concerns polymer electrolytes of poly(ethylene oxide) (PEO) containing multivalent cations and monovalent anions. In addition to their importance in furthering the understanding of the fundamental properties of polymer electrolytes such as identifying the factors controlling ion mobility, they also offer alternative technological applications, for example, in electrochemical displays and sensors where lower cost materials and less reactive metals than lithium may be desirable.

To date, there have been of the order of a dozen divalent and trivalent cation-based polymer electrolytes studied.[1-5] In general, multivalent cation polymer electrolytes are much more difficult to prepare: factors such as the process of film formation, thermal history, low molecular weight solvent and water residues are critical and it is often difficult to reproduce conditions where morphology and ionic mobility data are directly comparable.

In this paper, we present results obtained for a number of anhydrous polymer electrolytes: $Ca(ClO_4)_2$, $Cu(CF_3SO_3)_2$, $Hg(ClO_4)_2$ and $La(ClO_4)_3$. The electrical properties of the $Hg(ClO_4)_2$ system have been discussed elsewhere[6] and will not be dealt with here. In addition to considerations of thermal and electrical properties, the effects of preparation techniques on multivalent cation-based systems are highlighted.

EXPERIMENTAL

Poly(ethylene oxide) (BDH Polyox WSR-301 MW 4×10^6) and copper trifluoromethane sulphonate (Fluka) were dried by heating under vacuum. Calcium, lanthanum and mercury perchlorates (Alfa Products) were purchased as hydrated salts. The water of crystallisation was removed from the perchlorate salts by dissolving in acetonitrile and refluxing the solvent through a bed of molecular sieves.[7] A solution of the polymer–salt mixture was prepared by mixing acetonitrile solutions of appropriate quantities of dried PEO and anhydrous perchlorate salt. Excess solvent was removed in a flow of dry nitrogen to give a viscous solution which was suitable for solvent casting. CHN microanalysis confirmed the absence of residual solvent after the films were vacuum dried at 70°C for 48 h. PEO–Cu(CF$_3$SO$_3$)$_2$ films were prepared by both solvent casting and vacuum drying or by the hot-pressing technique.[8]

Characterisation of the multivalent cation-containing polymer electrolytes was carried out by X-ray diffraction and differential scanning calorimetry. Electrochemical measurements were performed in an inert atmosphere using a two-electrode, constant volume cell with various blocking and non-blocking electrodes. AC impedance measurements were carried out with a Solartron 1255 or 1250 frequency response analyser and a 1286 electrochemical interface and transference number measurements were carried out using the DC polarisation method.[9]

PREPARATION TECHNIQUES

PEO–Ca(ClO$_4$)$_2$, PEO–La(ClO$_4$)$_3$ and PEO–Hg(ClO$_4$)$_2$

Acetonitrile is a commonly used casting solvent in polymer electrolyte film preparation and has been used here throughout. Water associated with the perchlorate salts was removed by dissolving the salt in acetonitrile and refluxing the solvent through molecular sieves, to avoid isolating the anhydrous perchlorates. However, metal ions may act as catalysts for the hydrolysis of acetonitrile to acetamide.[10] Mercury perchlorate was found to act as a catalyst for hydrolysis. The partial infra-red spectrum given in Fig. 1 shows a loss of water over time but, after 1 h refluxing, an NH$_2$ stretching peak at 3360 cm^{-1}, indicative of an increasing concentration of acetamide. The mechanism for the hydrolysis reaction proposed by Sze & Irish[10] is:

$$(CH_3CN)_x Hg^{2+} \xrightarrow[\text{slow}]{H_2O} CH_3\!-\!C\!=\!N \cdots Hg^{2+}(CH_3CN)_{x-1}$$

$$\xrightarrow{\text{fast}} CH_3\!-\!C\!=\!N \cdots Hg^{2+}(CH_3CN)_{x-1}$$

$$\rightleftharpoons CH_3\!-\!C\!-\!N\!-\!Hg^{2+}(CH_3CN)_{x-1}$$

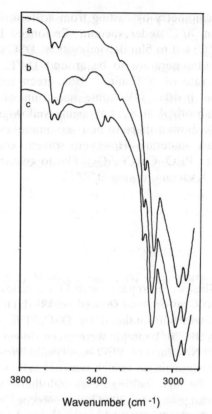

Figure 1. Partial infra-red spectrum showing: (a) purified acetonitrile; (b) acetonitrile + Hg(ClO₄)₂.3H₂O before refluxing; (c) after 1 h refluxing.

Calcium and lanthanum perchlorate did not show this effect and could thus be refluxed for appropriate periods of time. La(ClO₄)₃ solutions required 3 h refluxing to remove the water of crystallisation. Mercury perchlorate could be dried by equilibrating the salt solution with ion-exchanged molecular sieves,[6] thus avoiding the need to reflux.

PEO–Cu(CF₃SO₃)₂

PEO–Cu(CF₃SO₃)₂ films with O:Cu ratios of 50:1, 10:1 and 5:1 were prepared by the solvent-free hot-pressing technique and films with

50:1 and 5:1 stoichiometry by casting from acetonitrile solution and heating for 48 h at 70°C under vacuum. Prolonged heating at high temperatures (120°C) led to film decomposition. DSC studies revealed the decomposition temperature to be around 154°C, when samples were heated at a rate of 10°C min^{-1}. Blue-green solvent cast films showed absorption in the UV-visible spectrum at 750 nm while hot-pressed films absorbed at 825 nm, being yellow-green in colour. CHN microanalysis showed there to be trace quantities of acetonitrile in the solvent cast material. However, solvent could be readily removed from a 5:1 PEO–Cu(CF$_3$SO$_3$)$_2$ film to give the yellow-green colour after only 7 h vacuum drying at 70°C.

MORPHOLOGY

PEO–Ca(ClO$_4$)$_2$

PEO–Ca(ClO$_4$)$_2$ films were prepared with O:Ca ratios of 50:1, 20:1, 12:1, 8:1 and 4:1. X-ray analysis showed peaks which were indicative of PEO and this was substantiated by DSC. The recrystallisation kinetics of PEO in the 20:1 sample were much slower than found for the 50:1 material. No trace of PEO was found for a 12:1 stoichiometry but a crystalline phase, which was not salt, was present and which appeared to be high melting as no endotherm was detected in the temperature range up to 200°C. A eutectic, of composition between 20:1 and 12:1, was implied by the thermal traces. When the salt concentration was increased to 8:1, the X-ray diffraction photographs showed no trace of crystalline material. At 4:1, a new crystalline phase, again not pure salt, was observed and must also be high melting as no endotherm was detected in the DSC trace below 200°C. This suggests that there is a second eutectic with composition near 8:1 and with a melting temperature below 25°C.

PEO–La(ClO$_4$)$_3$

PEO–La(ClO$_4$)$_3$ films were prepared with the same compositions as the calcium perchlorate systems. The morphology of PEO–La(ClO$_4$)$_3$ based systems appears to be much simpler than the latter, with only one crystalline complex evident below an O:La ratio of 12:1.

Depression of the small PEO melting endotherm observed in the DSC traces indicated a eutectic point near an O:La ratio of 10:1, with a melting temperature around 50°C. For lower salt concentrations, only PEO is observed by X-ray and thermal analysis.

PEO–Cu(CF$_3$SO$_3$)$_2$

Films of PEO–Cu(CF$_3$SO$_3$)$_2$ with stoichiometries 50:1, 10:1 and 5:1 were analysed by X-ray and thermal techniques. Both solvent cast and hot-pressed films of O:Cu = 5:1 were prepared. Crystalline PEO is present in both 50:1 and 10:1 samples but the PEO melting point in the 10:1 sample was depressed by 12°C, indicating a eutectic point at a slightly higher salt concentration. A partial phase diagram for this system constructed by Passerini et al. [11] suggests the eutectic composition to have an O:Cu ratio of 9:1. No PEO was present at a mole ratio of 5:1 but evidence for a crystalline complex which melted between 80 and 90°C was obtained from both X-ray and thermal studies. The X-ray powder diffraction pattern also showed some pure salt to be present. This was not so for the solvent cast film, and suggests that polar solvents assist the salt solvation in the polymer.

IONIC MOBILITY

PEO–Ca(ClO$_4$)$_2$

AC impedance measurements were carried out on samples using stainless steel blocking electrodes. The bulk resistance and hence the DC conductivity was obtained from the impedance spectra in the usual manner.[12] Arrhenius type plots for various concentrations of this polymer electrolyte are given in Fig. 2. Hysteresis on cooling samples was found for 50:1, 20:1 and 12:1 molar ratios, indicating slow crystallisation rates which substantiated the results of the thermal studies on these systems. At 8:1 and 4:1 the conductivity–temperature behaviour is very strange and difficult to explain at this stage. It cannot be due to a decomposition reaction as this would have been observed by DSC and the temperature dependence of the conductivity would not have shown reversibility. However, in this salt concentration region a crystalline complex forms different to that in

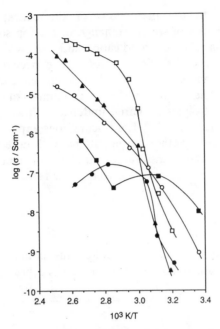

Figure 2. Temperature dependence of the conductivity for PEO–Ca(ClO$_4$)$_2$ electrolytes. O: Ca = 50:1. □; 20:1, ▲; 12:1, O; 8:1, ■; 4:1, ●.

the lower salt regions. It is possible that a temperature dependent phase change or salt redistribution occurs which could markedly affect the observed conductivity. A similar effect, although at much higher temperatures, has been observed by Farrington[13,14] in (PEO)$_8$NiBr$_2$ electrolytes. By heating and cooling or by hydrating and dehydrating again, the conductivity of PEO–NiBr$_2$ electrolytes increases dramatically, but also, the conductivity–temperature plot takes on a more 'normal' shape. This suggests some morphological redistribution has occurred. It is worth noting, however, that many transition metal ions can catalyse the decomposition of polymers.

Transference number measurements were carried out using the method of Evans et al.[9] on the assumption, at this stage, that the system behaved as a strong electrolyte. A 50:1 PEO–Ca(ClO$_4$)$_2$ electrolyte was used with calcium metal non-blocking electrodes and measurements were made at 75°C. A calcium transference number of approximately zero was obtained for this system. The immobility of

Ca^{2+} is contrary to the predictions of the concept of mobility paralleling the kinetics of water exchange in the ion solvent shell.[6,15] However, Ca^{2+} is a relatively hard cation and thus, with respect to the Hard Soft Acid Base Principle, would be expected to have low mobility. EXAFS studies on the local structure around Ca^{2+} ions in calcium halide complexes show a very large number of oxygen nearest neighbours for electrolytes prepared under anhydrous conditions and in the presence of moisture.[4,14] The calcium ions appear to be trapped within a cage formed by ether oxygen atoms, with no evidence for ion pairing observed. The transference number for anhydrous PEO–calcium halide films is also reported to be near zero.

PEO–La(ClO$_4$)$_3$

The conductivity measurements for anhydrous lanthanum perchlorate based electrolytes are summarised in Fig. 3. The presence of PEO is

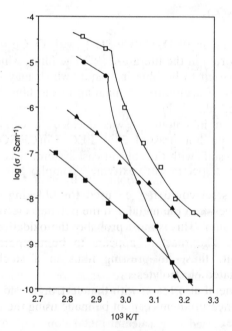

Figure 3. Temperature dependence of the conductivity for PEO–La(ClO$_4$)$_3$ electrolytes. O : La = 50 : 1, □; 20 : 1, ●; 12 : 1, ▲; 8 : 1, ■.

evident in a 20:1 electrolyte, but not at the higher concentrations measured, as would be expected from thermal and X-ray data. The low conductivity for 12:1 and 8:1 films may be due to ionic immobility but is probably the result of the materials comprising large quantities of high temperature melting crystalline complex.

AC impedance measurements were carried out under vacuum on an La $|P(EO)_{12}La(ClO_4)_3|$ La cell at 120°C and showed a high frequency bulk electrolyte semicircle and an inclined spike at lower frequencies, typical of blocking behaviour. On application of a 10 mV DC potential, the current decayed to zero. Thus La^{3+} is largely immobile in high molecular weight PEO. Again, from the predictions of the water exchange rate concept, some mobility of this ion would be expected. However, the high positive charge on the trivalent ion appears to have the greater influence and renders it essentially immobile.

PEO–Cu(CF$_3$SO$_3$)$_2$

Electrical studies on PEO–Cu(CF$_3$SO$_3$)$_2$ with O:Cu = 9:1 have been previously reported in the literature.[11,16] The films, which were solvent cast, were reported to be blue in colour, which may be indicative of trace quantities of acetonitrile remaining in the film. We have found that hot-pressed films, which have never been in contact with solvent, are yellow-green in colour. It was decided to carry out electrical measurements on hot pressed 50:1 PEO–Cu(CF$_3$SO$_3$)$_2$ electrolytes and compare results with the solvent cast system.

AC impedance spectra were extremely complex but indicated that:

1. Stainless steel could not be used for blocking electrodes as a reaction between the metal and the polymer electrolyte produced a surface film. This is most probably the oxidation of Fe to Fe^{2+} by Cu^{2+}. The reaction appears to be temperature and time dependent, thus, hot-pressing films in a steel die produces contaminated electrolytes.

2. Copper metal also reacts with the electrolyte and so transference number measurements cannot be made using the DC polarisation technique.

At present therefore, there exist no valid data on the electrical properties of this very interesting polymer electrolyte. We are presently investigating the true bulk properties of PEO–Cu(CF$_3$SO$_3$)$_2$ systems and these will be reported at a later date.

CONCLUSIONS

A number of multivalent-cation anhydrous polymer electrolytes have been investigated. Care must be taken in removing water of crystallisation from the acetonitrile solutions of salts as the cation may catalyse hydrolysis of the solvent.

REFERENCES

1. Yang, L. L., McGhie, A. R. & Farrington, G. C., Ionic complexes of PEO and MgCl$_2$. *J. Electrochem. Soc.,* **133** (1986) 1380.
2. Patrick, A., Glasse, M., Latham, R. & Linford, R., Novel solid state polymer batteries. *Solid State Ionics,* **18, 19** (1986) 1063.
3. Huq, R. & Farrington, G. C., Ion transport in divalent cation complexes of PEO. *Solid State Ionics,* **28–30** (1988) 990.
4. Cole, M., Latham, R. J., Linford, R. G., Schlindwein, W. S., & Sheldon, M. H., EXAFS of polymer electrolytes. Materials Research Society, Fall Meeting, 1988, Paper M6.1.
5. Vincent, C. A., Polymer electrolytes. *Prog. Solid State Chem.,* **17** (1987) 145.
6. Bruce, P. G., Krok, F., & Vincent, C. A., Preparation and characterisation of PEO–Hg(ClO$_4$)$_2$ complexes and some thoughts on ion transport in polymer electrolytes. *Solid State Ionics,* **27** (1988) 81.
7. Gray, F. M., Salt and polymer dehydration and the preparation of reproducible polymer electrolyte films. *Euro. Polym. J.,* **24** (1988) 1009.
8. Gray, F. M., MacCallum, J. R., & Vincent, C. A., PEO–LiCF$_3$SO$_3$–polystyrene systems. *Solid State Ionics,* **18, 19** (1986) 252.
9. Evans, J., Vincent, C. A., & Bruce, P. G., Electrochemical measurement of transference numbers in polymer electrolytes. *Polymer,* **28** (1987) 2324.
10. Sze, Y. K. & Irish, D. E., A Raman spectral study of hydrolysis of acetonitrile catalysed by Hg(II). *Can. J. Chem.,* **53** (1975) 427.
11. Passerini, S., Curini, R. & Scrosati, B., Characterisation of PEO–copper salt polymer electrolyte, *J. Appl. Phys. A.,* **49** (1989) 425.
12. Bruce, P. G., Electrical measurements in polymer electrolytes. In: *Polymer Electrolyte Reviews-1,* ed. J. R. MacCallum & C. A. Vincent. Elsevier Applied Science Publishers, London, 1987, p. 237.
13. Farrington, G. C. & Linford, R. G., PEO electrolytes containing divalent cations. In: *Polymer Electrolyte Reviews-2,* ed. J. R. MacCallum & C. A. Vincent., Elsevier Applied Science Publishers, London, 1989.

14. Huq, R. & Farrington, G. C., Solid polymeric electrolytes formed by PEO and a transition metal salt. *J. Electrochem. Soc.*, **135** (1988) 524.
15. Eigen, M., Fast elementary steps in chemical reaction mechanisms. *Pure Appl. Chem.*, **6** (1963) 97.
16. Bonino, F., Pantaloni, S., Passerini, S. & Scrosati, B., New PEO–Cu(CF₃SO₃)₂ polymer electrolytes. *J. Electrochem. Soc.*, **135** (1988) 1961.

Thermodynamics of Protonation and Cu(II)-Complex Formation with Polyelectrolytes Containing Different Functional Groups

Mario Casolaro & Rolando Barbucci

Dipartimento di Chimica, Università di Siena, Piano dei Mantellini 44, 53100 Siena, Italy

ABSTRACT

The thermodynamics of protonation and copper(II)-complex formation with synthetic polyelectrolytes, namely poly(amido-amine) containing amino acid residues and vinyl polymers carrying amido and amino or carboxylic groups as side-chain substituents, was studied in 0·1 M NaCl at 25°C. Spectroscopic investigations were performed before stability constant evaluation to determine the number and kind of species present in solution. The $\log \beta$ were calculated by a new method from potentiometric data. The properties in solution and in the solid state were compared. Each monomer unit of the polymer binds one copper(II) ion giving a complex species with clear-cut stoichiometry.

INTRODUCTION

The quantitative examination of heavy metal ion binding in polymer electrolyte systems underwent little development for many years.[1] The recent application of new physical methods and the study of the detailed thermodynamic properties of these complex reactions has provided much needed additional information.[2] Aqueous solutions of polyelectrolytes, and particularly the thermodynamics of Cu(II)-complex formation, can be better understood if different experimental techniques are employed in their investigation. Our contribution has

311

been to combine spectroscopic (ESR, UV-vis, FT-IR and CD), thermodynamic (potentiometry, calorimetry and viscometry)[3] and computational methods, with particular emphasis on stability constant evaluation.[4] The polymer electrolytes we review in this paper are either of poly(vinylic) structure or polymer containing amino acid residues in a poly(amido-amine) moiety. Some are able to form solid Cu(II)-complexes with clear-cut stoichiometry. Their physico-chemical properties were compared with those of the corresponding Cu(II)-complexes in aqueous solution.[2,5]

SYNTHESIS AND STRUCTURE OF POLYMERS

Poly(Amido-Amino)Acids

Linear polymers with a regular sequence of amido groups and amino acid residues are obtained by a Michael-type hydrogen transfer addition of simple α- or ω-amino acids to bis-acrylamides:[6]

$$x CH_2{=}CH{-}CO{-}\underset{\underset{R_1}{|}}{N}{-}R_2{-}\underset{\underset{R_3}{|}}{N}{-}CO{-}CH{=}CH_2 \ + \ x NH_2{-}R_4{-}COO^-$$

$$\left[-CH_2CH_2{-}CO{-}\underset{\underset{R_1}{|}}{N}{-}R_2{-}\underset{\underset{R_3}{|}}{N}{-}CO{-}CH_2CH_2{-}\underset{\underset{\underset{COO^-}{|}}{R_4}}{N}{-} \right]_x$$

This procedure is a general synthetic route for the class of poly(amido-amine)s[6] even though the presence of triethylamine is necessary to free the amino group from the zwitter form for polyaddition. In Table 1 is reported some examples of poly(amido-amino)acids. Low molecular weight products are obtained for natural glycine and α- or β-alanine, as showed by the intrinsic viscosity values.

Vinyl Polymers

The polymerization of vinyl monomer precursors can be readily accomplished with radical initiators:[7,8]

$$x\,CH_2{=}CH \quad \xrightarrow[60°C]{AIBN} \quad \left[\begin{array}{c} -CH_2-CH- \\ | \\ C{=}O \\ | \\ NH \\ | \\ R \end{array}\right]_x$$

High molecular weight products are formed for both acidic and basic polymers, the structures of which are listed in Table 2.

PROPERTIES OF POLY(AMIDO-AMINO)ACIDS

The protonation and Cu(II)-complex formation of a number of poly(amido-amino) acids were studied in aqueous solution by thermodynamic (viscometry, potentiometry, calorimetry) and spectroscopic (UV-visible, ESR, FT-IR, CD) techniques.[3,4,9-11]

Protonation Studies

All the polymers listed in Table 1 show typical polyelectrolyte behaviour since the basicity constants of both basic groups (the tertiary amino nitrogen and the carboxylate group) of each monomeric unit depend on the overall degree of protonation (α) of the whole macromolecule,[4,12] giving 'apparent' $\log K$ values. Both protonation constants of each polymer follow the modified Henderson–Hasselbalch equation[7]

$$\log K_i = \log K_i^\circ + (n_i - 1) \log [(1 - \alpha)/\alpha] \qquad (1)$$

in a wide range of α values. The large size of the stiff diacylpiperazine group is not enough to shield the charges present in different monomeric units,[10] as occurs in related poly(amido-amine)s[2] in which only 'real' basicity constants were obtained. The presence of carboxylate groups, as a pendant to the main chain, leads to loss of independence of the monomeric units. The charged COO^- groups evidently influence the electrostatic field force created on the chain even if the enthalpy changes are not so sensitive as the $\log K$ values. In fact, the $-\Delta H_i^\circ$ are 'real' and the variations of the $-\Delta G_i^\circ$ reflect only entropy effects (ΔS°) since the liberation of water molecules on protonation involves not only the hydration shell of the unit that is

Table 1. Some Examples of Poly(Amido-Amino) acids

No.	Structure of the repeating unit	$[\eta]$ (dl g^{-1})[a]	Ref.
1		0·12	4
2		0·12	4
3		0·13[b]	10
4		0·18	4
5		0·17	12
6		0·17	4

[a] In aqueous 0·1 M NaCl at 30°C where not otherwise indicated.
[b] In aqueous methanol (90/10 : v/v%).

Table 2. Some Examples of Vinyl Polymers

No.	Structure of the repeating unit	$[\eta]\,(dl\,g^{-1})^{a}$	Ref.
1	$\left[CH_2-CH\right]_x$ with $C=O$ attached to piperazine N, N–H	0.24^{c}	17
2	$\left[CH_2-CH\right]_{x'}$ with $C=O$ attached to piperazine N, N–CH_3	0.28^{d}	17
3	$\left[CH_2-CH_2\right]_x$ $O=C-NH(CH_2)_2NH_2$	2.4^{b}	18
4	$\left[CH_2-CH\right]_x$ $O=C-NH(CH_2)_3NH_2$	2.6^{b}	18
5	$\left[CH_2-CH\right]_x$ $O=C-NH-CH_2-COOH$	0.66	8
6	$\left[CH_2-CH\right]_x$ $O=C-NH(CH_2)_5COOH$	0.58^{e}	8
7	$\left[CH_2-C(CH_3)\right]_x$ $O=C-NH-CH_2-COOH$	0.27	21
8	$\left[CH_2-C(CH_3)\right]_x$ $O=C-NH(CH_2)_2COOH$	0.20	21

[a] In DMF at 30°C where not otherwise indicated.
[b] In aqueous methanol (90/10: v/v%)
[c] In methanol;
[d] In CHCl$_3$.
[e] Inherent viscosity.

Table 3. *Protonation* of Poly(Amido-Amino) Acids and Corresponding Non-Macromolecular Models: Thermodynamic Values at 25°C in 0·1 M NaCl

Compound	log $K_i^{\circ a}$	n_i^a	$-\Delta G_i^{\circ b}$	$-\Delta H_i^{\circ b}$	$\Delta S_i^{\circ c}$	Ref.
1	8·30	1·07	11·32	7·62	12·4	4, 9
	2·01	0·80	2·74	0·60	7·2	
2	8·52	1·14	11·62	8·64	10·0	4, 9
	3·57	1·23	4·87	2·57	7·7	
3	8·52	1·12	11·62	7·5	13·8	10
	2·34	0·57	3·2	0·6	9	
4	8·47	1·10	11·55	8·83	9·1	4, 9
	4·21	1·12	5·74	0·25	18·4	
5	8·49	1·31	11·58	8·2	11·3	12
	5·18	0·87	7·06	−0·38	25·0	
	3·1	1·0	4·2	0·1	13·7	
6	8·50	1·16	11·59	9·17	8·1	4, 9
	4·28	1·08	5·84	0·07	19·4	
M1	8·19	—	11·17	7·53	12·2	4, 9
	1·90	—	2·59	0·72	6·3	
M2	8·23	—	11·23	8·68	8·6	4, 9
	3·58	—	4·88	2·52	7·9	

[a] Log K = log K° + $(n - 1)$ log $(1 - \alpha)/\alpha$.
[b] kcal mol^{-1}.
[c] cal mol^{-1} K^{-1}.

being protonated, but the whole macromolecule. Table 3 summarizes the thermodynamic results of the polymers and two corresponding non-macromolecular models. The latter were purposefully synthesized from N-acryloylmorpholine and the appropriate amino acid in a 2:1 molar ratio:[4]

$$O \diagdown N—CO—CH_2CH_2—N—CH_2CH_2—CO—N \diagup O \quad M_1 \, (m = 1)$$
$$\underset{\underset{COOH}{|}}{\overset{|}{(CH_2)_m}} \quad M_2 \, (m = 2)$$

The log K_1 and log K_2 for models M_1 and M_2, obtained with the program Superquad,[13] are in agreement with those of the polymers. Polymers containing α-amino acid residues behave in a peculiar way during the protonation of the carboxylate groups.[4,10,12] The protonation of this group leads to easier proton accessibility which is reflected in a value of n_2 less than 1. Of all the series of polymers listed in Table

1, compact conformation of the zwitterionic form was only attributed to that containing α-amino acid residues and easier protonation, i.e. $n_2 < 1$ suggests a behaviour which can be generalized because it is found in all cases where the COO^- group being protonated is in the α-position with respect to other basic groups (1, 3 and 5 of Table 1). The n less than 1 can be considered diagnostic of a tightly compact conformation which uncoils during protonation. The lengthening of the aliphatic chain, i.e. the hydrophobic character, leads to n_i values which approach one and the monomer units behave independently of each other like small molecular weight molecules. Viscometric titrations show the effect of electrostatics, a decrease in the reduced viscosity leading the extended macromolecule to assume a coil-like conformation upon protonation until the zwitterionic form. Hence even the observed decrease in viscosity from the expanded polymer in the anionic form to the more compact coil in the neutral zwitterionic form does not alter the symmetry of the monomeric unit. Circular dichroism measurements on polymer 3, containing L-alanine residues, show a conformational transition in the alkaline region (Fig. 1). The

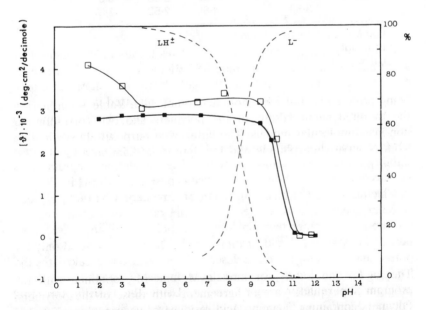

Figure 1. pH-induced conformational transition of polymer 3 in 0.1 M NaCl (■) and in salt-free water (□) solutions. Molar ellipticity [θ], at 215 nm, is plotted versus pH (full lines). Distribution curves of L^- and LH^+ species calculated from the log K_i in 0·1 M NaCl (dashed lines).

CD pattern of 3 is typical of polypeptides in a random coil conformation with a positive Cotton effect at 215 nm assignable to the $n - \pi^*$ amide transition. The molar ellipticity $[\theta]$ at 215 nm plotted against pH, shows the conformational destruction of an ordered coil form by the complete deprotonation of the tertiary amino nitrogen.[10] The increase in pH gives rise to other free coil forms. A locally ordered structure, even in the coil form, is hypothesized to occur between the zwitterionic form of the residue with the amido C=O groups of the backbone. The optically active centre is thus involved in the dissymmetrical perturbation and this agrees with previous results obtained by theoretical calculation for poly(amido-amine)s.[14] The infrared spectral data further show that the interaction between the C=O of the amidic group and the protonated aminic group in the main chain occurs by shifting the Amide I band to higher frequencies at low pH.[10] The strong band recorded for polymer 3 and due to the stretching of the NC=O group falls to 1675 cm^{-1} at high pH, but shifts to 1610 cm^{-1} at low pH, thus explaining the interaction.

Copper(II) Complex Formation Study

Stability constants for polymeric electrolytes towards heavy metal ions are usually evaluated by the modified Bjerrum method.[15] The modification was necessary because the polyelectrolytes show 'apparent' basicity constants. Even so, the log β values reported in the literature for polymeric electrolytes with Cu(II), Ni(II), etc., do not completely describe the complexing behaviour. To elucidate this phenomenon, i.e. to evaluate log β as a function of pH, we recently developed a new stability constant calculation method.[4] The calculation program takes into account the variation of the basicity constants with pH in all cases in which eqn (1) holds. A limitation of this method is that only one complex species can be present in the pH range considered, because the presence of more than one complex species leads to a divergence between observed pH and calculated pH. The program, used only for potentiometric data, is briefly described. In particular it calculates the log K_i for each value of observed pH by means of eqn (1) considering as input the values of log $K°$ and n. Using the basicity constants calculated in this way and the stability constant to be refined, together with the analytical data, a system of mass balance equations is solved for each titration point by the Newton–Raphson method, giving the equilibrium concentrations of all the species present in solution and in particular a calculated value of pH (pH_{calc}). The program refines the $\log \beta$ of the species considered with respect to a previous point, by choosing pH as titre parameter and iteratively solving the equation:

$$pH_{obs} = pH_{calc} + (\partial pH / \partial \beta) \Delta \beta \qquad (2)$$

in the form:

$$\Delta \beta = (pH_{obs} - pH_{calc}) / (\partial pH / \partial \beta)$$

where β is the stability constant to be refined and the derivative is evaluated numerically.

Several examples of complex systems with polymer electrolytes were analysed with the above method. The polymers 1, 2 and 3 of Table 1 are complexing agents towards heavy metal ions since the amino acid residues are able to form stable complexes with the copper(II) ion in acidic conditions due to the closer vicinity of the two different basic groups, i.e. the tertiary nitrogen and the carboxylate groups. The $\log \beta$ values are reported in Table 4 and refer to the $Cu^{2+} + L^- \rightleftharpoons CuL^+$ equilibrium (L^- is the anionic monomer unit). The presence of glycinate and β-alaninate residues strongly coordinates the metal ions and their $\log \beta$ values show a decreasing trend with increasing pH.[4] Polymer 1 has a higher stability constant than polymer 2 even if the mean values in both cases resemble the $\log \beta$ of the corresponding non-macromolecular analogues. The presence of the methyl group in the α-alaninate residue affects both the $\log \beta$ and the spectroscopic data.[11] The stability constant for polymer 3 does not seem to be

Table 4. *Cu(II)-Complex Formation* with Poly(Amido Amino) Acids and Corresponding Non-Macromolecular Models: Thermodynamic Values at 25°C in 0·1 M NaCl (reaction $Cu^{2+} + L^- = CuL^+$)

Compound	pH range	$\log \beta$	$-\Delta G^{\circ a}$	$-\Delta H^{\circ a}$	$\Delta S^{\circ b}$	Ref.
1	2·2–4·5	9·0–8·3	12·3–11·3	7·2(5)	17·1–13·8	3, 4
2	4·0–6·4	6·5–5·8	9·1–7·9	3·8(5)	17·8–13·8	3, 4
3	2·3–5·4	8·6(1)	11·7(1)	5·9(5)	20(2)	11
M1	2·3–3·8	9·98(2)	13·60(3)	7·2(5)	22(1)	3, 4
M2	3·3–7·6	6·05(6)	8·25(8)	4·1(5)	14(1)	

[a] kcal mol^{-1}.
[b] cal mol^{-1} K^{-1}.
Values in parentheses are standard deviations.

Table 5. ESR and Electronic Spectral Data for the CuL$^+$ Complex Species with Poly(Amido Amino) Acid Ligands

Compound	$g_{\parallel}$	$g_{\perp}$	$10^4 A_{\parallel}$ (cm^{-1})	$10^4 A_{\perp}$ (cm^{-1})	ε (dm^3 mol^{-1} cm^{-1})	λ_{max} (nm)
1	2·335	2·080	177·7	10·7	36	735
2	2·350	2·075	181·0	10·7	53	745
3	2·298	2·058	162·0	14·0	94	725

pH-dependent unlike that of related polymers 1 and 2. The difference was attributed to the methyl group which determines a more hydrophobic environment affecting electrostatic interactions. The same effect was found in the protonation process because hydrophobic polymers decrease polyelectrolyte behaviour. Finally, the methyl group in polymer 3 affects the symmetry of the monomeric unit once coordinated and determines a more distorted geometry of the complex species leading to an increase in ε value and a decrease in the ESR parameters with respect to polymers 1 and 2 (Table 5). A square-pyramidal structure of the Cu(II)-complex with polymer 3 was therefore suggested, unlike the octahedral tetragonally-distorted structure attributed to the complex species with polymers 1 and 2 and poly(amido-amine) giving complexes with copper(II) ion.[2] In all cases complex species of the general formula CuL were found in aqueous solution with 1:1 stoichiometry and this was also found in solid Cu(II)-complexes prepared with polymers and corresponding non-macromolecular models.[2] Elemental analysis was consistent with the formula CuL(NO$_3$)$_2$.xH$_2$O and the electronic spectra of the solid complexes were found to be practically identical to those of the same complexes in aqueous solution.[5] Contrary to all previously described complexes of metal ions with polymeric ligands, this means that the poly(amido-amine) complex compounds have well defined stoichiometry.

A linear relationship of molar ellipticity [θ] to Cu(II)/L molar ratios at 202 and 220 nm was found for polymer 3 by Circular Dichroism investigations.[11] The [θ] values increased with the amount of Cu(II) until the Cu(II)/L molar ratio approached one, beyond which the ellipticity remained constant even with a large excess of Cu(II) ions (Fig. 2). Thus saturation is reached when only one Cu(II) is bound for each monomer unit of the polymer involving also the interaction of the

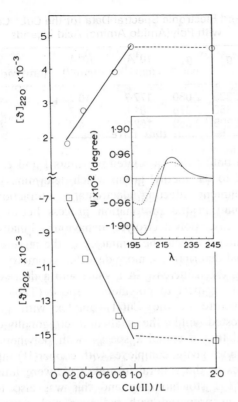

Figure 2. Molar ellipticity [θ], at two different wavelengths, for the CuL⁺ complex species of polymer 3 in aqueous solution versus Cu(II)/L molar ratios and at pH 3·5. (Inset: circular dichroism spectra of the complex species at two different Cu(II)/L molar ratios and at pH 4·2 (– – – – –, 1/2; ———, 1/1)].

amido C=O groups which play an important role as they do in the protonation process. These data are supported by the calorimetric results (Table 4) especially the $-\Delta H°$ values, which are 'real' in all cases and in good agreement with those of the corresponding non-macromolecular models. The amido C=O groups also play a role in the coordination process with the copper(II) ion. They not only act as a chromophore in the $n - \pi^*$ amide transition, but contribute to the slight exothermic process.[11,16]

SOME DATA ON VINYL POLYMERS

Vinyl polymers are more flexible than the previously reported poly(amido-amino acid)s and the absence of the stiffer bis(amido) group leads to a higher cooperativity between monomeric units. In fact, in all compounds listed in Table 2 the basicity constants relative to the protonation of the carboxylate groups or amino nitrogen are more α-dependent, and thus have greater polyelectrolyte behaviour than the polymers of Table 1. The polyelectrolyte character decreases with the hydrophobic nature of the polymer, and this is reflected by a lower value of n (Table 6). In some vinyl compounds the $-\Delta H°$ of protonation are also α-dependent and in most cases are 'real'. The rigid lateral chain in polymer 1 enhances H-bonding interactions between monomeric units which are absent in polymer 2 for steric reasons.[17] The reduced viscosity always bears a linear relationship to protonation.[18,19] The latter is entropy driven and the $\Delta S°$ values suggest high entropy contribution for the dehydration occurring on neutralization of carboxylate groups (polymers 5–8 of Table 6). Due to the high flexibility of the polymers, the rationalization of the copper(II)-complexation process becomes more difficult since cooperativity between monomeric units can easily coordinate metal ions of different stoichiometries. In the case of polymer 3 only one complex species is formed with a CuL stoichiometry.[20] Log β was evaluated with our calculation method and a 'real' stability constant found in the range of pH 4–6. The spectroscopic and thermodynamic data were compatible with the independence of the monomeric units because polymer 4 does not complex with Cu(II) ions.[20] The coordination to the copper(II) ion involves both the amido C=O and the primary amino groups of the same monomeric unit. Deprotonation of the amido group occurs at high pH since λ_{max} shifts to a shorter wavelength. The ability to form Cu(II)-complexes was demonstrated for polymers 5, 7 and 8. In these cases complex species are formed in the alkaline regions[21] and the lengthening of the methylenic chain does not lead to failed coordination to the copper(II) ion as in polymer 4. Only for longer aliphatic chains (polymer 6) does the coordination process fail because highly hydrophobic character leads to precipitation since charge neutralization renders the complex species insoluble. The same process occurs for the protonation of polymer 6 beyond $\alpha = 0.65$.[19] The smaller charge density is not enough to maintain the macromolecule in solution. The evaluation of log β for all these

Table 6. Vinyl Polymers: Thermodynamic Values of Protonation and Cu(II)-Complex Formation at 25°C in 0.1M NaCl

Compound	Protonation					Cu(II)-complex formation					
	$\log K^{\circ a}$	n^a	$-\Delta G^{\circ b}$	$-\Delta H^{\circ b}$	$\Delta S^{\circ c}$	Ability to form CuL complex species in solution	$\log \beta$	$-\Delta G^{\circ b}$	$-\Delta H^{\circ b}$	$\Delta S^{\circ c}$	Ref.
1	7.13	1.40	9.72	9.29[d]	1.44[d]	No					17
2	6.09	1.27	8.30	6.78[d]	5.10[d]	No					17
3	8.23	1.48	11.22	11.2	0.0	Yes	4.83	6.59	2.3	14	18
4	9.31	1.37	12.70	12.7	0.0	No					18
5	4.32	1.50	5.89	-0.3	20.7	Yes		—			19, 21
6	5.42	1.31	7.39	-0.5	26.5	No		—		—	19
7	4.26	1.40	5.81	0.3[d]	18.5	Yes		—		—	21
8	4.92	1.61	6.71	0.4[d]	21.2	Yes		—		—	21

[a] See footnote of Table 3.
[b] kcal mol⁻¹
[c] cal mol⁻¹ K⁻¹
[d] Values calculated at $\alpha = 0.5$.

polymeric systems (polymers 5, 7 and 8) is in progress and will be
reported in forthcoming papers.[21]

ACKNOWLEDGEMENT

The authors wish to thank the *Italian Ministry of Education* for
financial support.

REFERENCES

1. Marinsky, J. A., *Coord. Chem. Rev.*, **19** (1976) 125.
2. Ferruti, P. & Barbucci, R., *Adv. Polym. Sci.*, **55** (1984).
3. Barbucci, R., Campbell, M. J. M., Casolaro, M., Nocentini, M.,
   Reginato, G. & Ferruti, P., *JCS Dalton Trans.* (1986) 2325.
4. Barbucci, R., Casolaro, M., Nocentini, M., Corezzi, S., Ferruti, P. &
   Barone, V., *Macromolecules,* **19** (1986) 37.
5. Barbucci, R., Barone, V., Oliva, L., Ferruti, P., Soldi, T., Pesavento, M.
   & Riolo, C. B., In: *Polymeric Amines and Ammonium Salts,* ed. E. J.
   Goethals. Pergamon Press, New York, 1980.
6. Danusso, F. & Ferruti, P., *Polymer,* **11** (1970) 88.
7. Barbucci, R., Casolaro, M., Corezzi, S., Nocentini, M. & Ferruti, P.,
   Polymer, **26** (1985) 1353.
8. Barbucci, R., Casolaro, M., Magnani, A., Roncolini, C. & Ferruti, P.,
   Polymer, **30** (1989) 1751.
9. Barbucci, R., Casolaro, M., Ferruti, P. & Nocentini, M.,
   Macromolecules, **19** (1986) 1856.
10. Barbucci, R., Casolaro, M., Di Tommaso, A. & Magnani, A.,
    Macromolecules, **22** (1989) 3138.
11. Barbucci, R., Casolaro, M. & Magnani, A., *Polymer J.,* **21** (1989) 915.
12. Fini, A., Casolaro, M., Nocentini, M., Barbucci, R. & Laus, M.,
    Makromol. Chem., **188** (1987) 1959.
13. Gans, P., Sabatini, A. & Vacca, A., *Inorg. Chim. Acta,* **79** (1983) 219.
14. Barbucci, R., Casolaro, M., Ferruti, P., Barone, V., Lelj, F. & Oliva, L.,
    Macromolecules, **14** (1981) 1203.
15. Gregor, H. P., Luttinger, L. B. & Loebl, E. M., *J. Phys. Chem.,* **59**
    (1955) 34.
16. Barbucci, R., Casolaro, M., Barone, V., Ferruti, P. & Tramontini, M.,
    Macromolecules, **16** (1983) 1159.
17. Barbucci, R., Casolaro, M., Ferruti, P., Tanzi, M. C., Grassi, L. &
    Barozzi, C., *Makromol. Chem.,* **185** (1984) 1525.
18. Barbucci, R., Casolaro, M., Nocentini, M., Reginato, G. & Ferruti, P.,
    Makromol. Chem., **187** (1986) 1953.
19. Barbucci, R., Casolaro, M. & Magnani, A., *Makromol. Chem.,* **190**
    (1989) 2627.
20. Casolaro, M., Nocentini, M. & Reginato, G., *Polym. Comm.,* **27** (1986)
    14.
21. Barbucci, R. & Casolaro, M., *Macromolecules,* submitted.

ORMOLYTEs (ORganically MOdified ceramic electroLYTEs): Solid State Protonic Conductors by the Sol–Gel Process

**H. Schmidt,[a] M. Popall,[a] F. Rousseau,[a,b]
C. Poinsignon,[b] M. Armand[b] & J. Y. Sanchez[b]**

[a] *Fraunhofer-Institut für Silicatforschung, Neunerplatz 2, D-8700 Würzburg, FRG*
[b] *Laboratoire d'Ionique et d'Electrochimie du Solide de Grenoble, ENSEEG, BP 75, 38402 St Martin d'Hères Cedex, France*

ABSTRACT

Protonic conductive polymers have been synthesized by the sol–gel process. the materials are obtained by hydrolysis and condensation of amino alkyl trimethoxysilane, with dissolved CF_3SO_3H as proton donor. The influence of the alkylamine chain on hydrolysis and condensation and on the properties of the final material has been studied. These materials are obtained in the form of transparent films, which are amorphous, hard and non-porous, with densities around $1\cdot3\,g\,cm^{-3}$. The materials are stable in air up to 220°C. The ionic conductivity reaches from about $3 \times 10^{-5}\,\Omega^{-1}\,cm^{-1}$ at room temperature to about $10^{-2}\,\Omega^{-1}\,cm^{-1}$ at 120°C with a $1\cdot2\,V$ stability window.

INTRODUCTION

The main drawback of ionic conductive polymers such as PEO complexes is the formation of crystalline phases at room temperature, leading to a conductivity decrease.[1] Ravaine *et al.*[2] and Charbouillot[3] have shown the possibility of obtaining amorphous solvating polymers

325

by using the sol–gel route. Moreover the sol–gel process presents special advantages in comparison to conventional processing techniques:[4] besides high reactivity, homogeneity and purity of the precursors, it allows special shaping techniques such as coating films or powder formation.

A suitable route to sol–gel materials is the hydrolysis and the condensation of alkoxides, e.g. $Si(OR)_4$ or $R'_nSi(OR)_{4-n}$, R' representing an organofunctional grouping and R an alkyl group. The hydrolysis and the condensation of $R'_nSi(OR)_{4-n}$ can be schematically described by eqns (1)–(3):

Hydrolysis:

$$\equiv Si\text{---}OR + H_2O \rightleftharpoons \equiv Si\text{---}OH + ROH \tag{1}$$

Condensation:

$$\equiv Si\text{---}OH + RO\text{---}Si\equiv \rightleftharpoons \equiv Si\text{---}O\text{---}Si\equiv + ROH \tag{2}$$

$$\equiv Si\text{---}OH + HO\text{---}Si\equiv \rightleftharpoons \equiv Si\text{---}O\text{---}Si\equiv + H_2O \tag{3}$$

Condensates with one non-hydrolysable ligand can be described by a silicate network with functional organic substituents R' (Fig. 1). The inorganic backbone provides mechanical strength and an amorphous structure at room temperature. Organic groups can provide solvating properties to ionic compounds (acids, salts).

In this study the influence of the organofunctional grouping on the hydrolysis and condensation reactions will be shown and resulting material properties are investigated. The understanding of hydrolysis and condensation is necessary to control structure and properties of the material.

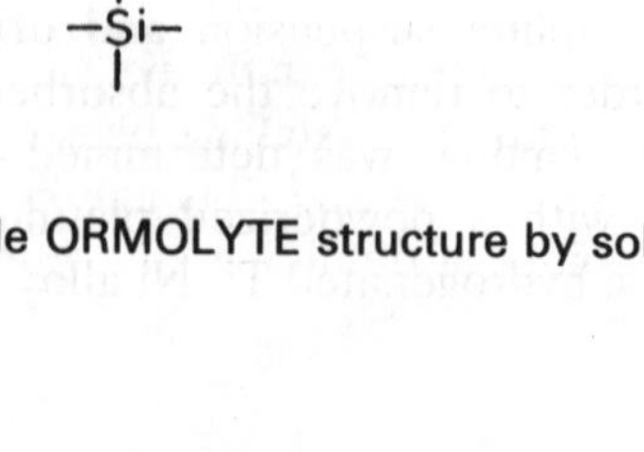

Figure 1. Possible ORMOLYTE structure by sol–gel process.

EXPERIMENTAL

The materials are synthesized by mixing at room temperature H_2O, CH_3OH and $(CH_3O)_3SiR'$ in a $1\cdot5:9\cdot5:1\cdot0$ molar ratio. The solution is stirred for 5 h and CF_3SO_3H is added as a proton donor before removing the excess of solvent by evaporation. The viscous solution obtained is cast on a Teflon plate and dried at 50°C. Gelation with minimum of shrinkage or cracking takes place rapidly during drying. In order to determine the influence of the organic group on hydrolysis and condensation, three variations of $(CH_3O)_3SiR'$ as precursor have been used:

amo: $R' = (CH_2)_3NH_2$;

diamo: $R' = (CH_2)_3NH(CH_2)_2NH_2$

triamo: $R' = (CH_2)_3NH(CH_2)_2NH(CH_2)_2NH_2$.

The CF_3SO_3H concentration is expressed as the molar ratio x of CF_3SO_3H to the total amount of available amino nitrogen, $[CF_3SO_3H]:[N]$. x has been varied between 0 and $0\cdot3$.

Kinetic measurements (gas chromatography, water titration by Karl Fischer method, FT–IR spectroscopy) have been performed by analysing the reaction solution (same conditions were used as for the synthesis). Gas chromatography (Carbowax 1500 column for methanol and an AK 30 000 column by Perkin Elmer for silanes) was used for monitoring the changes of the concentrations of methanol and monomer during the hydrolysis and condensation process. FT–IR spectroscopy has been carried out with the circle cell equipment (multiple internal reflection device) using a ZnSe crystal. The thermochemical behaviour has been investigated by differential thermal analysis and differential gravimetric analysis coupled with mass spectrometry using a heating rate of $10 \, K \, min^{-1}$. To obtain information about the amorphous or crystalline character of the materials X-ray diffraction analysis has been performed at room temperature.

AC complex impedance spectroscopy was used to measure the ionic conductivity of the samples. Flat samples are covered on both sides with a conductive graphite suspension and dried under vacuum at 120°C for 24 h in order to remove the absorbed carbon dioxide and water. The stability window was determined using three-electrode cyclic voltammetry, with a powderized platinum working electrode (area about $7 \, mm^2$), a hydrogenated Ti–Ni alloy electrode as reference and a stainless steel counter electrode. The experiments were carried out under vacuum at 100°C by starting on the cathodic side.

RESULTS AND DISCUSSION

Hydrolysis and Condensation Reaction

As shown for amo on Fig. 2 the water disappearance occurs with a nearly total depletion of the monomer within 15 min, indicating the catalytic effect of the amino groupings on the hydrolysis. Gas chromatography can only detect unhydrolysed monomers. After the unhydrolysed monomer has disappeared, hydrolysis still pro-

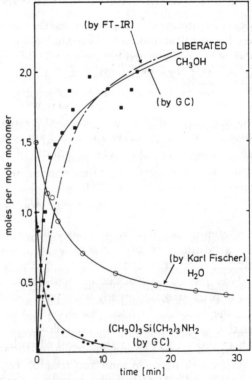

Figure 2. Hydrolysis of amo.

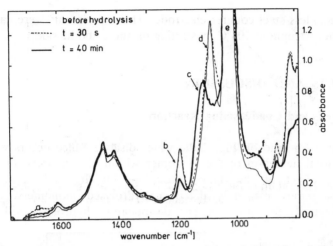

Figure 3. The FT–IR absorbance spectra of the amo system: before, 30 s and 40 min after water addition. Deformation of molecular water: peak (a) around 1650 cm^{-1}; Si—OCH$_3$ rocking and stretching vibrations: peaks (b) at 1190 cm^{-1} and (d) at 1080 cm^{-1}; Si—O—Si stretching vibration: peak (c) at 1130 cm^{-1}. Another peak at 1040 cm^{-1} corresponding to the silicate network, too, is superposed by the peak (e) (C—O stretching of methanol). The shoulder (f) at 950 cm^{-1} (SiO$^-$ SiOH stretching[14]) indicates the formation of silanols groups.

ceeds by hydrolysing mixed monomers (HO(CH$_3$O)$_2$SiR′ or (HO)$_2$(OCH$_3$)SiR′) or oligomers already formed.

The evolution of CH$_3$OH can be determined by gas chromatography as well as indirectly calculated by the disappearance of the Si—OCH$_3$ bond in the FT–IR spectrum (Fig. 3). The two curves in Fig. 2 do not match completely, probably due to the higher experimental error of the gas chromatography. The ratio of the H$_2$O decay to the evolution of CH$_3$OH, which is close to two, indicates an almost complete condensation immediately after hydrolysis. Continuous evaluation of the type of peaks indicated in Fig. 3 is shown in Fig. 4 for amo compared to triamo. It confirms the findings of Fig. 2 for amo, that the formation of SiOSi bonds ($\equiv$ condensation) starts at a very early stage of the reaction. The low concentration of SiOH groups within the first 60 min supports the interpretation that SiOH groups formed either condense by reaction (2) or (3) immediately after formation. Figure 4 shows clear differences between amo and triamo. Together with the H$_2$O consumption and CH$_3$OH deliberation kinetics in Fig. 5, the

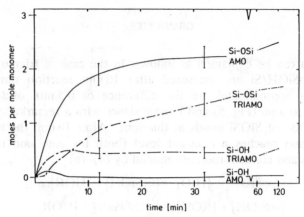

Figure 4. Formation of SiOH and SiOSi for amo and triamo by hydrolysis and condensation; FT–IT data.

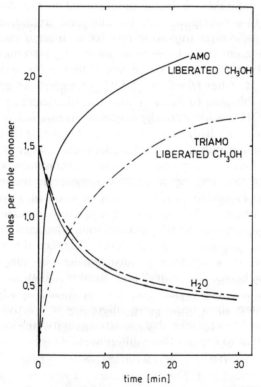

Figure 5. H_2O consumption and CH_3OH formation for amo and triamo; CH_3OH determined by GC; H_2O determined by Karl Fischer titration; ——, amo; · — · —·, triamo.

differences can be explained as follows. In the case of triamo, about 0·5 mol $\equiv$SiOH/Si are measured after 10 min reaction time, corresponding approximately to the difference of 0·5 mol of CH_3OH compared to amo (Fig. 5) and in accordance with a remarkably lower concentration of SiOSi bonds at this time. After 10 min the $\equiv$SiOH concentration reaches a constant level (both for amo and triamo). Hydrolysis and condensation are related by (4)–(6)

$$[\equiv SiOR] + [H_2O] \rightarrow [\equiv SiOH] + [HOR] \tag{4}$$

$$[\equiv SiOH] + [ROSi\equiv] \rightarrow [\equiv SiOSi\equiv] + [HOR] \tag{5}$$

$$\text{or } 2\,[\equiv SiOH] \rightarrow [\equiv SiOSi\equiv] + [H_2O] \tag{6}$$

((6) does not change the overall balance). For c_{SiOH} = constant the formation of $\equiv$SiOSi$\equiv$ is linear proportional to the decay of c_{H_2O}. Since the H_2O consumption rate for amo and triamo are the same (Fig. 5), the $\equiv$SiOSi$\equiv$ formation rate (condensation rate) has to be the same for amo and triamo, too, in the c_{SiOH} = constant region. From 0 to 10 min the condensation rate is higher for amo as one can see from Fig. 4. After 60 min the SiOH concentrations are increasing slightly, probably due to the fact that stiff oligomers are formed by time with SiOH groups sterically hindered from condensation. Amo and triamo are both hydrolysing fast at the beginning of the reaction. While amo immediately starts condensation, the condensation reaction of triamo is slower, thus producing higher concentrations of $\equiv$SiOH. The reason of this may lay in an increasing steric hindrance by the —$(CH_2)_3NH(CH_2)_2NH(CH_2)_2NH_2$ chain or even in a weak internal coordination of the —NH_2 group to the Si, thus blocking a reactive site.[5] With increasing SiOH concentration the condensation rate increases too, reaching a constant level. Whether the condensation occurs via (2) or (3) cannot be distinguished by this experiment. According to Livage & Henry[6] and Brinker,[7] with ortho esters a condensation via (2) should be the preferred reaction. Si—C-substituted esters show remarkable differences in reactivity compared to ortho esters,[8,9] suggesting that electronegativity considerations only are not sufficient to explain these differences.

Material Properties

The dried amo, diamo and triamo condensates are transparent and optically homogeneous. X-ray analysis shows the amorphous character for all types and all acid concentrations. The densities range around $1\cdot 3\ \text{g cm}^{-3}$. The materials are soluble in water (they can be recast from this solvent) and are non-porous.

The thermal stability in air was determined by thermal and thermogravimetric analysis coupled with mass spectrometry. These measurements for amo with $CF_3SO_3H:N = 0\cdot035$ are shown in Fig. 6. The endothermic peak (a) around 140°C can be clearly attributed to the evaporation of adsorbed water, carbon dioxide and nitrogen dioxide as shown by mass spectroscopic analysis. The exotherm peak (b) at 260°C combined with the weight loss of about 9% corresponds mainly to the formation of H_2O, a weak CO_2 formation and a yet unknown amount of alkyl groupings by degradation. The H_2O evolution can be explained by further condensation based on the increasing network mobility with temperature increase. As known

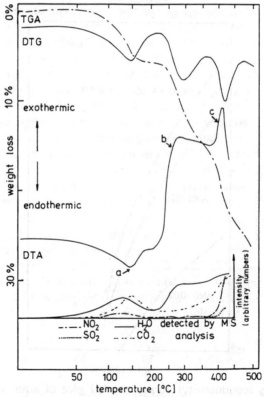

Figure 6. DTA, DTG, TGA and MS curves for amo 0·035 in air.

from IR investigations, after very long reaction times the ≡SiOH content in the condensates increases due to the steric hindrance-based slow down of the condensation rate. Considerable decomposition by oxidation only starts above 370°C (peak (c)). CO_2, H_2O and NO_2 are formed by oxidation of hydrocarbons and amines, SO_2 by reduction of CF_3SO_3H. Measurements carried out with diamo and triamo show the same phenomena and confirm a thermal stability of the materials at least up to 220°C.

Electrical Properties

The conductivity versus the reciprocal temperature exhibits non-Arrhenius behaviour in the 15–130°C range (Fig. 7) and it is

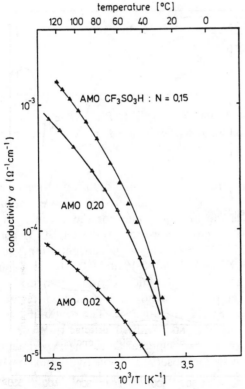

Figure 7. Log (conductivity) versus the $1/T$ plot of amo with different CF_3SO_3H contents.

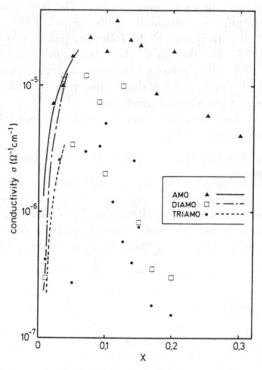

Figure 8. 25°C conductivities of amo, diamo and triamo with various ratios $x = CF_3SO_3H : N$.

characteristic for a free volume limited diffusion process, which can be expressed by the VFT (Vogel–Fulcher–Tamman) equation[10] (computer analysis yields the best-fit values for the parameters within this equation). The conductivity depends on the CF_3SO_3H concentration. It first increases with increasing acid concentration, but decreases between ratios of $CF_3SO_3H : N > 0.15$ and 0.20. Undoped amo shows a conductivity of about $10^{-8}\,\Omega^{-1}\,cm^{-1}$ at room temperature. Figure 8 shows the room temperature conductivities of amo, diamo and triamo with different $CF_3SO_3H : N$ ratios x. The optimum for all three compounds lies around $x = 0.1$ with a conductivity error of $\pm 20\%$ due to the scattering of the experimental data. The experiment also shows a slight difference in the observed conductivity maxima between the three types. To develop a structural model to explain this behaviour, more detailed investigations are necessary. The results for amo are comparable to those of Charbouillot *et al.*[11] The best conductivity at room temperature is obtained with amo 0·115: $\sigma = 3\cdot3 \times 10^{-5}\,\Omega^{-1}\,cm^{-1}$. It corresponds to the following parameters of the VTF equation: $-103°C$ for the ideal glass transition temperature and $9\cdot5 \times 10^{-2}\,eV$ for the pseudo-activation energy. According to Charbouillot,[3] the conductivity is due to the motion of the alkylamine chains, allowing an ion hopping mechanism.

As shown in Fig. 9(a) the cathodic wave corresponding to the evolution of hydrogen is stable:

$$H^+(ORMOLYTE) + e^-(Pt) \rightarrow \tfrac{1}{2}H_2 + ORMOLYTE$$

The anodic wave degrades the polymer and since the position of this peak is poorly defined, we have proceeded by gradually widening the studied domain. Between $0\cdot0\,V/Ref.$ and $1\cdot0\,V/Ref.$ the cyclic voltammogram is stable and the involved redox reactions do not affect the material (Fig. 9(b)). Over $1\cdot2\,V/Ref.$ the cycle is not stable and after

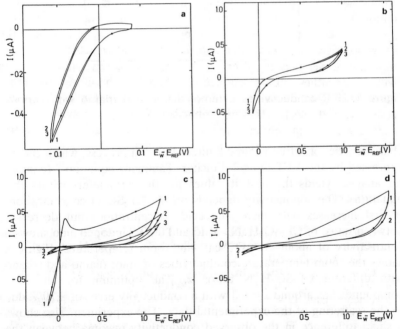

Figure 9(a–d). Cyclovoltammograms of amo 0·115. $dE/dt = 10\,mV/s$, $T = 100°C$, vacuum.

the first cycle, the cathodic wall is suppressed (Fig. 9(c)). This fact can be attributed to the degradation of the alkylamine chains, probably leading to an insulating layer. This is confirmed by cycles starting on the anodic side, where no cathodic wall is observed (Fig. 9(d)). In Fig. 9(c) due to the more negative potential applied, there is an enhanced production of hydrogen, which will be oxidized first in the first cycle resulting in the maximum in this figure. The stability window was found for diamo and triamo systems equal to 1·2 V, too.

CONCLUSION

The sol–gel process offers possibilities of preparing new materials with specific properties. The electrolytes obtained by this route present interesting properties: amorphous structure at room temperature, transparency, temperature stability and ionic conductivity.

For application of these new materials as conductors, which are comparable to the best polymer ionic conductors, it is of general importance to know the influence of the starting compounds and important reaction parameters on the polycondensation process and the gel properties. The knowledge and the control of structure by a more thorough understanding of the hydrolysis and condensation reaction will allow synthesis of ion conducting materials with tailored properties based on ORMOCERs (ORganically MOdified CERamics).[12] Forseeable practical electrochemical applications are electrolytes in solid state batteries, membranes in low temperature fuel cells, electrochromic displays and windows and hydrogen sensors.[13]

REFERENCES

1. Defendini, F., Thesis, Institut National Polytechnique, Grenoble, 1987.
2. Ravaine, D., Seminel, A., Charbouillot, Y. & Vincens, M., *J. Non-Cryst. Solids,* **82** (1986) 210–19.
3. Charbouillot, Y., Thesis, Institut National Polytechnique, Grenoble, 1987.
4. Schmidt, H., *J. Non-Cryst. Solids,* **100** (1988) 51–64.
5. Plueddemann, Edwin P. In: *Silane Coupling Agents.* Plenum Press, New York and London, 1982.
6. Livage, J. & Henry, M. In: *Ultrastructure Processing of Advanced Ceramics 183–195,* ed. J. D. Mackenzie & D. R. Ulrich. John Wiley, New York, 1988.

7. Brinker, C. J., *J. Non-Cryst. Solids*, **100** (1988) 31–50.
8. Schmidt, H., Scholze, H. & Kaiser, A., *J. Non-Cryst. Solids*, **63** (1984) 1–11.
9. Voronkov, M. G., Mileshkevich, V. P. & Yuzhelevskii, Yu. A., *The Siloxane Bond*. Consultants Bureau, New York, 1978.
10. Ratner, Mark A. In: *Polymer Electrolyte Reviews 173–236*. Elsevier Applied Science, London and New York, 1987.
11. Charbouillot, Y., Ravaine, D., Armand, M. & Poinsignon, C., *J. Non-Cryst. Solids*, **108** (1988) 325–30.
12. Schmidt, H. In: *Inorganic and Organometallic Polymers*. ACS Symposium Series 369, 1988.
13. Popall, M. & Schmidt, H. In: TEC 88, Grenoble, Materiaux. Proceedings (in press).
14. Orcel, G., Phalippou, J. & Hench, L. L., *J. Non-Cryst. Solids*, **88** (1986) 114–30.

Mixed Phase Solid Electrolytes Based on Poly(Ethylene Oxide) Systems

W. Wieczorek, K. Such, J. Płocharski & J. Przyłuski

Institute of Solid State Technology, Warsaw University of Technology, ul. Noakowskiego 3, 00-664 Warszawa, Poland

ABSTRACT

Properties of mixed phase electrolytes based on $(PEO)_{10}NaI$ system doped with various inorganic additives (SiO_2, aluminas, NASICON) are described. Addition of inorganic particles improved the mechanical and thermal stability of polymer electrolytes without impairing their ionic conductivity. With the correct amount of the added powders the conductivities of the studied systems exceeded 10^{-5} S/cm at room temperature. The relation between conductivity changes and structural parameters of the studied samples are discussed.

INTRODUCTION

During the last 10 years as a result of intensive research polymer solid electrolytes have become very promising substances. However, their ionic conductivities are still too low for technical applications and various ways to modify and improve these materials have been proposed. One of the proposals relies on dispersion of ceramic powder in a polymer matrix. In many cases increase of conductivity was observed due either to a percolation effect (conductive ceramics)[1] or to lowering crystallinity of a polymer complex (mainly inert ceramics).[2-4] In our opinion the literature is lacking in systematic investigations of relations between properties of ceramic additives and conductivity or other technically important parameters. In this paper we present some

339

collected results classifying an influence of some ceramic additives (Al_2O_3 and SiO_2 powders) on conductivity and phase structure of polymer electrolytes based on a $(PEO)_{10}NaI$ system.

EXPERIMENTAL

Polymer electrolyte samples were prepared by casting from a solution following the previously described procedure.[4] θ-alumina and silica were used as ceramic fillers. Alumina powder had a grain size less than $2\,\mu m$ and silica about $30\,\mu m$. In order to change the polarity of the SiO_2 surface the powder was treated with an aqueous alkaline solution or with a toluene solution of trimethylchlorosilane. In the former case the SiO_2 surface became rich in —OH groups whilst in the latter case $(CH_3)_3SiO$— groups of low polarity were created on the surface. These procedures, however, also affected the specific area of the powders. The values calculated from a BET equation were equal to $103\,m^2/g$ for the —OH rich powder and $24\,m^2/g$ for the one with the siloxy groups.

Ionic conductivity was evaluated from impedance spectra. The frequency range was $5\,Hz$ to $500\,kHz$ and stainless steel blocking electrodes were applied.

The thermal behaviour of the samples was studied by a DSC method and their phase structure by X-ray diffraction.

RESULTS AND DISCUSSION

In our previous papers we showed that addition of a correct amount of inert ceramic powder into a polymer electrolyte matrix lowered its crystallinity and therefore improved the ionic conductivity.[2-5] In Fig. 1 three sets of data are presented showing that the influence of the addition of the studied powders on crystallinity is the most pronounced for the case of SiO_2 powder with a highly polar surface. Both SiO_2 powders caused greater effects than θ-Al_2O_3 despite the fact that the latter substance had a finer granulation.

Figure 2 shows the results of ionic conductivity measurements on a $(PEO)_{10}NaI$ system filled with various amounts of SiO_2 powders of high and low surface polarity. The conductivity of the samples containing SiO_2 with hydroxy groups is about one order of magnitude

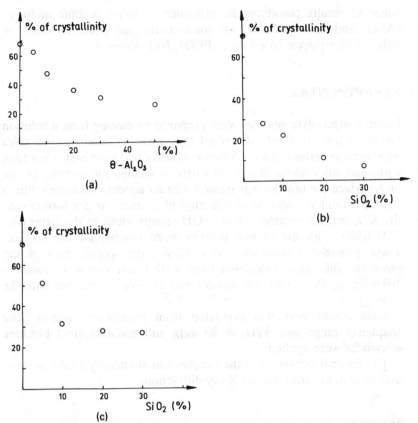

Figure 1. Degree of crystallinity versus ceramic powder concentration for: (a) PEO–θ-Al$_2$O$_3$ system; (b) PEO–SiO$_2$ system (powder with high surface polarity); (c) PEO–SiO$_2$ system (powder of low surface polarity).

higher than the conductivity of the samples with low polar SiO$_2$. Explanation of this fact may be related to a higher specific area of the highly polar powder. Another explanation, however, should also be considered. Grains of the modified powder of lower polar surface may not be as good nucleation centres as SiO$_2$ grains with hydroxy groups. In this situation the crystallisation process may take part mainly in the bulk of the polymer matrix lowering in consequence the final concentration of amorphous phase and the conductivity.

In order to distinguish between these two possibilities we compared the above described results with values of ionic conductivity for an

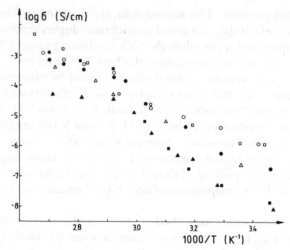

Figure 2. Bulk conductivity versus reciprocal temperature for the (PEO)$_{10}$NaI–SiO$_2$ systems:
○, 5 wt%
□, 10 wt% —SiO$_2$ containing polar surface groups;
△, 20 wt%
●, 5 wt%
■, 10 wt% —SiO$_2$ containing surface groups of low polarity.
▲, 20 wt%

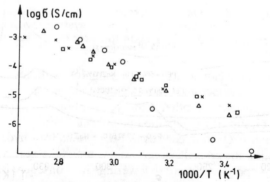

Figure 3. Bulk conductivity versus reciprocal temperature for the systems: ×, (PEO)$_{10}$NaI (NASICON)$_{0.5}$; □, (PEO)$_{10}$NaI–θ-Al$_2$O$_3$; △, (PEO)$_{10}$NaI–SiO$_2$; ○, (PEO)$_{10}$NaI.

alumina filled polymer. The related data are depicted in Fig. 3. The alumina filled electrolyte exhibited considerably higher conductivity in the low temperature region than the SiO_2 containing sample. It should be stressed that for this comparison the best SiO_2 containing material was chosen. The increase of conductivity caused by alumina powder was high despite the fact that its ability to hinder the crystallization of pure PEO powder was much weaker than that of the SiO_2 powders. However, the specific area of the θ-Al_2O_3 powder (exceeding $700\ m^2/g$) was higher than for the silica. Ambient temperature conductivities of both mixed phase systems mentioned above are much higher than those obtained for pristine $(PEO)_{10}NaI$ electrolyte but slightly lower than measured for samples containing highly conductive NASICON powder.

All the observations described above suggest that perhaps two mechanisms are responsible for increase of ionic conductivity caused by ceramic additive. The first one relies clearly on increasing the total volume of the amorphous phase in a polymer electrolyte matrix. The second one is less evident and may be attributed to some changes of polymer structure resulting from interaction with a surface of ceramic grains. Thus the basicity of the ceramic surface may also play an important role.

The changes of glass transition temperature evidencing partial immobilisation of polymer chains (see Fig. 4 and Table 1) may serve as

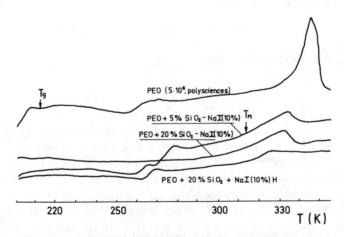

Figure 4. DSC traces for $(PEO)_{10}NaI$–SiO_2 systems: H–SiO_2 with low polar surface groups.

Table 1. DSC Data for $(PEO)_{10}NaI$–SiO_2 Systems (Data for Polar Powder)

C (wt%)	T_g (°C)	T_m (°C)	T_{mc} (°C)
0	−59	68	—
5	−22	65	133
10	−14	53	127
20	−9	49	115

T_g, glass transition temperature; T_m position of the low temperature endothermic peak; T_{mc}, position of the high temperature endothermic peak; C, inorganic powder concentration.

an argument for the second mechanism. The lower flexibility of a polymer resulting from addition of ceramics is also confirmed by NMR experiments.

Many inorganic crystalline and glassy superionic conductors obey the so called Meyer–Neldel rule assuming a linear relation between the logarithm of σ_0 and the activation energy of the ionic conductivity in an Arrhenius type equation:[6,7]

$$\sigma = \sigma_0 T \exp\left(-E_a/kT\right)$$

This rule has a simple mathematical form:

$$\ln \sigma_0 = \alpha E_a + \beta$$

where α and β are empirical constants. Since for some of our mixed phase electrolytes their temperature dependence of ionic conductivity satisfies the Arrhenius relation we have plotted the values of $\ln \sigma_0$ against E_a. The results are shown in Fig. 5 and it is evident that the Meyer–Neldel rule holds also for the $(PEO)_{10}NaI$ system with ceramic additives. This fact demonstrates that the number of charge carriers does not change and the ceramic grains influence only their mobility.

Ceramic additives also improved the mechanical stability of the polymer electrolytes studied.

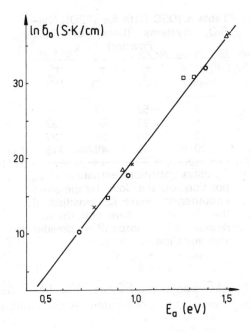

Figure 5. Meyer–Neldel dependence for: O, (PEO)$_{10}$NaI–10 wt% θ-Al$_2$O$_3$ system (powders of different grains sizes); □, (PEO)$_{10}$NaI–10 wt% SiO$_2$ system (powders of different grain sizes); △, (PEO)$_{10}$NaI–SiO$_2$ system (different polar powder concentration); ×, (PEO)$_{10}$NaI–20 wt% of SiO$_2$ (powders with different surface groups).

CONCLUSIONS

Addition of ceramic powder to a polymer electrolyte matrix increases its ionic conductivity. This effect is due to a reduction of crystallinity of the polymer and perhaps to some transformations of polymer electrolyte structure resulting from interaction with a surface of ceramic grains. Ionic conductivity of the modified electrolytes exceeds 10^{-5} S/cm at ambient temperature. The relation between σ_0 and activation energy of ionic conductivity obeys the Meyer–Neldel rule for polymer electrolytes.

ACKNOWLEDGEMENTS

This work was financially supported by the Ministry of Education according to the CPBP 01·15 research programme.

REFERENCES

1. Stevens, J. R. & Mellander, B. E., *Solid State Ionics*, **21** (1986) 203–6.
2. Płocharski, J. & Wieczorek, W., *Solid State Ionics*, **28–30** (1988) 979–82.
3. Wieczorek, W., Such, K., Wyciślik, H. & Płocharski, J., accepted for publication in *Solid State Ionics*.
4. Płocharski, J., Wieczorek, W., Przyłuski, J. & Such, K., *Appl. Phys. A*, **49** (1989) 55–60.
5. Przyłuski, J. & Wieczorek, W., *Proc. II Int. Symp. on Solid State Ionic Devices*, ed. B. V. R. Chowdari & S. Radhakrishna. World Scientific Publishing Co., Singapore, 1988, pp. 475–80.
6. Almond, D., & West, A., *Solid State Ionics*, **18, 19** (1986) 1105–9.
7. Garbarczyk, J. E., Kurek, P., Nowiński, J. L. & Jakubowski, W., accepted for publication in *Solid State Ionics*.

Lithium Electrode Behavior in Polymer Electrolyte Cells

A. Bélanger, M. Gauthier, M. Robitaille & R. Bellemare

Institut de recherche d'Hydro-Québec (IREQ), 1800 montée Ste-Julie, Varennes, Québec, Canada J3X 1S1

ABSTRACT

The work done at IREQ on the behavior of lithium and lithium-alloy electrodes in thin SPE batteries is reviewed. While lithium alloy composites suffer from utilization decrease during cycling, the cyclability of metallic battery-grade lithium was found to be very good owing to the stability of the lithium/electrolyte interface, to the use of low cycling current densities and low positive electrode loadings. The SPE characteristics allow for a wide range of operating temperatures: 0–100°C. A new molten-lithium printing technique to produce thin supported lithium electrodes has been developed and tested.

INTRODUCTION

Extensive work has been carried out for about 10 years to develop an all-solid-state battery based on the use of thin plastic films acting as ionic conductors, as proposed by Armand.[1,2] Compared with the liquid electrolyte technology, this approach offers clear advantages: all-solid-state, non-corrosive medium, high automation potential for the electrode preparation and cell and assembly techniques, broad operating temperature range, no separator, no filling procedure, intrinsic safety and others.

The use of intercalation positive electrodes combined with a thin solid electrolyte appeared to be very promising for rechargeable batteries. Hydro-Québec selected polyethers and lithium ion conducting electrolytes for its polymer electrolyte battery project (ACEP

347

project). The choice of Li ion conducting polymers was based on many considerations. First, they have a high energy content which is related to the lightness and the high electropositive character of lithium. Second, lithium offers apparent compatibility with organic electrolytes in general and with solid polymer electrolytes (SPE) in particular, due to the formation of favourable passivating films (SEI). Third, the electrochemical kinetics and reversibilities at both electrodes ensure fast ionic transport in the anodic SEI films and in the positive insertion material. In addition, lithium alloys (ductile or brittle) can also be used as reversible anodes. Lithium is not without its drawbacks, however. It has a high molecular volume ($13 \, cm^3/mol$), it is highly reactive and easily contaminated, and it is difficult to handle as very thin films. Under certain experimental conditions, it may suffer some dendrite growth problems. Lithium production at the mil level ($25 \, \mu m$ or less) is difficult and expensive, while storage over long periods is hazardous. Lastly, the conductivity of polyether–lithium salt complexes is not as high as that obtained with other alkali metal salts.

When IREQ started working on SPE batteries, 'battery grade' extruded or rolled lithium was used as the anode. At that time most of the electrolytes were prepared from commercial polyethylene oxide (PEO) with high O/Li ratios (4–10) of $LiClO_4$ and $LiCF_3SO_3$. As we know by now, these electrolytes, unless operated at high temperature, are inhomogeneous, multiphased and lack good wetting properties. The end result is resistive and passivated anodic interfaces often accompanied by dendrite growth problems. A typical example of the effect of the temperature on cycling is shown in Fig. 1 where the presence of dendrites affects the recharge efficiency when the operating temperature is lower than 90°C.

The use of lithium alloys in the form of composite electrodes was proposed essentially to alleviate the dendrite problem but also to standardize the electrode preparation techniques: the slurry processes.

The system most often studied is lithium–aluminum which was the focus of earlier work at Exxon.[3] This alloy system was also used in molten LiCl–KCl batteries where 50–50 LiAl electrodes perform more than adequately. Other binary systems including LiSn and LiMg have also been evaluated at IREQ. Since then, and in parallel with the work on alloys, the behavior of pure lithium anodes has been much improved as a result of a better understanding of the many factors governing the SPE cell electrochemistry and technological improvements in electrode preparation and assembly techniques.

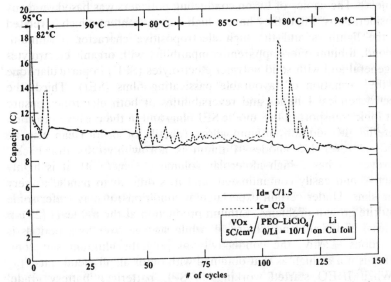

Figure 1. Effect of the temperature on the occurrence of current leakages and dendrite growth during recharge of a VO_x–Li cell using a PEO-based electrolyte with a high O/Li salt ratio.

LITHIUM ALLOYS IN SPE CELLS

Lithium–Aluminum Alloys

The lithium–aluminum binary alloy system is quite complex and different intermetallic compounds, each possessing its own reversibility, have been reported on the lithium-rich side. Depending on the cycling conditions, an overall electrode behavior yielded by these compounds is often satisfactory in terms of battery performance.

Two types of LiAl alloys were studied at IREQ:

—composite LiAl alloys,

—compact LiAl alloys obtained from chemically or electrochemically lithiated aluminum foils.

(a) Composite LiAl Alloy Electrodes

These electrodes make use of brittle LiAl intermetallics which can be processed in powder form. It precludes the use of very rich Li alloys which are too ductile to be ground. Making electrodes out of fine, reactive LiAl powders has proven difficult. First attempts yielded very electrically-resistive electrodes due to the multiplication of contacts between the passivated LiAl grains. This problem was solved by adding graphites and carbon blacks and by optimizing the electrode composition: LiAl particle size and volume fractions of black and binder electrolyte. These electrode materials can be extruded or slurry-processed using the 'doctor blade' technique onto a suitable current collector such as nickel, copper, iron or steel.

Unfortunately, most PEO-based composite LiAl electrodes exhibit a rapid decline of anode material utilization when cycled at 100°C. The cycling of these anodes seems to cause a pulverization of the LiAl powder followed by an encapsulating mechanism which renders most of the material unusable (Fig. 2). This interpretation was confirmed by post-mortem analysis of exhausted anodes which showed that the remaining chemical activity of the lithium in the alloy was exactly the same as that of a fresh uncycled anode. Another limitation of the LiAl composite electrode is its rather low surface electrode loading: $1.22 \, \text{mAh/cm}^2$ compared to $4.1 \, \text{mAh/cm}^2$ for pure lithium for $20 \, \mu\text{m}$ thick electrodes.

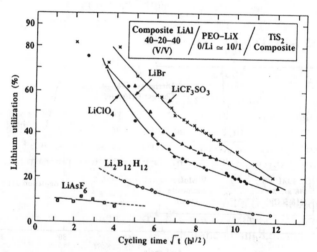

Figure 2. Rapid decline in cell utilization observed during cycling of anodes made of 50 at% LiAl powder in composite form against a TiS_2 composite electrode. After 6 days, the anode utilization is down to about 20% and the drop is almost independent of the salt used in the electrolyte.

(b) Dense LiAl Alloy Electrodes

To solve the above-mentioned particle-insulating problem, a new concept was developed in which a compact LiAl alloy electrode makes contact with the electrolyte on a single plane. In one scenario, the positive material was supplied in the form of a pre-inserted cathode of the type $Li_{1+x}TiS_2$, while a simple aluminum foil was used on the anodic side. This all-discharge-state battery avoided handling highly reductive lithium species at the assembly stage. In another scenario, the aluminum metallic foil was first lithiated by a proprietary chemical process to about 45 at% Li and used directly in a cell assembly. The latter type of electrode mounted counter to a V_6O_{13} electrode has undergone many hundreds of cycles in deep-discharge conditions.[4] It has been observed experimentally that the anode must be in excess with respect to the positive electrode as illustrated in Fig. 3 in which a compact LiAl electrode (18·5 C) is cycled against a MoO_2 composite cathode (14 C). The small excess in lithium (about 30%) is sufficient to reach over 80 cycles before any decrease in cell utilization occurs. With a full lithium excess, i.e. 28 C, that particular cell would attain many hundreds of cycles but with no excess lithium, cycling of an anode-limited cell causes a rapid decrease in cell utilization controlled

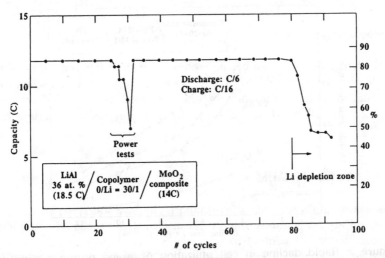

Figure 3. Cycling of a 36 at% Li compact LiAl electrode against a MoO_2 composite cathode at 100°C. The small excess in lithium (4·5 C) versus the positive electrode capacity improves drastically the cyclability of the cell.

Table 1. Comparison of Different Lithium Anodes

Anode type	Li (d = 0·53)	LiAl dense 30 at.% (d = 2·2)	LiAl dense 50 at.% (d = 1·94)	LiAl composite 50 at.% (d = 1·65) 40-20-40
Cathode Faraday (g)	TiS$_2$	TiS$_2$	TiS$_2$	TiS$_2$
(cm^3)	112	112	112	112
	34·8	34·8	34·8	34·8
Anode Faraday (g)	7	70	34·0	72·5
(cm^3)	13·21	31·8	17·5	43·9
Specific energy (Wh/kg)	451	236	294	232
Volumetric energy (Wh/litre)	1121	644	820	545
Capacity (mAh/cm^2) at 20 μm	4·1	1·7	3·1	1·2

by the anode depletion rate. Table 1 summarizes the resulting characteristics of a TiS$_2$ cell using various anodes.

Other Lithium Alloys

The thermodynamic and cycling behavior of the LiSn and LiMg alloy systems have been studied on specimens prepared by electrochemical lithium injection in the corresponding metallic foils. The behavior of dense LiMg alloys when cycled against a lithium electrode is depicted in Fig. 4.

METALLIC LITHIUM

As shown in Table 1, metallic lithium is the best source of lithium for thin battery applications, mainly because of its higher gravimetric and volumetric energy contents.

Most of the problems associated with passivation or dendrite formation have been associated with technological difficulties—lithium purity, surface uniformity, stability and wetting property of the electrolyte—and have been largely resolved. Optimization of the

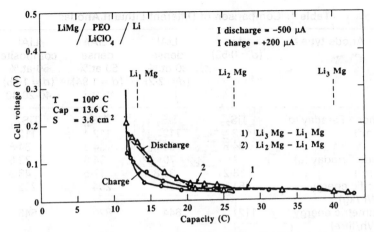

Figure 4. Cycling of LiMg compact electrodes which were obtained by electrochemical alloying with lithium thin magnesium foil. The first injected lithium is totally irreversible.

cycling conditions (current densities, voltage limits, temperature, etc.) has also improved the lithium electrode behavior.

In practical cells, however, several subjects need to be addressed:

—processing of thin lithium films at the mil level,
—handling and storage,
—assembly techniques and cycling conditions.

Thin Lithium Production
One of the main characteristics of SPE cells is their high surface-to thickness ratio which compensates for the relatively low conductivity of polymer electrolytes, especially at ambient temperature. In a typical example a cell using a 1 mAh/cm² TiS₂ composite electrode requires a lithium thickness of only 15 μm (2 Li excesses).

Normally lithium is produced either by extrusion (>100 μm) or by lamination down to about 40 μm, depending on the film width. The latter process is usually slow and expensive, however.

In order to adapt lithium needs to SPE technology requirements alternative techniques for producing thin lithium films had to be evaluated including electron-beam deposition and improved lamination, but finally the development of new techniques appeared necessary in order to be able to produce thin supported lithium electrodes

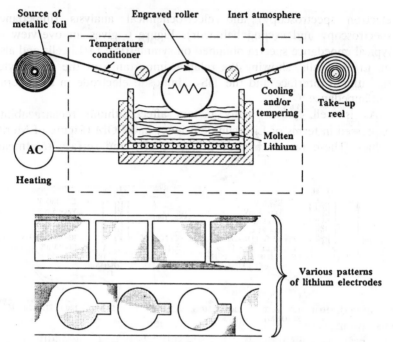

Figure 5. Schematic representation of a printing technique using molten lithium to prepare thin supported lithium anodes in the range from 1 to 50 μm.

from 1 to 50 μm thick at speeds compatible with those used to prepare the electrolyte and positive electrode. The apparatus that served to deposit thin molten lithium films on various metallic foils is schematized in Fig. 5. This 'printing' technique can also produce various patterns of lithium electrodes. The choice of optimized anodic current collectors is also of prime importance since, in certain cases, it may affect seriously the overall battery cost.[5]

Lithium Characterization and Rechargeability

The importance of lithium characterization is to understand better the different mechanisms responsible for the anode modifications during cycling. The techniques currently used for that purpose, either before cycling or in post-mortem analysis, include: physical examination by profilometry and micrographic analysis, chemical analysis (chemical activity of lithium before and after cycling, atomic absorption, Auger electron spectroscopy) and electrochemical analysis (impedance spectroscopy and cycling behavior). Figure 6 gives an overview of typical impedance spectra obtained on symmetrical Li/Li cells and also on Li/TiS$_2$ cells, showing the relative importance of the electrolyte, the lithium interface and the TiS$_2$ composite electrode in the overall cell impedance.

As is well known many factors affect lithium rechargeability expressed in terms of Li cycling efficiency and FOM (Figure of Merit) values. These factors include lithium purity and surface treatment,

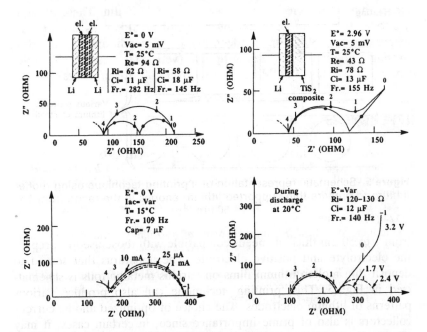

Figure 6. Impedance spectra of Li/Li symmetrical cell and Li/TiS$_2$ cell. (a) The first spectrum shows the separation of the large circle into two smaller circles corresponding to the two lithium interfaces (c. 60 ohms). (b) The same circle in the TiS$_2$ cell that also corresponds to the lithium interface (78 ohms) while the 45° straight line is characteristic of Warburg diffusion process in the composite electrode. (c) The Li/Li cell behavior under various AC perturbation currents: 25 μA, 1 mA and 10 mA showing the stability of the interface without any diffusion type limitations. (d) The Li/TiS$_2$ cell impedance at different stages during a discharge cycle. Most of the changes seem to be observed in the positive electrode while the anode remains unchanged.

nature of the polymer and lithium salt, positive electrode capacity, current densities and temperature.

CONCLUSIONS

Over the years, lithium quality and electrode preparation techniques have improved significantly and are now very well adapted to the particular requirements of SPE cells. Improved lamination and molten lithium coating processes have been developed and allow the production of thin lithium electrodes in the range of 1–50 μm. These lithium electrodes show very good cycling behavior in SPE cells from room temperature to 80°C under deep-discharge conditions. Electrodes carrying a nominal 2 excesses of lithium (versus the positive electrode capacity) have shown FOM values well over 100 (Li turnovers).

REFERENCES

1. Armand, M. B., PhD Thesis, University of Grenoble, 1978.
2. Armand, M. B., Chabagno, J. M., & Duclot, M. In: *Fast Ion Transport in Solids,* ed. P. Vashishta, J. N. Mundy & G. K. Shenoy. North Holland, New York, 1979, p. 131.
3. Rao, B. M. L., Francis, R. W. & Christopher, H. A., *J. Electrochem. Soc.,* **124** (1977) 1490.
4. Gauthier, M., *Third International Seminar on Lithium Battery Technology and Applications,* ed. S. P. Wolsky and N. Marincic. Shawmco, Inc., Tulsa, USA.
5. Owens, B. B., Technology Assessment of Ambient Temperature Lithium Secondary Batteries Technology: Electric Utility and Vehicle Applications. EPRI Final Report EPRI-5218, Project 0370-30, June 1987.

The Cathode/Polymer Electrolyte Interface

Peter G. Bruce, Eileen McGregor

Polymer Electrolyte Group Scotland, Department of Chemistry, Heriot-Watt University, Riccarton, Edinburgh EH14 4AS, UK

&

Colin A. Vincent

Polymer Electrolyte Group Scotland, Department of Chemistry, University of St Andrews, St Andrews, Fife KY16 9ST, UK

ABSTRACT

AC impedance methods are employed to study the interface between V_6O_{13} cathodes and lithium ion conducting polymer electrolytes. Data are presented for single crystal and polycrystalline cathodes in contact with both low molecular weight, liquid, and high molecular weight, solid, poly(ethylene oxide) based electrolytes. $LiClO_4$ and $LiCF_3SO_3$ are introduced into the polymer to form the electrolyte. The addition of impurities to the pure, liquid electrolyte produces a rapidly growing interfacial impedance which is similar to that observed with the unpurified, solid, polymer electrolyte, suggesting that impurities dominate the interfacial impedance of such systems. A brief description of possible models for the polymer electrolyte/intercalation electrode interface is also given.

INTRODUCTION

A major technological application of polymer electrolytes in combination with intercalation cathodes is to be found in the field of all-solid-state lithium batteries. For high temperature rechargeable batteries, polymer electrolytes consisting of high molecular weight poly(ethylene oxide)(PEO) containing salts such as $LiCF_3SO_3$ and intercalation cathodes such as V_6O_{13} are being extensively investigated. To achieve high cycling rates it is essential that the electrolyte film possesses a low resistance ($<10^3 \, \Omega \, cm$); this has been the driving force behind extensive research programmes aimed at the development of polymer electrolytes with a high conductivity. However, high rate capability also demands low impedances at the interface between the electrolyte and the lithium and V_6O_{13} electrodes, it is therefore apparent that the nature and stability of these interfaces is of paramount importance.

It is already established that in non-aqueous liquid and solid polymer electrolytes a layer grows on the lithium surface, passivating it and possibly leading to inferior cell performance. The current state of understanding of the lithium electrode in polymer electrolyte cells is summarised in the paper by Bélanger *et al.* in this volume. Until recently work on intercalation compounds has concentrated on their bulk properties—particularly structure and lithium ion diffusion rates. Work on the electrolyte/intercalation electrode interface *per se* is scarce. One of us has investigated the high voltage cathode $Li_{(1-x)}CoO_2$ in contact with propylene carbonate based electrolytes[1] and we have already published results on the interface between powdered V_6O_{13} and solid PEO:$LiCF_3SO_3$ or $LiClO_4$.[2,3] In this paper we extend our investigation of this interface to include single crystal V_6O_{13} electrodes and also describe work on electrolytes based on a low molecular weight liquid polyether used as a model solvent for the solid polymer.

EXPERIMENTAL

The manipulation of air sensitive material was carried out in a Millar-Howe high integrity glove box with an H_2O level of <7 ppm.

Electrolyte Preparation

PEO (Union Carbide, MW = 100 000) was purified by dissolution in water followed by pressure filtration in an attempt to remove ionic impurities. A solution of PEO in freshly distilled and deionised water was passed through a Sartorius Ultrasart Cell 50 fitted with a porous membrane possessing a 10 000 MW cut off. The sample of PEO was washed three times with the distilled water before being collected from the filter and dried. Initial drying was carried out by heating the solution to 55°C under dynamic vacuum. In a further chemical dehydration process suggested by Armand,[4] the remaining solid was dissolved in dry acetonitrile and an excess of trimethoxymethane $(CH_3O)_3CH$ was added. This compound reacts with the remaining water to yield volatile products:

$$(CH_3O)_3CH + H_2O \longrightarrow H-C\overset{\displaystyle O}{\underset{\displaystyle OCH_3}{\big<}} + 2CH_3OH$$

This appears to be a very efficient method of removing water when distillation is not a possible alternative. The acetonitrile and products of the reaction were removed by pumping on the mixture until a solid mass of PEO remained. An FTIR spectrum of a thin slice of the PEO indicated the absence of water in the solid. The efficiency of this form of chemical dehydration was demonstrated in a separate experiment when an excess of trimethoxymethane was added directly to untreated $CH_3-(OCH_2CH_2)_8-OCH_3$ (Fluka, purum), a short chain liquid poly-ether, and the resulting solution stirred for 12 h, then pumped for 5 h at room temperature. FTIR spectra before and after the drying procedure indicated that a significant quantity of water present initially had been removed; again no trace of O–H stretching peaks could be detected in the final product.

LiCF$_3$SO$_3$ (Aldrich, 97%) and LiClO$_4$ (Alfa, 99·5%) were dried by heating under dynamic vacuum at 120°C and 175°C respectively. The dried PEO was dissolved in anhydrous acetonitrile and combined with the appropriate salt, LiCF$_3$SO$_3$ or LiClO$_4$, also in acetonitrile. The electrolyte was then cast in Teflon moulds and the majority of the solvent was removed by flowing dry Ar gas over the solution until soft tacky films remained. The residual acetonitrile was removed by heating the films to 70°C under dynamic vacuum. The resulting films were typically several hundred μm thick. Electrolyte compositions of 20 ether oxygens per Li$^+$ ion were used throughout.

Liquid electrolytes were prepared as follows. Low molecular weight, liquid, tetraethylene glycol dimethyl ether, $CH_3-(OCH_2CH_2)_4-OCH_3$, hereafter referred to as tetramer, was purified by addition of LiAlH$_4$, primarily to remove peroxides and then vacuum distilled at 81°C to produce a dry, high purity, solvent. The dryness of the tetramer was again demonstrated by FTIR. LiClO$_4$ and LiCF$_3$SO$_3$, dried as indicated above, were added to the purified and unpurified tetramer respectively, to yield an electrolyte composition of 20:1. In certain experiments impurities were deliberately added to the liquid electrolyte.

V$_6$O$_{13}$ Electrodes

Polycrystalline cathodes were prepared by pressing the V$_6$O$_{13}$ powder onto 13 mm diameter stainless steel grids. Single crystal cathodes were prepared by selecting suitable needle-shaped crystals of approximate dimensions $1 \times 1 \times 4$ mm which were then suspended by a thread in a cylindrical cup of 10 mm diameter and the crystals potted in resin. When hard, the disks were removed from the cup and their flat surfaces polished, first on a diamond ring until the crystal faces were exposed on each of the flat surfaces, and then on successively finer grades of SiC powder and diamond paste. Final polishing was carried out with 0·25 μm diamond paste. A layer of gold was evaporated onto one of the flat surfaces to facilitate good electronic contact when mounting the electrode in the cell. V$_6$O$_{13}$ is an anisotropic intercalation compound with channels capable of accepting Li$^+$ ions lying in the [010] crystallographic direction. Crystals were potted such that in some the intercalation channels lay parallel, and in others perpendicular, to the axis of the disks.

Three-Electrode Cells

Cylindrically shaped Teflon cells containing stainless steel current collectors were used for all the electrochemical measurements. In the case of the solid polymer electrolyte cells a lithium disk counter electrode followed by two electrolyte films and then the mounted, single crystal working electrode were placed between the stainless steel end pieces. A third lithium ring reference electrode was placed between the two polymer films. In the case of liquid electrolytes the films were replaced by Whatman glass fibre pads soaked in electrolyte.

For measurements on the polycrystalline V_6O_{13} electrodes impregnated pressed powder grids described above replaced the potted single crystals.

Cells were contained in gas-tight stainless steel chambers under vacuum which were in turn mounted in tube furnaces controlled to $\pm 1°C$ for the elevated temperature measurements. AC impedance measurements were carried out using a computer controlled Solartron 1250 Frequency Response Analyser and 1286 Electrochemical Interface.

MODELS OF THE POLYMER ELECTROLYTE/INTERCALATION ELECTRODE INTERFACE

The equivalent circuit shown in Fig. 1(a) represents a simple model of the interface between a homogeneous electrolyte and a smooth planar intercalation electrode. C_e represents the electrical double layer of the electrode, R_e the process of Li^+ ion transfer across the interface, Z_d the diffusion of Li^+ ions within the electrode and R_b the bulk electrolyte resistance. Figure 1(b) represents the complex impedance plot anticipated on the basis of Fig. 1(a), where the semicircle arises from the combination of R_e and C_e. Even if we retain a smooth electrode surface this model may be inadequate as several more complex processes may occur which are now briefly considered. Extensive ion pairing in polymer electrolytes would result in a charge density very much less than that anticipated on the basis of a fully dissociated electrolyte, and hence to a diffuse double layer at the interface. A further complication in the non-Faradaic process may occur due to specific adsorption of the polymer, cation or anion onto the electrode surface. In particular it is known that the anions are not strongly bound to the polymer chains and thus may be particularly available for adsorption. Faradaic processes may also be more complex than a one step charge transfer mechanism followed by bulk diffusion within the intercalation host. In a similar fashion to the mechanism commonly accepted for deposition on a simple elemental metal surface, the Li^+ ions may become associated with the surface of the intercalation electrode diffusing along it before being incorporated into the bulk electrode. This adion mechanism is distinct from that occurring in liquid electrolytes where the adion diffuses with a partial

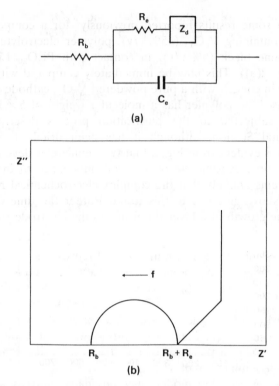

Figure 1. (a) Equivalent circuit for a simple model of the polymer electrolyte/intercalation electrode interface. (b) Complex impedance plot arising from the circuit shown in (a).

solvent coordination sephere, only losing this completely at the point of final incorporation into the lattice. With a solid polymer electrolyte it is not possible for an electroactive ion to surface diffuse together with a solvent sheath. Further complications may arise due to surface roughness, particularly with compacted powder electrodes. Roughness generally has the effect of broadening the interfacial semicircles, as was discussed in more detail in Ref. 1.

RESULTS AND DISCUSSION

Since it is the aim of this paper to consider the processes occurring at a practical solid polymer electrolyte/V_6O_{13} interface it is instructive to begin with some results reported previously[2] for a composite V_6O_{13} cathode, containing V_6O_{13} (45% v/v), polymer electrolyte (50% v/v) and acetylene black (5% v/v), in contact with $PEO_{20}:LiCF_3SO_3$ at 100°C (Fig. 2(a)). This may be immediately compared with the same electrolyte in contact with a pure powdered V_6O_{13} cathode (Fig. 2(b)). In both cases the polymer has a molecular weight of 5×10^6 and has not been subjected to the purification process described in the Experimental Section, although it has been dried at 55°C under vacuum. It is evident from Fig. 2 that two semicircles develop with the passage of time causing the overall cell impedance to increase with time. It seems unlikely that the complex electrochemical mechanisms described above give rise to this trend. Rather the time dependence suggests the growth of a layer on or across the electrode surface. This

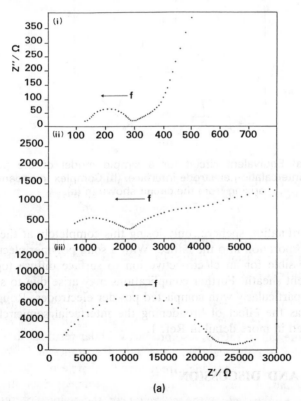

(a)

Figure 2. (a) The $(PEO)_{20}:LiCF_3SO_3/V_6O_{13}$ (composite cathode) interface, at 100°C: (i) 2, (ii) 200, (iii) 320 h after cell assembly. PEO MW = 5×10^6. Electrode area = 1 cm^2.

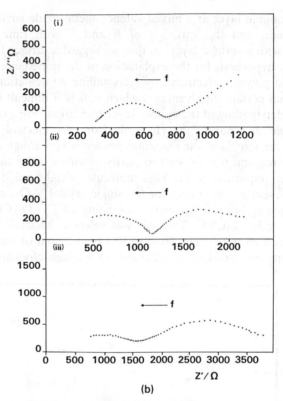

Figure 2. (b) The $(PEO)_{20}:LiCF_3SO_3/V_6O_{13}$ (polycrystalline cathode) interface, at 100°C: (i) 2 h, (ii) 90 h, (iii) 220 h after cell assembly. PEO MW = 5×10^6. Electrode area = 1 cm².

assumption is further reinforced by the fact that the interfacial impedance is significantly reduced immediately after the passage of a constant current across the interface,[3] consistent with the removal, or reduction in the extent, of a surface layer. Broadly similar results are obtained with electrodes which have been chemically intercalated with lithium in the range $Li_xV_6O_{13}$, $0 < x < 6$, prior to cell assembly. Changing the salt to $LiClO_4$ also produces similar results although now both semicircles are present from the start. Resistance and capacitance values associated with the high frequency semicircle of Fig. 2(b) were extracted, and it was noted that the resistance rises in inverse proportion to the fall in the capacitance; the values of this latter parameter (8×10^{-8} to $3 \times 10^{-7} \, F \, cm^{-1}$) are too small for a typical electrode double layer at a mixed valence metal oxide such as V_6O_{13}. These values, and the variation of R and C with time are again consistent with a surface layer, so that we regard a surface layer as the most likely hypothesis for the explanation of the impedance results.

The solid polymer electrolyte/polycrystalline intercalation electrode interface has certain disadvantages which makes it difficult to obtain a deeper understanding of the surface layer formation. For example, it is not possible to separate the electrode from the electrolyte without disrupting the interface, the electrode surface is very rough and thus of unknown area and it is difficult to purify or characterise the type and level of the impurities in the high molecular weight polymer. Therefore we present in Fig. 3 data for the single crystal V_6O_{13} electrode in the high purity liquid electrolyte, consisting of $CH_3O\text{-}(CH_2CH_2O)_4\text{-}CH_3$ with added $LiClO_4$. This salt was selected because it could be obtained in high purity and dried to less than 2 ppm of water. Figure 3(a) exhibits the well-known behaviour of a rough blocking electrode

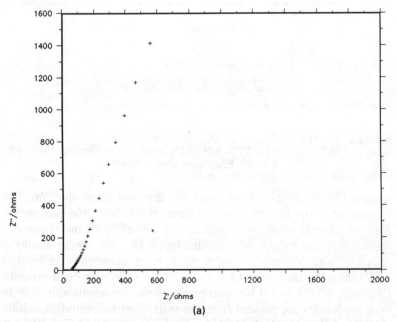

(a)

Figure 3. (a) The (pure tetramer)$_{20}$:$LiClO_4$/V_6O_{13} (single crystal) interface at room temperature. The single crystal is orientated with the diffusion channels perpendicular to the direction of current flow in the electrolyte, i.e. the blocking orientation. Electrode area = 4 mm^2.

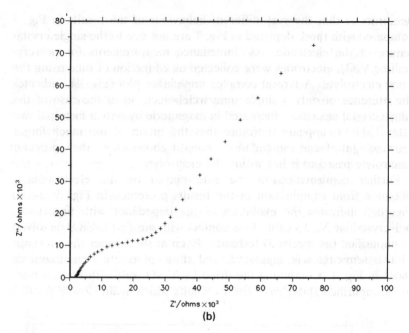

Figure 3. (b) The (pure tetramer)$_{20}$:LiClO$_4$/V$_6$O$_{13}$ (single crystal) interface at room temperature. The single crystal is orientated with the diffusion channels parallel to the direction of current flow in the electrolyte, i.e. the non-blocking orientation. Electrode area = 1 mm^2.

with an inclined spike at high frequencies turning towards a steeper gradient in the low frequency region.[5] These results confirm that lithium intercalation does not occur in crystallographic directions perpendicular to [010]. In strong contrast, in Fig. 3(b) where the diffusion channels are parallel to the field direction in the electrolyte a high frequency semicircle is seen. DC polarisation results with this orientation of crystal further confirm that Li$^+$ ions may be easily inserted and removed. A small amount of lithium, sufficient to achieve an average composition Li$_{0.01}$V$_6$O$_{13}$, was inserted galvanostatically prior to the AC measurements thus ensuring that the electrode was fully poised. The impedance of this electrode was monitored for more than 400 h without detecting an increase in impedance of more than a factor of two. The capacitance associated with the semicircle is ~10 μF cm^{-1} which is typical of an electrode double layer. It would thus appear that a surface layer is not present in this case. To demonstrate that the very different behaviour of the results in Fig. 3 compared with those depicted in Fig. 1 are not due to the single crystal nature of the electrode, AC impedance measurements for polycrystalline V_6O_{13} electrodes were collected as a function of time using the same electrolyte. A typical complex impedance plot (Fig. 4), indicates the presence of only a single semicircle which, as in the case of the single crystal electrode, increased in magnitude by only a factor of two after 134 h. It appears therefore that the origin of the much larger growing interfacial impedance noted above for the polymer electrolyte-based cells lies within the electrolyte.

Further confirmation of the role played by the electrolyte is obtained from examination of the results presented in Figs 5 and 6. Figure 5 indicates the evolution of the impedance with time for a polycrystalline V_6O_{13} cathode in contact with purified $LiClO_4$ dissolved in unpurified (as received) tetramer. Even at short times the presence of two semicircles is suggested, and these plots are reminiscent of those in Fig. 1. Combining the dried $LiCF_3SO_3$ salt with the tetramer in its unpurified state, gave rise to results which again closely parallel

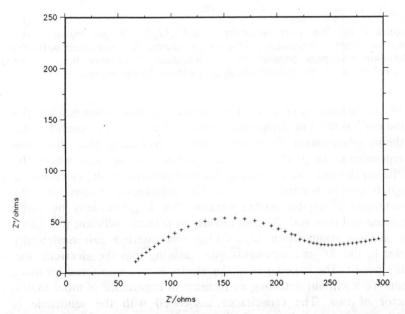

Figure 4. The (pure tetramer)$_{20}$:$LiClO_4$/V_6O_{13} (polycrystalline cathode) interface at room temperature, after 134 h. Electrode area = 0·24 cm².

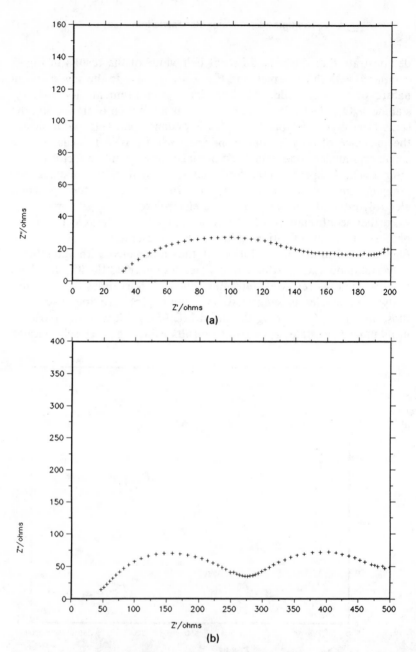

Figure 5. The (unpurified tetramer)$_{20}$:LiClO$_4$/V$_6$O$_{13}$ (polycrystalline cathode) interface at room temperature: (a) 2 h, (b) 191 h after cell assembly. Electrode area = 0·12 cm^2.

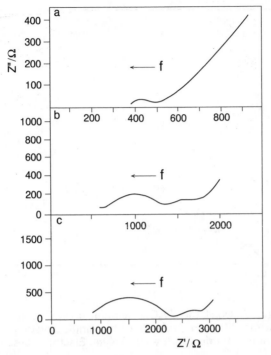

Figure 6. The (unpurified tetramer)$_{20}$:LiCF$_3$SO$_3$/V$_6$O$_{13}$ (polycrystalline cathode) interface at room temperature: (a) 0·5 h, (b) 24 h, (c) 72 h after cell assembly. Electrode area = 1 cm^2.

those of the solid polymer electrolyte containing LiCF$_3$SO$_3$, Fig. 6. It therefore appears that an impurity or impurities present, primarily in the ether-based solvent, are responsible for the surface layer formation.

Commercial high molecular weight PEO is a complex material which inevitably contains a variety of impurities. Among these are included Ca^{2+} residues from the polymerisation process, SiO$_2$ which is added as an aid to assist the flow of the powder during manufacture. Low MW cyclic ethers may also be present.[6] Furthermore, the polyether is known to age by oxidation and other processes, producing a variety of organic products. We have not carried out a detailed study of the impurities present in PEO but in order to evaluate the influence of some of the possible contaminants on the interfacial impedance we have added several different substances to the carefully purified liquid

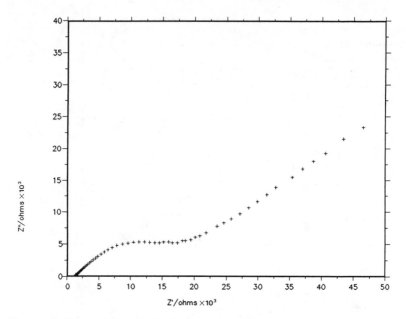

Figure 7. The PEO_{20} : $LiClO_4/V_6O_{13}$ (single crystal) interface, MW = 100 000, with the diffusion channels of the single crystal parallel to the direction of current flow in the electrolyte. Electrode area = 1 mm².

electrolyte. Four impurities were added at two concentrations, 100 ppm and 500 ppm, to the pure liquid electrolyte and the impedance monitored over a 48-h period.

At the lower level of 100 ppm, two semicircles were evident in the impedance plots of the cells containing H_2O_2, C_2H_5OH or H_2O. In the case of the peroxide the two semicircles are well resolved and the high frequency semicircle is clearly seen to grow with time. Because of the poor resolution in the plots for electrolytes with added C_2H_5OH and H_2O it is only possible to conclude that the overall impedance increases with time. In contrast, impedance plots of the electrolyte containing $Ca(ClO_4)_2$ as a source of the Ca^{2+} impurity, closely resembled those of the pure electrolyte (Fig. 4).

At the higher level of 500 ppm, the growth in the impedances is even more marked. For the cells with added H_2O, C_2H_5OH or H_2O_2, the impedances are all higher at this concentration. In the case of Ca^{2+} addition, the cell impedance showed a broad arc (probably encompassing two semicircles) which grew with time. It appears therefore that possible impurities in PEO itself may have a major effect on the impedance of the cathode, and that the organic impurities tested in this work are more active in this regard than Ca^{2+}.

Finally, AC impedance measurements were carried out on a cell containing a single crystal V_6O_{13} electrode in contact with the polymer electrolyte $PEO:LiClO_4$, which was prepared and the constituents purified as described in the Experimental Section. These results (Fig. 7), are reminiscent of those for the same electrode in contact with the purified liquid electrolyte at room temperature, with only one semi-circle present.

In conclusion, data have been presented for both single crystal and polycrystalline V_6O_{13} electrodes in contact with both solid and liquid polymer electrolytes, which suggests that impurities in the high molecular weight PEO are primarily responsible for the impedance at the electrode interface which grows with the passage of time.

ACKNOWLEDGEMENTS

The authors gratefully acknowledge the provision of single crystals of V_6O_{13} by B. B. Owens and M. Z. A. Munshi, and the Harwell Laboratory for supplying composite cathodes. P. G. Bruce and C. A. Vincent are indebted to the SERC for financial support.

REFERENCES

1. Thomas, M. G. S. R., Bruce, P. G. & Goodenough, J. B., AC impedance analysis of polycrystalline insertion electrodes: application to $Li_{(1-x)}CoO_2$. *J. Electrochem. Soc.*, **132** (1985) 1521–8.
2. Bruce, P. G. & Krok, F., Studies of the interface between V_6O_{13} and poly(ethylene oxide) based electrolytes. *Electrochimica Acta*, **33** (1988) 1669–74.
3. Bruce, P. G. & Krok, F., Characterisation of the electrode/electrolyte interfaces in cells of the type $Li/PEO–LiCF_3SO_3/V_6O_{13}$ by ac impedance methods. *Solid State Ionics*, **36** (1989) 171–4.
4. Armand, M. B., Private communication.
5. de Levie, R., The influence of surface roughness of solid electrodes on electrochemical measurements. *Electrochimica Acta*, **10** (1965) 113–30.
6. Dale, J., Borgen, G. & Daasvatn, K., The oligomerization of ethylene oxide to macrocyclic ethers, including 1,4,7-trioxacyclononane. *Acta Chem. Scand.*, **27** (1973) 378–9.

Polyheterocyclic Polymers as Positive Electrodes in Solid-State Polyethylene Oxide–LiClO$_4$ Lithium Rechargeable Batteries

C. Arbizzani & M. Mastragostino

Dipartimento di Chimica 'G. Ciamician' dell'Universita' di Bologna, Via Selmi 2, 40126 Bologna, Italy

ABSTRACT

Cyclability features of solid-state lithium–polypyrrole and lithium–polybithiophene batteries with PEO$_{20}$–LiClO$_4$ polymer electrolyte were investigated at 70°C working temperature. The cyclability performance of a lithium–polybithiophene battery is more promising than that of a lithium–polypyrrole battery. The shelf-life of these batteries was also investigated. The self-discharge rate is fast at 70°C but not appreciable at room temperature.

The preliminary results, which were obtained at room temperature using a new polymer electrolyte (PEO/SEO)$_{20}$LiClO$_4$ in the lithium–polybithiophene battery, are promising.

INTRODUCTION

Currently much attention is focused on the development of solid-state lithium rechargeable batteries. To this end studies have been made on the use of both intercalation compounds[1,2] and electronically conducting polymers[3-5] as positive electrodes together with polymer electrolytes based on lithium salts–polyethylene oxide (PEO) complexes.

The major limitation of the batteries using 'conventional' PEO-based solid electrolyte is their high working temperature, i.e. over

60°C (the conductivity of these electrolytes attains useful values for battery application after transition in the amorphous phase).

We are investigating electrochemically prepared polypyrrole (pPy) and polybithiophene (pBT) as positive electrodes in solid-state rechargeable lithium batteries with $PEO_{20}-LiClO_4$ as polymer electrolyte. Cyclability data of these batteries at 70°C working temperature as well as stability data will be presented. The conductivity of $PEO_{20}-LiClO_4$ at 70°C is higher than that of $PEO_8-LiClO_4$, the most widely tested polymer electrolyte in battery application.[6,7] The viability of pPy and pBT positive electrodes in rechargeable lithium batteries has already been proved when organic liquid electrolyte is used.[8]

During the charge and discharge processes of a pPy or pBT/Li battery

$$nx\text{Li} + \{Py^{+x}(ClO_4^-)_x\}_n \underset{\text{charge}}{\overset{\text{discharge}}{\rightleftharpoons}} nx\text{LiClO}_4 + \{Py\}_n$$

the polymer becomes p-doped and undoped; ClO_4^- ions are introduced into the polymer electrode during charge (to balance the injected charge) and diffuse back into the electrolyte on discharge; x is the doping level and its maximum value is 0·33 for pPy and 0·66 for pBT (this means one counterion for every three pyrrole or thiophene rings). This fact makes conducting polymers less promising than intercalation compounds from the standpoint of specific capacity and energy density as shown in Table 1, where Li/LiV_3O_8 battery data are also reported in comparison (for LiV_3O_8 the maximum reversible uptake of Li is 3 eq/mol[9]).

In addition, increasing the polymer electrode thickness to enhance the capacity of the batteries modifies the kinetics of the electrochemical doping–undoping process and, consequently, the performance of the batteries. The ionic movement into and out of the polymer structure controls charge–discharge rates except in the case of thin

Table 1

	x	Ah/kg	V	Wh/kg
Li–pdoped pPy	0·33	90	3	270
Li–pdoped pBT	0·66	77	3·4	262
Li–LiV_3O_8	3	260	2·8	730

films or low current densities. This is an intrinsic limitation of these electrode materials.

EXPERIMENTAL

The preparation of PEO and of PEO/SEO based polymer electrolytes is described in Refs. 6 and 10, respectively. The pPy and pBT films were galvanostatically grown on stainless steel electrodes by monomer oxidation: pPy in propylene carbonate (PC)–LiPF$_6$ 0·2 M M-pyrrole 0·2 M at 0·8 mA/cm^2 and pBT in acetonitrile (ACN)–LiClO$_4$ 0·5 M-bithiophene 20 mM at 1·6 mA/cm^2. The polymer electrodes were washed respectively with PC and ACN and vacuum-dried. The solid–state polymer cells were assembled in a dry-box by sandwiching an Li disk, a thin layer of electrolyte (200–50 μm) and pPy or pBT film. All the electrochemical tests were carried out with AMEL (Italy) instrumentation interfaced with personal computer.

RESULTS AND DISCUSSION

How the electrochemical kinetics of pPy and pBT vary with thickness was tested first in liquid organic electrolyte (PC–LiClO$_4$ 1 M) by cyclic voltammetry (CV). Figures 1 and 2 show the capacities, as evaluated by recovered charge during CV, of pPy and pBT electrodes of different thicknesses. Thickness is represented as the film formation charge. The cyclable doping level decreases as thickness increases (Figs 1, 2), and decreases even more as the sweep rate increases (Fig. 1). The pPy cyclability in liquid electrolyte is more promising than that of pBT (it is worth mentioning that cyclability performances of these polymer electrodes are strongly dependent on film electrosynthesis conditions).

Figure 3 shows CV of thin and thick pPy and pBT electrodes in solid-state cells at different sweep rates. When the polymer film is thin, the cyclability at 50 mV/s compares with that in the liquid cell; with increased film thickness, the cyclable charge in the solid-state at 50 mV/s is very low and only at 5 mV/s compares with the liquid cell.

The solid-state cells were tested by repeated charge–discharge galvanostatic cycles. Figure 4 shows the capacity data (evaluated by recovered charge during discharge) of Li–pBT batteries at three

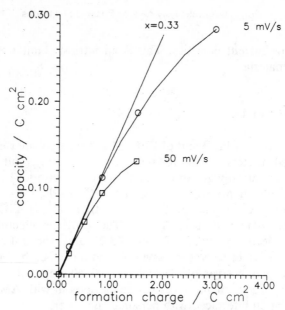

Figure 1. Capacities of pPy electrodes at different thicknesses evaluated by CV from 1·8 to 3·8 V versus Li, in PC–LiClO$_4$ 1 M.

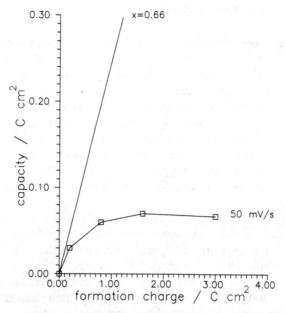

Figure 2. Capacities of pBT electrodes at different thicknesses evaluated by CV from 2·5 to 4·3 V versus Li, in PC–LiClO$_4$.

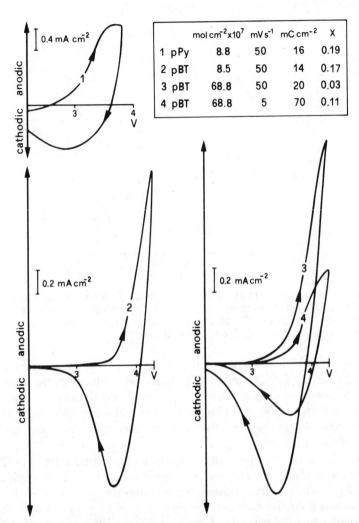

		mol cm^{-2}x10^7	mV s^{-1}	mC cm^{-2}	X
1	pPy	8.8	50	16	0.19
2	pBT	8.5	50	14	0.17
3	pBT	68.8	50	20	0.03
4	pBT	68.8	5	70	0.11

Figure 3. CV of pPy and pBT electrodes of different thicknesses in solid-state cells. The capacity values and their corresponding doping levels at different sweep rates are reported in the inset.

C. Arbizzani & M. Mastragostino

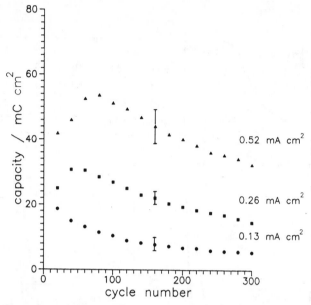

Figure 4. Capacity values at different numbers of galvanostatic cycles of Li/(PEO)$_{20}$LiClO$_4$/pBT batteries at three different pBT thicknesses: ●, $\Gamma = 8.4 \times 10^{-7}$ mol BT cm^{-2}; ■, $\Gamma = 34.8 \times 10^{-7}$ mol BT cm^{-2}; ▲, $\Gamma = 75.5 \times 10^{-7}$ mol BT cm^{-2}.

different pBT thicknesses up to the 300th cycle (cycles between 2·2 and 4·2 V at 0·13, 0·26 and 0·52 mA/cm^2 with efficiency 98–99%). Figure 5 shows the data of Li/pPy batteries at two thicknesses (cycles between 1·6 and 3·8 V at 0·13 mA/cm^2 with efficiency 98–99%). The patterns are the average of five battery runs; pPy data are more scattered than pBT data.

It is important to note that in assembling solid-state Li–pPy batteries with thick pPy electrodes, a small amount of PC was added to PEO (20% in weight); without this addition the cyclability of these batteries was poor. By contrast, the cyclability of Li–pBT batteries did not change when PC was added. Presumably the structure of pPy film is more compact than pBT film and the PC swells the pPy and facilitates the counterion movement into the polymer electrode; the PC effect will be on the polymer electrode more than on the conductivity of the polymer electrolyte.[2] Thus, from the standpoint of cyclability pBT is more promising than pPy as a positive electrode in solid-state batteries.

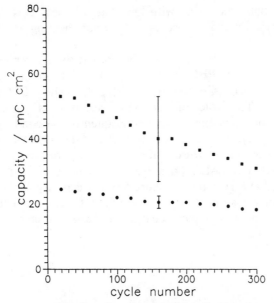

Figure 5. Capacity values at different numbers of galvanostatic cycles of Li/(PEO)$_{20}$LiClO$_4$/pPy batteries at two different pPy thicknesses: ●, $\Gamma = 8\cdot6 \times 10^{-7}$ mol Py cm^{-2}; ■, $\Gamma = 36\cdot6 \times 10^{-7}$ mol Py cm^{-2}.

As concerns the shelf-life of these solid batteries, the self-discharge is fast when the batteries are stored at 70°C. pPy loses 50% of charge over 4 days and pBT completely discharges over 1 day. In contrast, when the batteries are stored at room temperature, no significant self-discharge occurs over 1 month. The data of batteries charged at 70°C, stored at room temperature for different times, and heated to 70°C just 1 h before discharge (to assure equilibrium temperature) are shown in Table 2. The efficiency values of galvanostatic charge–discharge cycles at 0·26 mA/cm^2, after 1 h, 3 days, 1 week and 1 month

Table 2

	Q_f (C/cm^2)	Charge (mC/cm^2)	Discharge								
			Immediate		1 h		3 days		1 week		1 month	
			(V)	(η%)	(V)	(η%)	(V)	(η%)	(V)	(η%)	(V)	(η%)
pPy	0·20	16	3·60	99	3·48	89	3·49	90	3·45	89		
pBT	1·64	25	3·89	99	3·78	79	3·79	82	3·81	83	3·80	82

are substantially the same, with low charge losses. This demonstrates that the loss of charge occurs only when heating to 70°C before discharge.

Recently a new polymer electrolyte based on a complex of Li salts and composite of PEO and a styrenic macromonomer of PEO ($CH_2=CH-C_6H_4-CH_2O(CH_2CH_2O)_nCH_3$, called SEO, has been developed.[10] The role of the SEO addition is to enhance the amorphous phase in the system; consequently the conductivity of the PEO/SEO polymer electrolyte is higher than that of conventional PEO at room temperature. Our preliminary results on the use of (50/50 PEO/SEO)$_{20}$–LiClO$_4$ polymer electrolyte in Li–pBT batteries are reported in Figs 6 and 7. Figure 6 shows capacity data evaluated by discharge of galvanostatic cycles at 50°, 40°C and room temperature, and Fig. 7 shows a typical charge–discharge galvanostatic cycle at

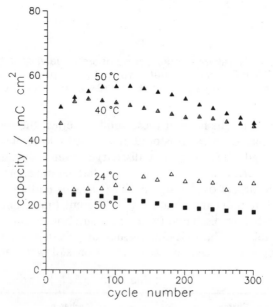

Figure 6. Capacity values at different numbers of galvanostatic cycles (between 2·2 and 4·2 V versus Li), at different temperatures, of Li/(PEO–SEO)$_{20}$LiClO$_4$/pBT batteries at two different thicknesses: ■, $\Gamma = 34\cdot8 \times 10^{-7}$ mol BT cm^{-2}; ▲, △, △, $\Gamma = 69\cdot5 \times 10^{-7}$ mol BT cm^{-2}. $I = 0\cdot13$ mA cm^{-2} at 50° and 40°C and $I = 26\ \mu$A cm^{-2} at 24°C. Efficiency values 99–100%.

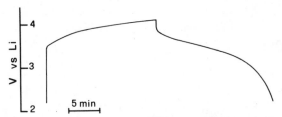

Figure 7. Charge–discharge galvanostatic cycle (300th between 2·2 and 4·2 V versus Li) at $I = 26 \, \mu A \, cm^{-2}$ of Li/(PEO–SEO)$_{20}$LiClO$_4$/pBT battery, at 24°C.

room temperature at $0.026 \, mA/cm.^2$ These preliminary results are very promising and tests are in progress.

ACKNOWLEDGEMENTS

The authors wish to thank ENI for the research grant, Professor P. Ferloni for kindly providing the PEO$_{20}$LiClO$_4$ film and Drs A. Guyot and T. Hamaide for kindly providing (PEO/SEO)$_{20}$LiClO$_4$.

REFERENCES

1. Gauthier, M. *et al.*, *J. Electrochem. Soc.*, **132** (1985) 1333.
2. Munshi, M. Z. A. & Owens, B. B., *Solid State Ionics*, **26** (1988) 41.
3. Panero, S. *et al.*, *Electrochimica Acta*, **32** (1987) 1461.
4. Novak, P. & Inganas, O., *J. Electrochem. Soc.*, **135** (1988) 2485.
5. Arbizzani, C. *et al.*, *Synth. Metals*, **28** (1989) 663.
6. Ferloni, P. *et al.*, *Solid State Ionics*, **18–19** (1986) 265.
7. Robitaille, C. D. & Fauteux, D., *J. Electrochem. Soc.*, **133** (1986) 315.
8. Corradini, A. *et al.*, *Synth. Metals*, **18** (1987) 625.
9. Bonino, F. *et al.*, *J. Electrochem. Soc.*, **135** (1988) 12.
10. Carre, C., *et al.*, *Brit. Polym. J.*, **20** (1988) 269.

Polymer Electrodes and Polymer Electrolytes in Li Secondary Batteries

O. Inganäs, X. Shuang, R. B. Bjorklund

Department of Applied Physics, IFM, University of Linköping, S-581 83 Linköping, Sweden

&

P. Reinholdsson

Department of Chemical Engineering II, Chemical Center, Lund Institute of Science and Technology, Box 124, S-221 00 Lund, Sweden

ABSTRACT

The combination of electroactive polymers with ion conducting polymers in composite materials offer opportunities for the construction of all solid state polymer electrochemical devices. We report materials development of polypyrroles in combination with poly(ethylene oxide) (PEO) electrolytes. Modification of the electrode/electrolyte interface through polymer exchange reactions has been demonstrated. We note that the normal anion doping of polypyrroles necessitates a salt reservoir in the electrolyte if combined with Li electrodes. This fact leads to a rather low theoretical upper limit to energy and charge capacity of polymer/lithium batteries. Improvement on this situation can be expected if a Li-intercalating polymer electrode is used. We suggest a way of accomplishing the incorporation of lithium, by using an anionic polyelectrolyte as an immobile dopant of polypyrrole, with lithium movement active in the electrochemistry. We also report initial studies of such materials, verifying this possibility.

INTRODUCTION

The development of electroactive polymers for use in electrochemical devices has already led to practical batteries, 10 years after the introduction of these materials. So far, the batteries produced are based on polymer electrodes combined with liquid organic electrolytes and lithium electrodes. The parallel development of polymeric electrolytes for thin film lithium batteries is also rapidly leading to practical devices. An attractive alternative for the future may be the combination of these two classes of materials in solid state polymer batteries. Work on such devices has recently appeared in the literature.[1-8] We have earlier reported a practical route towards such batteries based on composite electrodes with the electroactive polymer, polypyrrole, dispersed in the polymer electrolyte in the form of submicron particles.[1-3] The submicron particles of polymer materials are in principle advantageous, as charge and discharge of the small particles should be rapid, leading to good capacity utilization and high power densities. We would also expect to find an advantage in the use of small particles as we may be able to construct thin film electrochemical devices with less risk for electronic short circuit due to protruding particles with electronic conductivity.

The major problem with these batteries is their high rate of self-discharge. This self-discharge process has been studied in detail, and a suggestion for the mechanism of self-discharge has been put forward.[8] In essence, this model assumes the existence of some reducing species, possibly generated at the polymer electrolyte/Li interface, capable of diffusion across the electrolyte to reduce the oxidized polypyrrole electrode. On the positive side, this may mean that we can clean up the electrolyte to avoid the presence or generation of the culpable species. It is demonstrated that polypyrrole/lithium batteries in organic electrolytes can have a low and acceptable rate of self-discharge, 2% per day.[9] It may be difficult to reach that value for solid state polymer batteries, but in principle we do not see why it would be impossible. It may be that the high temperature at which we characterize the polymer battery is a major reason for the self-discharge. If so, we should expect to find better performance with the polymer electrolytes of higher conductivity now available.

Among other problems we identified in the first generation of these batteries was the low capacity utilization.[1] Early observations indicated that one cause of inefficient capacity utilization may be inefficient ion transfer at the interface between polymer electrode and polymer electrolyte. The first generation of material was prepared by polymerizing pyrrole to polypyrrole in the presence of a sterically stabilizing polymer, methylcellulose.[10] This led to the growth of particles 100–200 nm in diameter, which were stable in the form of a suspension. This suspension was then mixed with an aqueous solution of a polymer electrolyte, poly(ethylene oxide) (PEO) with $LiClO_4$. The interface between the polymer electrode and the polymer electrolyte should thus include a polymer layer of low electronic and ionic conductivity, methylcellulose. We thus set out to modify this interface by using new sterically stabilizing polymers in the synthesis of the polypyrroles. These polymers combined an anchoring function for adsorption onto the growing polypyrrole particle, and a poly(ethylene glycol) side chain, mimicking the polyethylene oxide electrolyte, and hopefully forming a good interface for ion transfer between polypyrrole and the polymer electrolyte.

But the major problem for development of polymer batteries is the fact that they are not very attractive when compared to other advanced battery systems now under development. This was emphasized by Passiniemi & Osterholm,[11] who showed that neither energy nor coulombic capacity for polymer batteries will be very impressive. We have since extended that analysis to secondary batteries incorporating polymeric electrolytes,[12] and can give upper limits on the energy and coulombic capacity of such systems. All of this analysis demonstrates the necessity of finding a lithium intercalating polymer electrode of sufficiently high potential.

Normally lithium intercalation is done on reduction of the polymer electrode, a process that is possible with polyacetylene and polyparaphenylene, but not with polypyrrole or polythiophene. In fact, in order to increase the voltage of such batteries, it is necessary to find polymers with a high potential for lithium intercalation, which will require a small bandgap material (if conductive conjugated polymers are the target material). The synthesis of such materials is possible, but few materials are yet available. We suggest and argue for a different way of using electroactive polymers in combination with charged polyelectrolytes in order to convert the electrochemical activity of the electroactive polymer to transport small cations.

Synthesis of materials for these purposes is reported, and we also demonstrate that they might be used in combination with polymer electrolytes.

EXPERIMENTAL

Materials Synthesis and Characterization

Pyrrole (Merck) synthesis quality, anhydrous ferric chloride, used as received. Stabilizer: The preparation of the comb-shaped stabilizer A, B, C, D is described by Wesslen & Wesslen.[13] The composition of the combshaped stabilizers is shown in Table 1.

Polypyrrole Synthesis

Colloidal solutions of polypyrrole were prepared by chemically oxidizing pyrrole monomer with ferric chloride in aqueous solutions of different stabilizers. The dispersion polymerization reactions were carried out in 100-ml flasks. Initially x g of the stabilizers and 0·75 g ferric chloride was dissolved in 50 ml millipore water. The pH was adjusted to 0 by adding HCl. To this solution was added 0·25 ml pyrrole monomer, and the mixture was stirred at room temperature (Table 2).

Table 1

Sample	Backbone polymer[a] (mole ratio)	Mn (GPC)	Grafted polymer[b] (% by weight)
A	PS/EHMA-56/44	11 600	76% MPEG 2 000
B	PS/EHMA-56/44	11 600	65% MPEG 2 000
C	PMMA	35 000	50% MPEG 5 000
D	DMA/PMMA 4/13	28 000	50% MPEG 2 000
E	MPEG 2 000		
F	Methocel 60 HG, PREM, 50 cps (methylcellulose)		

[a] PS = poly(styrene)
  EHMA = Poly(2-ethylhexyl methacrylate)
  PMMA = poly(methylmethacrylate)
  DMA = poly/dodecyl methacrylate).
[b] MPEG = poly(ethylene glycol) monoethyl ether.

Table 2

Stabilizer	Amount (g)	Particle size (nm)	Comment
A	0·1		Flocculation
B	0·1		Flocculation
C	0·1		Flocculation
D	0·1		Flocculation
E	0·1		Flocculation
F	0·05	20–40	

Within a minute of adding the pyrrole, the colour in the reaction mixture changed from weak orange to brown-black, which indicates that polymerization occurs. The product was centrifuged at 4000 rpm for about 1 h. This led to a black sediment and a pale green but transparent supernatant. The sediment was redispersed in millipore water, adjusted to pH 1 and centrifuged. This procedure was repeated three times to remove ferrous chloride.

Synthesis of Polypyrrole–Polyelectrolyte Complexes

Synthesis and characterization of polypyrrole–polystyrenesulphonate and polypyrrole–poly(vinylsulphate) grown by chemical polymerization is reported in detail in Ref. 14. Briefly, the sodium salt of the polyelectrolyte was exchanged with acid to protonate the polymer, a necessary step to put the system at the right pH level for synthesis of polypyrrole.[10,15] Thereafter exchange of some protons for Fe(III) was done. Both these ion exchange processes were performed by using ion exchange materials in the form of small beads, in contact with water solutions. Pyrrole was then added, and a polypyrrole–polyelectrolyte complex grew. The product was washed by adding water, centrifuging the product and discarding the aqueous supernatant. This was done up to five times to clean out the iron. Analytical tests indicated absence of iron after this cleaning procedure.

Particle Size Determination

The particle sizes were determined by transmission electron microscopy (TEM) on a JEOL 100 U instrument. Results are given in Table 2.

Electrode Formation

Electrochemical experiments were carried out as reported earlier.[1] The polymer electrodes were formed by mixing solutions of PEO and salt with a suspension of polypyrrole. In the case of the polypyrrole stabilized with comb co-polymers (samples A–D), no such mixing was done, and the materials were used as obtained from synthesis.

RESULTS AND DISCUSSION

Polypyrrole Stabilized by Comb Copolymers

The synthesis of polypyrroles in the presence of modified stabilizers has not led to a material of more efficient capacity utilization. Indeed, with all materials synthesized we have obtained much poorer capacity utilization than with the first generation of methylcellulose stabilized polypyrrole. As shown in Table 3, only polypyrrole polymerized in the presence of PEO or methylcellulose give capacity utilization of some significance. The chemical nature of the interface between the polypyrrole particle and the polymer electrolyte thus appears to be of great significance. We find that the comb copolymers chosen for the specific purpose of making an interface with ease of ion transport, give the worst results. The poor results merit an explanation.

Table 3 Estimates of Coulombic Capacity in Various Forms of Polypyrrole

Material	C/C_t (%)
Methylcellulose–polypyrrole	30–40
Polyvinylalcohol–polypyrrole	1–2
Polyethylene oxide–polypyrrole	10
Comb–EO–polypyrrole	1–2

Method of evaluation: Galvanostatic charge/discharge between 2·2 and 3·1 V versus Li, typically at 30 μA/cm^2.
Capacity is given in percentage of theoretical capacity, assuming a doping level of one electron for four pyrrole units.

The PMMA or PS/EHMA backbone chosen for adsorption on the polypyrrole particles is not a good ion conductor, and we might at first hand suggest that this layer is too thick for ion transfer. Why then is the methylcellulose, not known for its ion transport properties, the best material? We suggest that the process of forming the composite electrodes actually removes the methylcellulose from the interface, leading to a polypyrrole/PEO interface of the kind we desire.

Polymer Exchange Reactions

The formation of the composite electrodes, as reported in the experimental section, includes mixing of the suspensions of polypyrrole particles with a PEO/salt solution in water, before drying the composite material. It is known that suspensions stabilized by adsorption of non-ionic polymers undergo polymer exchange reactions in contact with polymer solutions.[16] We may thus expect that if methylcellulose is acting as a sterically stabilizing polymer clothing the polypyrrole particles, it will desorb if these particles are put in a solution of PEO, and that the PEO polymer will adsorb instead, leading in a short time to a more or less complete exchange of methylcellulose for PEO.

Experiments to test this possibility have verified the hypothesis. The polymer exchange was done by mixing a suspension of colloidal polypyrrole with a water solution of PEO, then centrifuging the suspension to collect the polypyrrole particles in the solid phase. This material was then redispersed in PEO solution, and the procedure was repeated. Finally, the pellet was redispersed in distilled water. Thin films of the virgin and the PEO-exchanged suspension of polypyrrole were deposited on gold surfaces, and FTIR spectroscopy was performed (Fig. 1). The results clearly show major loss of the methylcellulose material upon exchange, as the characteristic band at $c.$ $3400 \, \text{cm}^{-1}$ is lost in the exchanged sample. A band at $840 \, \text{cm}^{-1}$ (not shown), characteristic of polyethylene oxide only, is also apparent in the PEO-exchanged polypyrrole, and is absent in the methylcellulose polypyrrole, thus proving that PEO is adsorbed. We conclude that polymer exchange is an attractive route for modifying the polymer electrode/polymer electrolyte interface.

It should be noted that the electrochemical study of polypyrrole stabilized with comb copolymers used the homogeneous material,

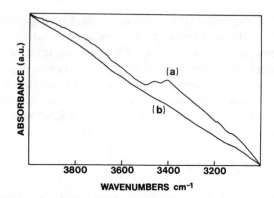

Figure 1. FTIR spectra of thin films of colloidal polypyrrole (a) as synthesized and purified and (b) after polymer exchange with polyethylene oxide. The band at 3400 cm^{-1} is characteristic of methylcellulose and does not exist in PEO.

without equilibration in PEO solutions. There would thus be no opportunity for polymer exchange reaction (except maybe over very long times, in contact with the polymer electrolyte).

Electrochemical Mechanisms in Polymer Electrodes: Anion and Cation Intercalation

The early identification of anions as active species in the oxidation/reduction reactions of polyheterocycles came partly from elemental analysis and partly from surface spectroscopy. Further evidence obtained since has complicated that picture, proving the transport of both anions and cations. Such evidence has been obtained by gravimetric electrochemical studies[17] in the case of polypyrrole, and by secondary ion mass spectroscopy for poly(3-methylthiophene)[18] It now appears clear that the current flowing during oxidation and reduction of (at least) some of the poly(heterocycles) is carried by both anions and cations, moving in opposite directions.

A parallel development of novel types of polypyrrole, by electrochemical synthesis in the presence of polymeric anions[19–25] have led to materials of high enough conductivity, some processability[19] and attractive mechanical properties.

The electrochemical study of these materials was initiated by the group of Shimidzu, who early realized their possibilities. With polypyrrole synthesized electrochemically from a solution of an anionic polyelectrolyte and pyrrole, they obtained what was named a pseudo-cationic doping mechanism (Fig. 2). Namely, on oxidation of the polymeric material, current was carried by small inorganic cations (Li, Na), and none whatsoever was carried by anions. The polymeric anions are hopelessly entangled in the polymeric material, and cannot diffuse in and out. All electrochemical current is carried as cations which leave the polymer material on oxidation, leaving a polycationic polypyrrole compensated by the polymeric anion, and on reduction small cations are included in the polymer, compensating the polymeric anions and leaving the electroactive polymer in the neutral state. Proof of this mechanism has been obtained with elemental analysis of electrodes and electrolytes. Further proof, obtained through a sensitive luminescence assay[27] has recently been presented. The use of this pseudo-cationic doping mechanism for electroactivated ion sorption[25] as well as metal ion-intercalating polymer electrodes in secondary batteries of aqueous type has been suggested.[26]

We propose to use this type of material in all solid state polymeric Li batteries. An example of the electrochemical performance of such a material in an all solid state polymer battery is given in Fig. 3, showing a voltammogram of a polypyrrole–poly(vinylsulphate) material in

Electrode mechanism in polymer secondary batteries
(a) Oxidation of poly(heterocycles)

$$Li \leftrightharpoons Li^+ + e^-$$
$$P^+(X^-) + e^- \rightleftarrows P^0 + X^-$$

$$P^+(X^-) + Li \leftrightharpoons P^0 + Li^+ + X^-$$

(b) Suggested mechanism of Li-intercalation during oxidation:

$$Li \leftrightharpoons Li^+ + e^-$$
$$P^+(A^-) + Li^+ + e^- \leftrightharpoons P^0(Li^+A^-)$$

$$P^+(A^-) + Li \leftrightharpoons P^0(Li^+A^-)$$

Figure 2. Traditional electrochemical mechanism of polymer oxidation through anion transport (a) and novel pseudo-cationic mechanism of polymer oxidation by cation exchange (b).

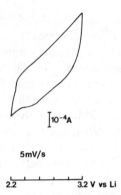

Figure 3. Voltammogram of polypyrrole–polyelectrolyte complex (polypyrrole–poly(vinylsulphate)) electrode (area 1·77 cm²) in contact with a $(PEO)_8(LiClO_4)$ electrolyte at 80°C. The counter and reference electrode is a Li button.

contact with a PEO electrolyte in a lithium battery. While the coulombic capacity of the present materials is not impressive, we expect to be able to optimize materials for these purposes and thereby increase energy and power density to more attractive levels.

The advantage of using a Li-intercalating polymer electrode is mainly that no storage of electrolyte salt is necessary, as only cations are active both at cathode and anode. When storage of salt is not necessary, we will be able to choose the thickness of electrolyte on engineering grounds, which may enable us to use a much thinner electrolyte film, enhancing charge and energy density, and reducing the amount of material. The thinner films will also let us use an electrolyte material of lower conductivity. The advanced battery concepts utilizing polymer electrolytes are all based on cation-intercalating inorganic electrodes.

Comparing the use of polymer electrodes based on cation inclusion during reduction of the polymer, for instance polyacetylene and poly(paraphenylene), we find that the potential of this process is much lower than that found for pseudo-cationic oxidation of polypyrrole. The reduction potential of polymer electrodes is basically identical to the conduction band edge of the materials, and we will have to design new materials to shift this band level. This is possible, but requires new small-bandgap materials not presently available.

An alternative synthetic route to cation-including materials is that of the so-called self-doping materials.[28-30] These include an ionic substituent on the electroactive polymer chain, which can dissociate to form an anionic group with concurrent transport of the cation. Such substituted polythiophenes are now available[30] and would essentially have the same electrochemical functions as the pseudo-cationic polymers. Comparison of the weight of the substituent on the electroactive polymer for the self-doped materials, and weight of the polyanion in the case of pseudo-cationic material, will decide which of these materials would be the better for application in Li batteries. Other aspects such as electronic and ionic conductivity and morphology will probably be of greater importance.

CONCLUSION

Systematic design of stabilizing polymers for use in the suspension polymerization of pyrrole to polypyrrole particles for use in solid state PEO-based batteries has been unsuccessful. Demonstration of simple polymer exchange reactions for modifying the polymer electrode/polymer electrolyte interface has opened up a new method for optimization of such interfaces. We argue the necessity of using a Li-intercalating polymer as the cathode in a polymer battery, and demonstrate the possibility of using polypyrrole–polyelectrolyte complexes for PEO-based batteries.

ACKNOWLEDGEMENTS

This work was supported by the Swedish Board for Technical Development. We thank Dr B. Wesslen and Dr K. B. Wesslen, LTH, for discussions and the donation of amphiphilic polymer stabilizers. The infra-red studies were done in collaboration with Mr Luigi Bonfatti and Dr B. Liedberg. Discussions with Dr Petr Novak, Prague, are gratefully acknowledged.

REFERENCES

1. Novak, P., Inganäs, O. & Bjorklund, R., *J. Electrochem. Soc.*, **134** (1987) 1341.

2. Novak, P., Inganäs, O. & Bjorklund, R., *J. Power Sources*, **21** (1987) 17.
3. Inganäs, O., *Brit. Polym. J.*, **20** (1988) 233.
4. Pantaloni, S., Passerini, S. & Scrosati, B., *J. Electrochem. Soc.*, **134** (1987) 753.
5. Panero, S., Prosperi, P., Bonino, F. & Scrosati, B., *Electrochimica Acta*, **32** (1987) 1007.
6. Scrosati, B., *Progr. Solid State Chem.*, **18** (1988) 1.
7. Mastragostino, M., Marinangeli, A. M., Corradini, A. & Giacobbe, S., *Synth. Met.* **28** (1989) C663.
8. Novak, P. & Inganäs, O., *J. Electrochem. Soc.*, **135** (1988) 2485.
9. Münstedt, H., Köhler, G., Möhwald, H., Naegele, D., Bittihn, R., Ely, G. & Meissner, E., *Synth. Met.*, **18** (1987) 259.
10. Bjorklund, R. & Liedberg, B., *J. Chem. Soc. Chem. Commun.*, (1986) 1293.
11. Passiniemi, P. & Österholm, J.-E., *Synth. Met.*, **18** (1987) 637.
12. Passiniemi, P. & Inganäs, O., *Solid State Ionics*, **34** (1989) 225.
13. Wesslen, B. & Wesslen, K. B., submitted, *J. Pol. Sci. A. Pol. Chem.*
14. Bjorklund, R., Work in preparation.
15. Bjorklund, R. B., *J. Chem. Soc. Faraday Trans.* **1**, 83 (1987) 1507.
16. de Gennes, P. G., *Annali di Chimica*, **77** (1987) 389.
17. Kaufman, J. H., Kanazawa, K. K. & Street, G. B., *Phys. Rev. Lett.*, **53** (1984) 2461.
18. Chao, F., Baudoin, J. L., Costa, M. & Lang, P., *Macromol. Chem. Macromol. Symp*, **8** (1987) 173.
19. Bates, N., Cross, M., Lines, R. & Walton, D., *J. Chem. Soc. Chem. Comm.*, (1985) 871.
20. Ulanski, J., Glatzhofer, D. T., Przybylski, M., Kremer, F., Gleitz, A. & Wegner, G., *Polymer*, **28** (1987) 859.
21. Glatzhofer, D. T., Ulanski, J. & Wegner, G., *Polymer*, **28** (1987) 449.
22. Iyoda, T., Ohtani, A., Shimidzu, T., & Honda, K., *Chem. Lett. (Jpn)*, **687** (1986).
23. Shimidzu, T., Ohtani, A., Iyoda, T. & Honda, K., *J. Chem. Soc. Chem. Comm.*, (1987) 327.
24. Shimidzu, T., Ohtani, A., Iyoda, T. & Honda, K., *J. Chem. Soc. Chem. Comm.*, (1987) 1415.
25. Shimidzu, T., Ohtani, A., Iyoda, T. & Honda, K., *J. Electroanal. Chem.*, **224** (1987) 123.
26. Ohtani, A., Abe, M., Higuchi, H. & Shimidzu, T., *J. Chem. Soc. Chem. Comm.*, (1988) 1545.
27. Basak, S., Ho, Y.-H., Tsai, E. W. & Rajeshwar, K., *J. Chem. Soc. Chem. Comm.*, (1989) 462.
28. Patil, A. O., Ikenoue, Y., Colaneri, N., Chen, J., Wudl, F. & Heeger, A. J., *Synth. Met.*, **20** (1987) 151.
29. Patil, A. O., Ikenoue , Y., Wudl, F. & Heeger, A. J., *J. Am. Chem. Soc.*, **109** (1987) 1858.
30. Ikenoue, Y., Uotani, N., Patil, A. O., Wudl, F. & Heeger, A. J., *Synth. Met.*, **30** (1989) 305.

Characteristics of MHB* Polymer–Electrolyte Battery Technology

J. S. Lundsgaard, S. Yde-Andersen, R. Koksbang

Energy Research Laboratory (ERL), Vestergade 24,
5700 Svendborg, Denmark

&

D. R. Shackle, R. A. Austin, D. Fauteux

Mead Imaging, Miamisburg, Ohio, USA

ABSTRACT

The MHB polymer electrolyte is characterised by a conductivity exceeding 10^{-3} S/cm at temperatures in the range -20 to $45°C$, by good mass transport properties and by its transparency and mechanical stability. The importance of controlling the interfacial properties during processing of polymer electrolyte based devices is demonstrated as the interface impedances exceed the electrolyte impedance. Room temperature application of the polymer electrolyte is demonstrated in rechargeable lithium batteries, cycled at C/10 and C/2·5 rates.

INTRODUCTION

Since Armand conceived the idea of solid polymer electrolytes and published the first results on PEO-based systems in 1978,[1] significant improvements in room temperature performance have taken place. Simultaneously, a variety of applications, notably polymer electrolyte lithium batteries, has been investigated. The feasibility of rechargeable all solid state polymer electrolyte lithium batteries employing polymeric electrolytes has been demonstrated at temperatures exceed-

* MHB is a joint venture between ERL and Mead Imaging.

ing 80°C[2] and near room temperature.[3] The polymers used in these studies were of the poly(ethylene oxide) (PEO) type but the exact nature of the electrolyte used in the latter study was not disclosed. The most successful approach to increasing the ionic conductivity of the polymer electrolytes seems to be by addition of various plasticisers. Kelly *et al.*[4] added polyethyleneglycol-dimethylether (PEGDME) to PEO, resulting in an increase of the ionic conductivity from 3×10^{-7} to 10^{-4} S/cm at 40°C. Recently, Munshi & Owens[5] demonstrated the feasibility of incorporating up to 40% propylene carbonate (PC) in PEO. With this electrolyte it was possible to cycle a V_6O_{13}-based battery at low rate and modest capacity at room temperature. The best performance was achieved by addition of only 20% PC. In the case of PEO/PC, a serious loss of mechanical stability is experienced as the plasticiser level is increased, because PC (and other polar solvents) dissolve PEO. For polyvinylidene fluoride (PVdF) and poly-acrylonitrile (PAN)[6-9,12] to which a variety of liquid polar solvents was added, the ionic conductivity increased as the plasticiser content increased. Room temperature conductivities around 10^{-3} S/cm were achieved. The latest results include plasticising salts of the $Li(CF_3SO_3)_2N$ type, which together with PEO give rise to room temperature conductivities around 10^{-4} S/cm.[10]

The feasibility of polymer electrolyte technology in batteries for commercial applications requires high ionic conductivity in a broad temperature range. However, the capabilities of processing thin electrode and electrolyte layers and at the same time controlling the properties of the electrolyte/electrode interfaces are of equal impor-tance. The intention of this paper is to describe some typical characteristics of the MHB polymer electrolyte technology, that is, electrolyte conductivity, transport properties of the electrolyte and performance of the first generation of batteries.

MEASUREMENTS

The conductivity and the interfacial properties of both electrodes were investigated by the use of AC measurements on different cell configurations:

Li/electrolyte/composite cathode   (1)

Li/electrolyte/Li   (2)

Composite cathode/electrolyte/composite cathode   (3)

Usually cell configuration (1) included either a LiAl or a Li reference electrode[11] in order to study the behaviour of both electrode/electrolyte interfaces simultaneously while the two other cell types were used for studies of the respective interfaces. The transport properties were determined by measurements on type (2) cells.

The cell areas for these measurements were generally in the range 1-20 cm^2, while electrolyte and cathode thicknesses were less than 100 μm. For the measurements, a Solartron 1250 FRA connected to a Solartron 1286 Interface were used. Cycling of batteries was realised using a personal computer based data acquisition system developed at the Energy Research Laboratory (ERL).

ELECTRICAL PROPERTIES OF THE ELECTROLYTE

The as-prepared polymer electrolyte films are transparent, and optical microscopy revealed no evidence of crystallisation of salt complexes at room temperature. The amorphous nature of the electrolytes is a first indication of the improved conductivity, as the conductivity of polymer electrolytes generally is thought to be related to the non-crystalline phases. Also, the thin films (100 μm) were mechanically stable, i.e. self-supporting and flexible.

Conductivity

The conductivity dependence on salt concentration and temperature is shown in Fig. 1, and the room temperature conductivity is compared with the conductivities of other polymer electrolytes in Table 1.

The conductivity is in the range 10^{-4}–10^{-2} S/cm in the temperature range investigated (−20 to 45°C). The temperature dependence, when plotted in an Arrhenius type plot, shows definite curvature and has to be described within other models. Comparing the salt concentration dependence and temperature dependence of the electrolyte shows that conductivities greater than 10^{-3} S/cm are achievable at temperatures down to −20°C.

The comparison of room temperature conductivities given in Table 1, shows that the MHB polymer electrolyte is superior to other solid polymeric electrolytes, including polymers plasticised to as high a level as 40%.[7] Also, at high temperature, the MHB electrolyte is superior to other polymer electrolytes.

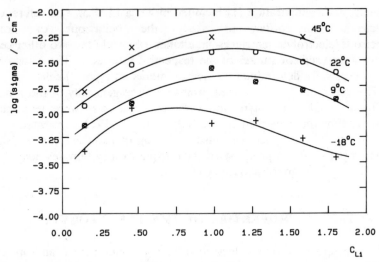

Figure 1. Conductivity versus salt concentration and temperature. The salt concentration is normalised with respect to the salt concentration corresponding to the highest conductivity at room temperature.

Table 1. Comparison of Room Temperature Conductivities of Various Polymer Electrolytes and Plasticised Polymer Electrolytes

Electrolyte	Conductivity (mS/cm)	Reference
PEO/PEGDME/LiI	0·001	4
PEEEVE[a]	0·01	23
MEEP[b]	0·02	24
PAN/EC/LiClO$_4$[c]	0·2	25
PVDF/PC/LiClO$_4$[d]	1	12
MHB	5	This work

[a] Polyethoxy-ethoxy-ethoxy-vinyl-ether/LiBF$_4$ (O/Li = 8).
[b] Poly(bis(methoxyethoxyethoxy)phosphazene)(LiCF$_3$SO$_3$)$_{0.167}$.
[c] Polyacrylonitrile doped with ethylene carbonate.
[d] Polyvinylidenefluoride doped with 40% propylene carbonate.

Furthermore, cyclic voltammetry shows that the electrolyte is electrochemically stable at potentials exceeding 4 V versus lithium, i.e. sufficiently stable for practical applications in batteries employing most of the known solid cathode materials.

Transport Properties

Although a high conductivity is needed for a satisfactory battery performance, the mass transport properties are of equal importance. Thus, the transport number of the positive ions as well as the salt diffusion coefficient were determined. The latter is a measure of the mass transport limitations which determine the battery performance, whether the limitations originate from cation or anion depletion on the electrodes.

The transport number was determined by the method developed by Evans et al.[13] After determination of the initial AC impedance (R_0), a small DC voltage (V) is applied to polarise the cell and the initial current (I_0) measured. The current was monitored as a function of time until the steady-state current (I_s) was reached. At this point the steady-state AC impedance (R_s) was measured. The transport number of the positive ions (t_+) was determined from the relation:

$$t_+ = \frac{I_s(V - I_0 R_0)}{I_0(V - I_s R_s)}$$

The transport number (t_+) of the MHB polymer electrolyte is in the range 0·5–0·6, i.e. the charge is mainly carried by positive ions. Comparison with other polymer electrolytes and liquid non-aqueous electrolytes (Table 2), shows that this is an improvement even compared with liquid electrolytes for which t_+ usually is less than 0·5.

Table 2. Comparison of the Transport Number of the Positive Ion of the MHB Polymer Electrolyte with those of other Polymer Electrolytes and Liquid Electrolytes

Electrolyte	t_+	$T(°C)$	Reference
$(PEO)_9LiCF_3SO_3$	0·46	90	13
PEEVE	0·33	48	23
MEEP	0·32	54	26
$PC/LiClO_4$	0·3–0·4	25	27
MHB	0·5–0·6	22	This paper

The salt diffusion coefficient was determined chronopotentiometrically by passing a constant current, sufficient to force the cell into diffusion limitation, through a Li/electrolyte/Li cell and monitoring the voltage as a function of time. Under these conditions the surface concentration of either positively or negatively charged species, reach zero after some time. The ion flux to the electrode surface is no longer sufficient to maintain the current and consequently the voltage increases sharply. The transition time (t, s) characterising the system, is the period from the time the current is applied, to when the voltage increase takes place. The salt diffusion coefficient (D, cm^2/s) is calculated from the transition time, the current density (I, A/cm^2) and the salt concentration in the bulk electrolyte by the use of Sand's equation:

$$D = \frac{t}{\pi}\left(\frac{2I}{nFC}\right)^2$$

where n and F have the usual meaning.

By this method, salt diffusion coefficients in the range 10^{-6}–10^{-5} cm^2/s were determined. This is in the range of the lithium salt diffusion coefficients of liquid electrolytes commonly used for primary and secondary lithium batteries.[14]

Thus, the high transport numbers and salt diffusion coefficients of the MHB polymer electrolyte show that the mass transport properties are sufficient to achieve a performance identical to that with liquid electrolyte lithium batteries.

INTERFACIAL CHARACTERISTICS

In spite of the high conductivity and the good mass transport properties of the MHB polymer electrolyte, limitations in performance can occur due to the interfacial properties of a complete cell laminate. In addition to the interfacial problems encountered in liquid electrolyte cells, polymer cells also suffer from contact problems associated with the solid/solid interfaces. However, when the contact problems are solved, other problems, such as formation of a passivating interface layer due to reaction between the electrolyte (or impurities) and the electrode materials, appear.

Figure 2 shows the AC impedance spectra of a full cell and the two electrode/electrolyte interfaces. The contribution to the cell im-

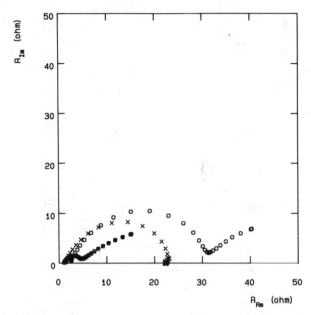

Figure 2. AC impedance spectra (65 kHz–0·1 Hz) of a cell equipped with LiAl reference electrode. Cell impedance (O), anode/electrolyte interface impedance (×) and cathode/electrolyte interface impedance (●).

pedance of the electrolyte is negligible as is the cathode/electrolyte interface impedance. The cell impedance appears to be solely dominated by the anode/electrolyte interface impedance.

Anode/Electrolyte Interface

In Fig. 3, the interface impedance is shown in an Arrhenius type plot for the temperature range −40 to +60°C. The activation energy of 0·67 + −0·04 eV, is close to those measured in both liquid electrolytes[15] and high temperature PEO-based polymer electrolytes.[16] Activation energies of both these types exceed 0·6 eV.

The size of the semi-circle observed in the AC diagram, decreases when a current is applied to the cell, i.e. it shows a behaviour typical of an electrode surface covered with a thin film due to reaction between electrolyte and electrode material.

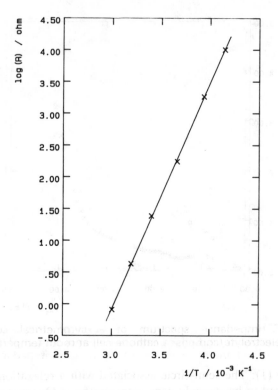

Figure 3. Log (R_i) versus 1/T for a symmetrical Li/electrolyte/Li cell.

Cathode/Electrolyte Interface

Although the cathode impedance is of apparently minor importance, some effort was devoted to characterisation of the cathode/electrolyte interfacial properties. A typical AC spectrum of a symmetrical cathode/electrolyte/cathode cell is shown in Fig. 4. Three semi-circles, with relaxation times corresponding to approximately 10^{-6} s(1), 10^{-4} s(2) and 10^{-2} s(3) can be resolved as indicated in the figure. The temperature dependence as well as the dependence on process parameters such as cathode composition and mixing, coating and lamination have been examined and are interpreted in the following where it should be noted that the semi-circle impedances are converted to conductivities.

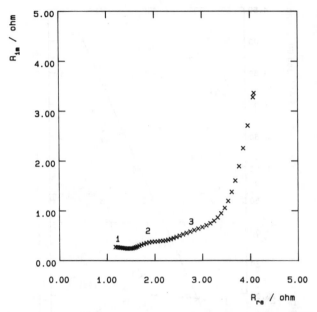

Figure 4. AC impedance spectrum of a symmetrical composite cathode/electrolyte/composite cathode cell at room temperature.

Semi-circle (1): The semi-circle associated with a relaxation time of 10^{-6} s appeared to be strongly dependent on the V_6O_{13} content in the cathode. In cells without V_6O_{13} in the cathode, i.e. only electrolyte and carbon present, this semi-circle is absent. Additionally, the impedance is temperature dependent, as shown by the Arrhenius type of behaviour in Fig. 5(a). This semi-circle is therefore assigned to lithium insertion in the V_6O_{13} in the composite cathode. The associated activation energy is 0.19 eV (± 0.08 eV), i.e. considerably lower than that of the surface film on the anode.

Semi-circle (2): The temperature dependence of the semi-circle associated with a relaxation time of 10^{-4} s shows a less regular variation (Fig. 5(b)). However a definite dependence on the total loading of the cathode with solid non-electrolyte material, i.e. carbon and V_6O_{13}, was observed. A high weight fraction of solids in the cathode is related to high conductivity and an almost linear temperature dependence. At medium and low weight fractions of solid materials, an optimum conductivity is observed around room temperature. At low temperature the three lines converge towards the high

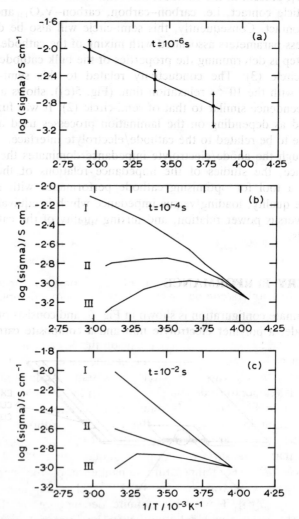

Figure 5. Impedance variation versus temperature and process parameters of (a) semi-circle (1), (b) semi-circle (2) and (c) semi-circle (3) as indicated in Fig. 4.

load line. As the behaviour is associated with the amount of carbon and V_6O_{13} present in the cathode, this semi-circle has been ascribed to interparticle contact, i.e. carbon–carbon, carbon–V_6O_{13} and V_6O_{13}–V_6O_{13} contact. Consequently, this semi-circle will also be dependent on process parameters associated with mixing of the cathode materials as this step is determining the properties of the bulk cathode.

Semi-circle (3): The conductivity related to this semi-circle associated with the 10^{-2} s relaxation time (Fig. 5(c)), shows a temperature dependence similar to that of semi-circle (2). It was furthermore identified as depending on the lamination processes used and seems therefore to be related to the cathode/electrolyte interface.

Although the anode/electrolyte impedance dominates the total cell impedance, the studies of the impedance relations of the cathode provide a tool for optimising cathode performance with respect to interface quality, loading versus impedance which is equivalent to an energy versus power relation, and mixing quality of the cathode raw materials.

BATTERY PERFORMANCE

The laminate configuration is shown in Fig. 6, and consists of a lithium foil anode, a polymer electrolyte film and a composite cathode on a

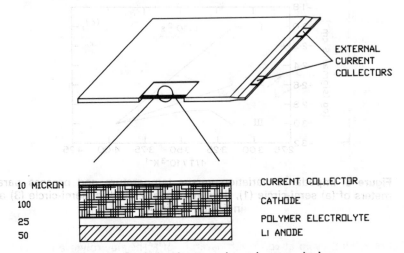

Figure 6. Basic laminate and test battery design.

current collector. The latter could be a metal foil, e.g. nickel as shown
in the figure, while the composite cathode consists of an intimate
mixture of MHB electrolyte, acetylene black and V_6O_{13}. Other active
cathode materials have been investigated but the results presented
here will be restricted to V_6O_{13} based batteries.

A typical test battery design is shown in Fig. 6 also. Metal current
collectors are attached to the laminate electrodes and the laminate is
placed in a flexible encapsulation which is evacuated and heat sealed
for protection of the battery laminate. This design provides easy access
to different battery configurations as it is easy to manufacture batteries
as both single cells and as bipolar cells having a higher voltage and
with areas in the range from one to a few hundred cm^2.

In the following, the utilisation of the active cathode material is
relative to 8 Li/V_6O_{13}, which is the maximum reversible capacity.[17-21].
The estimation of the utilisation is based on the charge passed through
the batteries and a chemical vanadium analysis of the cathodes.

The voltage versus time curves (first cycle) at room temperature and
at various rates, shown in Fig. 7, illustrate the discharge and charge
behaviour of typical first generation batteries. At the C/10 rate, 100%
utilisation of the active cathode material is achieved. The number of
plateaus, their relative sizes and voltage levels are identical to the

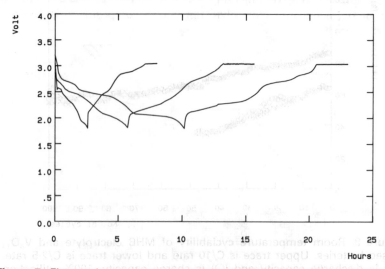

Figure 7. First cycle at room temperature of MHB electrolyte and V_6O_{13}
based batteries at different rates.

OCV curve[17] and show that the battery is close to equilibrium and that the overvoltage is negligible in both charge and discharge. Moreover, the voltage characteristics are identical to those observed in liquid electrolyte cells at room temperature[18–20] and at elevated temperature using conventional polymer electrolytes.[21] Similarly, the presence of plateaus during charge indicate equilibrium and good recharge properties. Even after 100 cycles the characteristic plateaus associated with the Li/V_6O_{13} system can be distinguished. The virtually unchanged shape of the voltage curves indicate that the integrity of the composite cathode is preserved and that no appreciable degradation of the vanadium oxide has taken place. The same general considerations apply to the rechargeability of the battery cycled at medium rate (C/5). At high rate (C/2·5) a voltage delay, presumably related to breakdown of the anode/electrolyte layer, is observed.

The rechargeability characteristics at C/10 and C/2·5 rates are illustrated in Fig. 8. At both rates, high material utilisation is apparent. The decaying capacity is associated with the inherent lithium insertion properties of V_6O_{13}[17–21] and inefficient recharge as the amount of charge used to recharge the battery is lower than the amount of charge extracted in the preceding discharge. Furthermore,

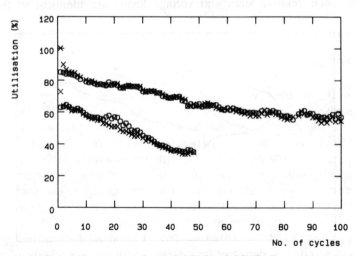

Figure 8. Room temperature cyclability of MHB electrolyte and V_6O_{13} based batteries. Upper trace is C/10 rate and lower trace is C/2·5 rate. (×) is discharge capacity and (○) is charge capacity. 100% utilisation corresponds to $8 Li/V_6O_{13}$.

100% utilisation is recovered by decreasing the current density by a factor of ten. Thus, the results are similar to those achieved with both liquid and polymer electrolytes.[17-21]

Comparison of room temperature and sub-room temperature discharge curves at low rates show that at low temperature, the overvoltage is slightly higher during discharge of the first 50% of the capacity. Then the overvoltage of the low temperature curve starts to increase. Full utilisation of the active material is still possible but only at discharge to 1·0 V (1·8 V at room temperature). This observation seems to be related to the lithium insertion properties of the vanadium oxide rather than to the electrolyte as both the electronic conductivity and the lithium ion diffusion coefficient of V_6O_{13} decrease as a function of the degree of insertion.[17,22]

The room temperature power capabilities are illustrated by the ability of a 16 cm^2 active area battery to power a motor, requiring 0·5 A. The battery was able to deliver a current density exceeding 30 mA/cm^2 continuously for more than 1 min. Similar experiments with smaller batteries show that the peak current density exceeds 60 mA/cm^2 at room temperature at shorter discharge times. For comparison, the peak current density, recorded under similar conditions, of a PVdF/PC/LiClO$_4$ electrolyte battery, using polyacetylene as both anode and cathode,[9] was 7·5 mA/cm^2.

CONCLUSION

The MHB polymer electrolyte is characterised by a conductivity greater than 10^{-3} S/cm at temperatures between −20 and 45°C and the transport properties are similar to those of typical liquid non-aqueous electrolytes used in lithium batteries. The cathode/electrolyte interface properties are determined by (1) lithium insertion in V_6O_{13}, (2) interparticle contact between V_6O_{13} and carbon and (3) the physical characteristics of the cathode/electrolyte interface, i.e. the quality of the lamination of the two parts. However, the cell impedance seems to be dominated by the anode/electrolyte interface.

High utilisation of the active material is possible at room temperature and at current densities which ensure the commerciality of the battery. Similarly, the charge retention during cycling is as good as that observed in both liquid electrolyte and high temperature polymer electrolyte batteries operated under similar conditions.

A typical battery with an area of $16\,cm^2$ is able to deliver $0.5\,A$, corresponding to more than $30\,mA/cm^2$, for approximately 1 min.

ACKNOWLEDGEMENT

The authors wish to express their acknowledgement to all their colleagues at the Energy Research Laboratory and at Mead Imaging who have contributed to this project.

REFERENCES

1. Armand, M. B., Chabagno, J. M. & Duclot, M., Second International Meeting on Solid Electrolytes, St Andrews, Scotland, 20–22 September 1978, extended abstract no 6·5.
2. West, K., Zachau-Christiansen, B., Landeira, M. J. & Jacobsen, T., In: *Transport-Structure Relations in Fast Ion and Mixed Conductors*, ed. F. W. Poulsen, N. Hessel Andersen, K. Klausen, S. Skaarup & O. Toft Sorensen. *6th Riso International Symposium on Metallurgy and Materials Science*. Riso National Laboratories, Denmark, 1985, pp. 265–72.
3. Gauthier, M., Fauteux, D., Vassort, G., Belanger, A., Duval, M., Ricoux, P., Gabano, J.-M., Muller, D., Rigaud, P., Armand, M. B. & Deroo, D., *J. Electrochem. Soc.*, **132** (1985) 1333–40.
4. Kelly, I. E., Owen, J. R. & Steele, B. C. H., *J. Power Sources*, **14** (1985) 13–21.
5. Munshi, M. Z. A. & Owens, B. B., *Solid State Ionics*, **26** (1988) 41–6.
6. Ohno, H., Matsuda, H., Mizoguchi, K. & Tsuchida, E., *Polym. Bull.*, **7** (1982) 271–5.
7. Tsuchida, E., Ohno, H. & Tsunemi, K., *Electrochim. Acta*, **28** (1983) 591–5.
8. Tsunemi, K., Ohno, H. & Tsuchida, E., *Electrochim. Acta*, **28** (1983) 833–7.
9. Nagatomo, T., Ichikawa, C. & Omoto, O., *J. Electrochem. Soc.*, **134** (1987) 305–8.
10. Armand, M., Gorecki, W. & Andreani, R., *Second International Symposium on Polymer Electrolytes*, ed. B. Scrosati. Elsevier Science Publishers, London, 1990, pp. 91–7.
11. Koksbang, R., Fauteux, D., Norby, P. & Nielsen, K. A., *J. Electrochem. Soc.*, **136** (1989) 598–605.
12. Nagatomo, T., Kakehata, H., Ichikawa, C. & Omoto, O., *Jap. J. Appl. Phys.*, **24** (1985) L397–L398.
13. Evans, J., Vincent, C. A. & Bruce, P. G., *Polymer*, **28** (1987) 2325–9.
14. Venkatasetty, H. V. (ed.), *Lithium Battery Technology*, John Wiley, New York, 1984, p. 39.

15. Geronov, Y., Schwager, F. & Muller, R. H., *J. Electrochem. Soc.*, **129** (1982) 1422–9.
16. Fauteux, D., PhD Thesis, INRS-Energie, Varennes, Canada 1986, pp. III-14–III-82.
17. West, K., Zachau-Christiansen, B. & Jacobsen, T., *Electrochim. Acta*, **28** (1983) 1829–33.
18. Murphy, D. W., Christian, P. A., DiSalvo, F. J. & Carides, J. N., *J. Electrochem. Soc.*, **126** (1979) 497–9.
19. Murphy, D. W., Christian, P. A., DiSalvo, F. J., Carides, J. N. & Waszczak, J. V., *J. Electrochem. Soc.*, **128** (1981) 2053–60.
20. Abraham, K. M., Goldman, J. L. & Dempsey, M. D., *J. Electrochem. Soc.*, **128** (1981) 2493–501.
21. West, K., Zachau-Christiansen, B., Jacobsen, T. & Atlung, S., *J. Power Sources*, **14** (1985) 235–45.
22. West, K., Zachau-Christiansen, B., Ostergaard, M. J. L. & Jacobsen, T., *J. Power Sources*, **20** (1987) 165–72.
23. Pantaloni, S., Passerini, S., Croce, F., Scrosati, B., Roggero, A. & Andrei, M., *Electrochim. Acta*, **34** (1989) 635–40.
24. Blonsky, P. M., PhD Thesis, Northwestern University, Evanston, Illinois, 1986, p. 8.
25. Watanabe, M., Kanba, M., Nagaoka, K. & Shinohara, I., *J. Polym. Sci., Pol. Phys. Ed.*, **21** (1983) 939–48.
26. Blonsky, P. M., PhD Thesis, Northwestern University, Evanston, Illinois, 1986, p. 48.
27. Venkatasetty, H. V. (ed.), *Lithium Battery Technology*. John Wiley, New York, 1984, p. 34.

Lithium Thiochromite All-Solid-State Secondary Cell

Y. Geronov, B. Puresheva, P. Zlatilova

Central Laboratory of Electrochemical Sources,
Bulgarian Academy of Sciences, Sofia 1040, Bulgaria

&

P. Novak

J. Heyrovsky Institute of Physical Chemistry and
Electrochemistry, Czechoslovak Academy of Sciences,
102 00 Prague 10, Czechoslovakia

ABSTRACT

The electrochemical performance of the $Li/(PEO)_8LiClO_4/Li_xCr_{0.9}V_{0.1}S_2$ all-solid-state secondary cell is studied.

It is shown that utilization between 50 and 90% with respect to the theoretical capacity is achieved at 140°C at a wide range of current densities.

Galvanostatic pulse and AC impedance measurements are performed for elucidating the reasons for the cell capacity decline during prolonged cycling at 140°C.

Both methods give strong evidence that the increase of the ohmic drop resistance as a result of the degradation of the polymer electrolyte, and the cathode impedance due probably to deterioration of the contacts between the cathode particles, mainly contribute to the decrease of the cell discharge capacity.

INTRODUCTION

At the present time several all-solid-state rechargeable cells with polymer electrolytes are not far from practical application.[1-3] Such

411

cells include a Li or Li-based negative electrode and a composite positive one containing vanadium oxides of TiS_2.[1-4] As electrolytes, a mixture of poly(ethylene oxide) and a salt ($LiClO_4$ or $LiCF_3SO_3$) working at 70–150°C are commonly employed.[1-3]. There are no available data for other compounds tested under these conditions as cathode active materials.

It was recently shown[5,6] that sodium thiochromites $Li_xNa_{0.1}Cr_{0.9}V_{0.1}S_2$ behave excellently in Li cells with aprotic liquid electrolytes.

This study is aimed at the elucidation of the compatibility of sodium thiochromite composite cathodes with a polymer electrolyte in a cell working at temperatures above 100°C.

EXPERIMENTAL

All electrochemical experiments were carried out in stainless steel cells with a configuration described in Ref. 7. In the glove box freshly scraped Li foil was pressed to the bottom and covered with one or two sheets of the solid electrolyte. The cathode was then attached to the top of the package and spring loaded at $2\,kg\,cm^{-2}$. The cell was hermetically sealed with a PTFE ring and heated in an oven out of the glove box.

The cathode active material $Na_{0.1}Cr_{0.9}V_{0.1}S_2$ was synthesized according to procedures described in Refs 5 and 6. The composite cathode contained 40 wt% active material, 40 wt% acetylene black and 20 wt% $(PEO)_8LiClO_4$. The cathodes were prepared as follows: the acetonitrile was evaporated from the slurry of $NaCrVS_2$, acetylene black and $(PEO)_8LiClO_4$ at 25°C. The resulting mixture was heated at 120°C under vacuum for 3 h and homogenized. The dry mixture was pressed at 600 MPa in a special die. The cathode was dried at 120°C for 4 h and stored in a dry-box prior to use. The theoretical capacity of the cathodes used in the cells was $1\cdot3$–$1\cdot5\,mAh\,cm^{-2}$. The geometric surface area of the cathode was $1\cdot8\,cm^2$. The cycling was performed automatically between $1\cdot70$ and $3\cdot10$ V.

Foils of $(PEO)_8LiClO_4$ polymer electrolyte with a thickness of about $0\cdot1$ mm and a specific conductivity of 10^{-3}–$10^{-4}\,S\,cm^{-1}$ were prepared as described in Ref. 8.

The galvanostatic pulse method measurements were performed according to the method widely described in the literature.[9,10]

The electrochemical impedance was measured by the standard procedure in the potentiostatic mode using 1 mV AC perturbation with a Solartron 1250 Frequency Response Analyzer coupled to a Solartron 1286 Electrochemical Interface.

RESULTS AND DISCUSSION

Figure 1 presents typical discharge curves of a $Li/(PEO)_8LiClO_4/Na_{0.1}Cr_{0.9}V_{0.1}S_2$ cell at different current densities (CD) and 140°C. The charge current -0.11 mA cm^{-2} was the same for all cycles. The discharge voltage–time curves are very similar to those obtained in a solution of $LiClO_4$ in PC/DME. The lower specific conductivity of the polymer electrolyte with respect to the liquid one and some diffusion limitations in the electrolyte and composite cathode are probably the reasons for the lower (by 0.1 V) mid-discharge voltage. On the basis of a voltage of 2.4 and a CD of 0.27 mA cm^{-2} a specific energy of 290 Wh kg^{-1} for the $Li/Na_{0.1}Cr_{0.9}V_{0.1}S_2$ cell was calculated or 100 Wh kg^{-1} including all material weights without the metal case.

The effect of temperature and discharge CD on the utilization of the cathode active material is shown in Fig. 2. At 140°C utilization between 50 and 80% with respect to the theoretical capacity, C_t, was achieved at a wide range of CD. However, at 115°C the cathode

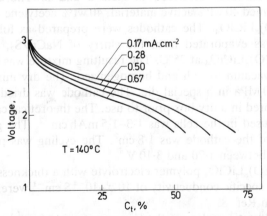

Figure 1. Typical discharge curves at different current densities; $T = 140$°C, $i_c = 0.11$ mA cm^{-2}.

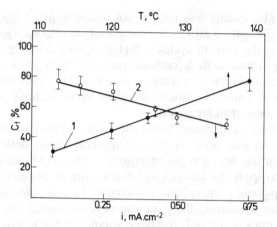

Figure 2. Effect of temperature and current density on the cathode utilization: 1, at $i = 0.27$ mA cm^{-2}; 2, at $T = 140°C$.

utilization drops to 30% (at $i = 0.27$ mA cm^{-2}). This value is twice as low than that found by Gauthier *et al.*,[1] for TiS$_2$. On the other hand the results in Fig. 2 are better than those given by Hooper & North[2] for V$_6$O$_{13}$ at 140°C. It is difficult at this stage to decide whether the reasons for the lower utilization at 115°C are related to the polymer electrolyte or to the properties of the composite thiochromite cathode.

Figure 3 presents the discharge capacity of a typical cell of Li/(PEO)$_8$LiClO$_4$/Na$_{0.1}$Cr$_{0.9}$V$_{0.1}$S$_2$ as a function of the cycle number

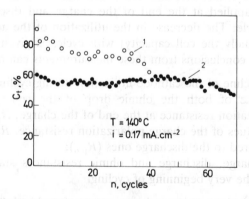

Figure 3. Utilization of the active cathode material upon cycling: $T = 140°C$; $i_c = i_d = 0.17$ mA cm^{-2}: 1, Na$_{0.1}$Cr$_{0.9}$V$_{0.1}$S$_2$; 2, Li$_x$Cr$_{0.9}$V$_{0.1}$S$_2$.

(curve 1). The cycling was carried out at 140°C at $i = 0.17$ mA cm^{-2} between 1.70 and 3.10 V. As compared to V_6O_{13} (Ref. 2) the utilization in the first 40 cycles is higher. Curve 2 in Fig. 3 presents preliminary results with a cathode prepared with an $Li_xCr_{0.9}V_{0.1}S_2$ active material obtained by direct synthesis.[11] This material is promising, giving a cathode utilization of 50–60% without appreciable decline for more than 80 cycles.

An attempt to understand better the cycling behaviour of a polymer electrolyte cell was made by the galvanostatic pulse method (GPM). As shown in our previous investigations[9,10] the GPM can be successfully used to study the kinetics and mechanism of the electrochemical processes in some primary[9] and secondary[10] Li cells. Here the method is applied for measuring the polarization resistance of the electrodes, R_p, ohmic drops in the cell, R_e (the resistance of the electrolyte, metal conductors and other contacts) and high frequency capacitance, C, during cycling. It was very difficult to build in our cell construction[7] a reliable reference electrode, working at temperatures above 100°C. Therefore some approximations are necessary because the overvoltages were measured between the two working electrodes. In such a case the R and C values of the anode and the cathode cannot be determined separately. The cell resistance R includes the polarization resistance of the cathode and anode and the ohmic drop resistance in the electrolyte: $R = R_{p,a} + R_{p,c} + R_e$.

The results obtained by using the galvanostatic pulse method during prolonged cycling of a cathode limited $Li/(PEO)_8LiClO_4/LiCr_{0.9}V_{0.1}S_2$ cell with $i = 0.17$ mA cm^2 and 130°C are shown in the Fig. 4. The pulses were applied at the end of the charge and discharge of the respective cycle. The decrease in the utilization of the active cathode material (actually the cell capacity) with cycles is given in the same figure. A few conclusions from these measurements can be drawn:

(i) the decline of the cathode utilization during cycling follows the increase of both the ohmic drop of the cell, R_e, and the polarization resistance at the end of the charge, $R_{p,c}$;

(ii) the values of the charge polarization resistance, $R_{p,c}$ are higher compared to the discharge ones $(R_{p,d})$;

(iii) the charge, discharge and ohmic resistances start increasing from the very beginning of cycling.

These conclusions suggest that a reason for the drop of the cell capacity could be probably related to some degradation of the polymer

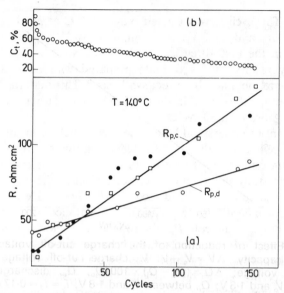

Figure 4. Change of the polarization resistance, ohmic drop and the cathode utilization during cycling of a Li/(PEO)$_8$LiClO$_4$/Li$_x$Cr$_{0.9}$V$_{0.1}$S$_2$ cell with $i_c = i_d = 0.17$ mA cm^{-2}: (a) ●, R_e; □, $R_{p,c}$; ○, $R_{p,d}$; data obtained by GPM; (b) percentage cathode utilization.

electrolyte at the high cathodic potential resulting in a decrease in its conductivity. From Fig. 4 it is seen that after 150 cycles R_e and $R_{p,c}$ increase more than four times compared to the 10th cycle, reaching values of 120 and 140 ohm cm^2 respectively. With the sum of these values and $i = 0.17$ mA cm^{-2} a loss in the charge voltage of 42 mV is estimated. Figure 5 presents the effect of the reduction of the charge voltage cut off below 3·10 V on the discharge capacity of a polymer electrolyte cell between the 10th and 30th cycles. In this range of cycling and voltage limits of 1·80–3·10 V the cathodic utilization is found to be constant. In the same figure ΔV is defined as $V_o - V_i$ where V_o is the charge cut-off voltage of 3·10 V and V_i the respective one below 3·10 V, and $\Delta Q = (Q_o - Q_i) \times 100/Q_o$ where Q_o is the discharge capacity at a cut-off voltage of 3·10 V and Q_i is the discharge capacity at the respective charge cut-off voltage. From this relationship it is seen that a reduction in the charge cut-off voltage of 50 mV corresponds to a 6% decrease in the discharge capacity.

If the same dependence of the cell capacity on the discharge cut-off voltage is assumed a decrease in the cell capacity of 12% should be

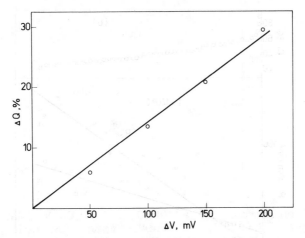

Figure 5. Effect of reduction of the charge cut-off voltage on the discharge capacity: $\Delta V = V_o - V_i$; V_o, charge cut-off voltage = 3·1 V, V_i, respective voltage; $\Delta Q = (Q_o - Q_i) \times 100/Q_o$; Q_o, discharge capacity between V_o and 1·8 V; Q_i, between V_i and 1·8 V; $i_c = i_d = 0.17$ mA cm^{-2}.

expected after 150 cycles. From Fig. 4 one can see that the discharge capacity drops by 50% at the 150th cycle as compared to the 10th cycle during cycling in the voltage limits between 1·80 and 3·10 V. Therefore some other additional limitations should be responsible for this larger decrease in the discharge capacity.

Impedance measurements were performed for further elucidation of the phenomena responsible for this process.[12] Figure 6 presents impedance plots at 120°C of a Li/(PEO)$_8$LiClO$_4$/Na$_{0.1}$Cr$_{0.9}$V$_{0.1}$S$_2$ cell before cycling, and after the 2nd and 32nd discharges at $i =$ 0·27 mA cm^{-2}. At the very beginning of the cycling the impedance diagrams have a hyperbolic shape, where two Warburg linear portions could be considered as asymptotes. From the intersection of the impedance plots with the real axis the sum of ohmic resistance could be estimated. For the fresh cell this value is low (about 10 ohms) but rises with the cycle number (about 5 times at the 32 cycle). As was shown before, approximately the same values of R_e were obtained by the GPM method (see Fig. 4).

After 32 cycles the shape of the impedance plots changes and a transition from capacitive to more resistive behaviour is observed, both at the end of charge and discharge. The change in the shape of the impedance plots and the increase of the impedance with cycling fit

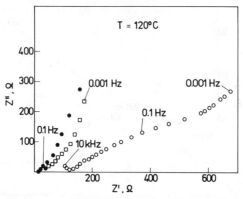

Figure 6. Impedance plots of a Li/(PEO)$_8$LiClO$_4$/Li$_x$Cr$_{0.9}$V$_{0.1}$S$_2$ at different cycle numbers; T = 120°C; i = 0·27 mA cm^{-2}: ●, before cycling; □, 2nd cycle; ○, 32nd cycle.

quite well with the results presented in Fig. 4. Both the GPM and impedance measurements were performed without a reference electrode. Therefore it is difficult from the results presented in Figs 4 and 6 to determine unambiguously the contribution of the impedance of the cathode/electrolyte and the lithium/electrolyte interface to the overall cell impedance of a cell containing polymer electrolyte.

The impedance plots at the 45th cycle of a Li/LiCr$_{0.9}$V$_{0.1}$S$_2$ cell with exactly the same construction but with a Li reference electrode and containing a liquid aprotic solution of THF/2-MeTHF −1·5 LiAsF$_6$ are presented in Fig. 7. The impedance of the Li cell in this electrolyte does not undergo any essential changes during cycling. However some substantial differences in the impedance plots between the cells with a liquid and a polymeric electrolyte are obvious (see Figs 6 and 7). The negligible value of the cathode impedance (Fig. 7(a)) of a cell containing a liquid electrolyte compared to the overall cell impedance implies that the latter is determined basically by the impedance of the Li electrode. On the other hand the theory of the composite cathodes containing a polymer electrolyte for the case of an electroactive material with large grains[4,13] predicts a Warburg behaviour followed by a capacitive behaviour at low frequencies. The curves in Fig. 6 show a similar shape. The higher $R_{p,c}$ values measured by the GPM of the charged cells compared to the discharged ones, $R_{p,d}$ (see Fig. 4) suggest as well that the Li electrode in a polymer electrolyte containing cell could not be responsible for such impedance behaviour

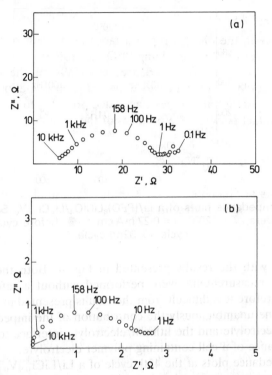

Figure 7. Impedance plots of a Li/2MeTHF-THF-1·5MLiAsF$_6$/ Li$_x$Cr$_{0.9}$V$_{0.1}$S$_2$ cell at 45th cycle: (a) between the cathode and the anode; (b) between the cathode and Li reference.

because its surface area increases during the cathodic deposition. For example, R_p values of secondary cells with liquid electrolyte after charge have never exceeded those after discharge.[14,15]

CONCLUSION

All results presented provide strong evidence that the main contribution to the overall impedance of a Li/(PEO)$_8$LiClO$_4$/Na$_{0.1}$Cr$_{0.9}$V$_{0.1}$S$_2$ cell comes from the impedance of the positive composite electrode which is probably responsible for the decrease in the cell capacity with cycling. Both methods used for the present investigation of a polymer electrolyte Li cell appear to show that the reasons for the capacity decrease during cycling are:

—increase in the ohmic drop resistance as a result of the degradation of the LiClO$_4$ containing PEO electrolyte;

—increase in the cell impedance, especially of the charged cell which could be attributed to some diffusion limitations and/or an increase in the cathode resistance due probably to deterioration of the contact between the particles of the cathode.

ACKNOWLEDGEMENT

The authors are indebted to Dr Moshtev for useful discussion of this paper and for kindly presenting them with samples of NaCr$_{0.9}$V$_{0.1}$S$_2$ and LiCr$_{0.9}$V$_{0.1}$S$_2$.

REFERENCES

1. Gauthier, M., Fauteux, D., Vassort, G., Bélanger, A., Duval, M., Ricoux, P., Chabagno, J., Muller, D., Rigaud, P., Armand, M. & Deroo, D., *J. Electrochem. Soc.*, **132** (1985) 1333.
2. Hooper, A. & North, J., *Solid State Ionics*, **9/10** (1983) 1161.
3. Gauthier, M., *Proc. 3rd Int. Seminar on Li Battery Technology and Applications*, 1987, Deerfield, Florida.
4. Scrosati, B., *Brit. Polym. J.*, **20** (1988) 219.
5. Moshtev, R., Manev, V., Puresheva, B. & Pistoia, G., *Solid State Ionics*, **20** (1986) 259.
6. Moshtev, R., Manev, V., Nassalevska, A., Gushev, A. & Pistoia, G., *J. Power Sources*, **26** (1989) 285.
7. Novak, P., Inganas, O. & Bjorklund, R., *J. Power Sources*, **21** (1987) 17.
8. Weston, J. & Steele, B. C. H., *Solid State Ionics*, **7** (1982) 81.
9. Moshtev, R., Geronov, Y. & Puresheva, B., *J. Electrochem. Soc.*, **128** (1981) 1851.
10. Geronov, Y., Schwager, F. & Muller, R. H., *J. Electrochem. Soc.*, **129** (1982) 1422.
11. Moshtev, R. V., unpublished results.
12. Novak, P., Geronov, Y., Puresheva, B., Podhajecky, P., Klapste, B. & Zlatilova, P., *J. Power Sources*, **28** (1989) 279.
13. Owen, J., Drennan, J., Lagos, G. E., Spurdens, P. & Steele, B. C., *Solid State Ionics*, **5** (1981) 343.
14. Geronov, Y., Zlatilova, P., Puresheva, B., Pasquali, M. & Pistoia, G., *J. Power Sources*, **26** (1989) 585.
15. Geronov, Y., Puresheva, B., Zlatilova, P., Pasquali, M. & Pistoia, G., *J. Electrochem. Soc.* (in press).

Manganese Dioxide as a Rechargeable Cathode in a High Temperature Polymer Electrolyte Cell

R. J. Neat, W. J. Macklin & R. J. Powell

Solid State Chemistry Group, MDD, Harwell Laboratory, Oxfordshire, OX11 0RA, UK

ABSTRACT

This paper discusses the use of manganese dioxide, a cheap readily available primary active cathode material, as a rechargeable cathode in a lithium anode polymer electrolyte cell operating at 120°C. We have discovered that an in-situ transformation of partially lithiated manganese dioxide occurs during the initial discharge process. The resultant material is shown to be the spinel $Li_{1+x}Mn_2O_4$, a well-known reversible intercalation cathode material. Test cells (40 cm^2) constructed from battery grade MnO_2 are highly reversible and exhibit no loss in capacity over the first twenty cycles. This discovery may have a significant effect on the success of the polymer electrolyte battery in the cost-sensitive traction application.

INTRODUCTION

The original work of Armand *et al.*,[1] who suggested the use of polymer electrolytes for high energy density batteries, has led to a concerted research effort in the field of solid-state lithium polymer batteries. Although recently much of the attention has focused on the development of room temperature polymer electrolytes, there have been signficant improvements in the performance of the high temperature (100–140°C) polymer electrolyte battery system.[2] This battery system (Fig. 1), utilising a PEO–LiX electrolyte, a lithium anode and

421

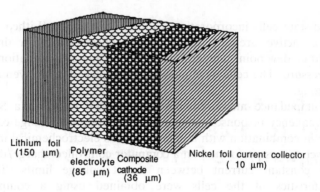

Figure 1. Configuration of polymer electrolyte battery.

a composite cathode incorporating an intercalation compound, now has considerable potential for electric vehicle traction and general space power applications.

Traction batteries in particular, provided they meet the required performance criteria, are very cost sensitive. The majority of the high temperature polymer cells built at Harwell are based on the relatively expensive cathode material V_6O_{13}. In this paper we report recent work on alternative cathode materials, and in particular the use of manganese dioxide as a rechargeable intercalation material in a high temperature polymer electrolyte cell. Manganese dioxide is a cheap, readily available cathode material widely used in primary liquid electrolyte lithium cells. Its use in a polymer electrolyte traction battery would have significant cost benefits.

EXPERIMENTAL

A composite cathode containing MnO_2 (battery grade), ketjenblack carbon, PEO (Union Carbide MW 4 000 000) and $LiClO_4$ (Aldrich) was prepared via doctor blade casting from the appropriate solvent slurry onto a nickel current collector.[3] The resulting cathode had the composition 45 vol% MnO_2, 5 vol% carbon and 50 vol% PEO–$LiClO_4$ ([EO units]/[Li] = 12) with a capacity of $\sim 1 \cdot 0$ mA h cm^{-2} based on a value of 308 mA h g^{-1} for MnO_2. Sheets of the electrolyte PEO–$LiClO_4$ ([EO units]/[Li] = 12) were cast from acetonitrile solution onto silicone release paper. This electrolyte composition was previously found to have the highest ionic conductivity at 120°C.

Solid-state cells incorporating a lithium foil anode (Lithco 150 μm) with an active area of 40 cm^2 were constructed in a dry room ($T = 20°C$, dew point temperature $-30°C$) using a combination of heat and pressure. The cells were packaged and placed in an oven at 120°C for testing.

AC impedance measurements were carried out with a Solartron 1254 frequency response analyser over the frequency range 65 kHz to 0·1 Hz in combination with a Solartron 1286 electrochemical interface. Cell cycling was performed on a computer controlled charge/discharge rig at constant current between preset voltage limits. The rate characteristics of the cells were obtained using a computerised constant current source. X-ray diffraction patterns were obtained using a Philips PW 1050 goniometer employing CuK$_\alpha$ radiation and incorporating pulse height discrimination and a curved graphite secondary monochromator. Data collection and processing over the range $4° \leq 2\theta \leq 70°$ was controlled by a PDP11/23 plus computer. Cathode films from dismantled cells were mounted inside an air sensitive chamber incorporating a beryllium window and a hot stage. Samples were heated above 70°C to melt any crystalline PEO present within the composite cathode.

RESULTS AND DISCUSSION

AC impedance data obtained after 2 h at 120°C (Fig. 2), indicated the very low overall impedance of the Li/PEO/MnO$_2$ cell (~28 Ω cm^2).

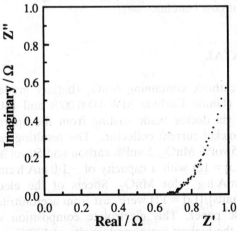

Figure 2. AC impedance of Li/PEO/MnO$_2$ cell at 120°C.

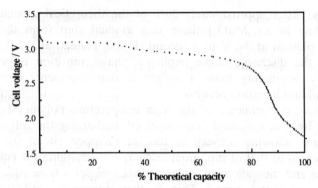

Figure 3. First discharge (C/10) of MnO_2 in a high temperature polymer electrolyte cell.

The electrolyte impedance (series resistance) is comparable to that observed in the V_6O_{13} cells,[2] reflecting both the high ionic conductivity of the electroyte ($\sim 10^{-3}$ S cm^{-1}) and the adequate electronic conductivity of the composite cathode. The interfacial resistance is very low, again similar to that in V_6O_{13} cells.

The first discharge of a Li/MnO_2 high temperature polymer cell at the 10-h rate (C/10) is shown in Fig. 3, and is essentially the same as that observed in a typical MnO_2 liquid electrolyte primary cell (Fig. 4). However, the discharge curve in Fig. 3 shows two distinct voltage

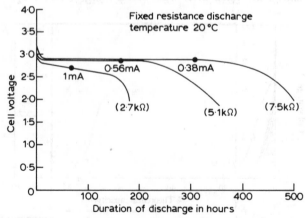

Figure 4. Typical discharge curves of Li/MnO_2 button cells under various loads.

plateaux. After approximately 30% of the theoretical capacity (cor-responding to $Li_{0.3}MnO_2$) there is a gradual step from the initial voltage plateau at 3·1 V to a second sloping plateau at ~2·8 V. This step in the discharge curve implies a phase transition is occurring (possibly originating from a structural rearrangement) during the initial lithium insertion process.

The rate performance of the high temperature polymer Li/MnO_2 cell has been investigated. This involved discharging the cell at the C (1-h) rate, allowing the cell to recover on open circuit for a time period equal to that of the initial discharge, discharging the cell at the C/2 rate and then allowing the cell to recover for a time equal to the total experimental time.[4] This is then repeated until the C/16 discharge has been completed. In this method the accumulative capacity of each discharge gives an indication of expected rate performance at a given rate of discharge, and has the advantage of only using one cell. Figure 5 shows the discharge curves at various discharge rates during the rate experiment, and Fig. 6 shows the capacity that may be utilised versus the rate. These results illustrate the excellent rate capability of the $Li/PEO/MnO_2$ cell with ~70% of the theoretical capacity being removed at the C rate.

On recharging at the C/10 rate, the discharged cell accepted a large degree of charge and was able to deliver ~58% of the initial theoretical capacity on the second discharge. Subsequent cycling indicated no loss in discharge capacity over the next 20 cycles (Fig. 7). Figure 8 shows the third discharge/charge cycle. These results clearly

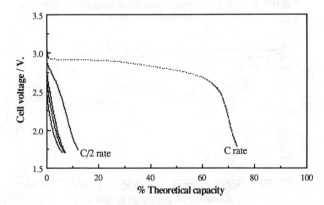

Figure 5. Discharge curves at various rates for a $Li/PEO/MnO_2$ cell during the rate experiment.

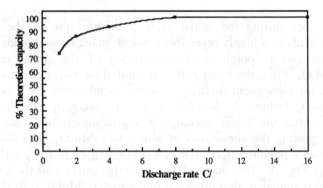

Figure 6. Available capacity from the Li/PEO/MnO₂ cell versus the rate.

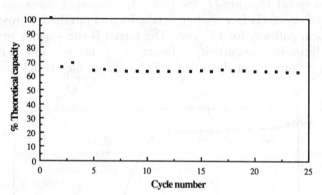

Figure 7. Cycle performance of a Li/PEO/MnO₂ cell.

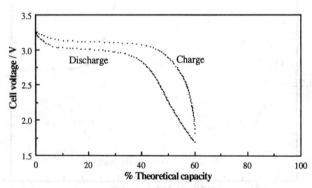

Figure 8. Third discharge and charge of a Li/PEO/MnO₂ cell.

indicate that during the initial cycle a modification of the MnO_2 cathode leads to a highly reversible form of 'manganese dioxide'.

Several factors point to the generation of the spinel material $Li_{1+x}Mn_2O_4$. First, the latter half of the initial discharge curve and the second and subsequent discharge curves resemble the OCV data for $Li_{1+x}Mn_2O_4$ (where $0 < x < 1.2$) (Fig. 9). Secondly, the capacity obtained after the initial discharge is approximately half that of the first; suggesting the conversion of MnO_2 to $LiMn_2O_4$. The discharge curve from a polymer electrolyte cell containing a $LiMn_2O_4$ cathode is shown in Fig. 10. The shape of the discharge curve and the capacity obtained is similar to that observed in the cycled 'MnO_2' cells (Fig. 7), consistent with the generation of $Li_{1+x}Mn_2O_4$ *in situ*.

In the spinel $Li_{1+x}Mn_2O_4$ the $[Mn_2]O_4$ framework possesses a 3-D interstitial space via face sharing octahedra and tetrahedra providing a conduction pathway for Li^+ ions. The mixed B-site valency results in good electronic conductivity.[5] Insertion of lithium into $LiMn_2O_4$ induces a Jahn–Teller distortion which reduces the crystal symmetry

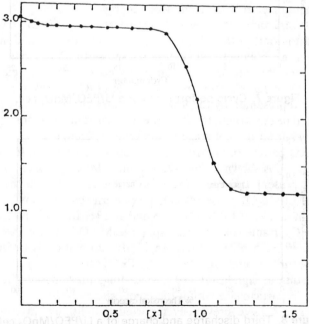

Figure 9. OCV versus composition for $Li_{1+x}Mn_2O_4$.

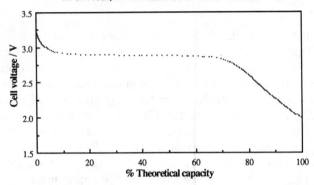

Figure 10. Typical discharge (C/10) of LiMn$_2$O$_4$ in a high temperature polymer cell.

from cubic in LiMn$_2$O$_4$ to tetragonal in Li$_{1+x}$Mn$_2$O$_4$, resulting in a two-phase electrode with a flat OCV at ~2·9 V (Fig. 9). Cyclic voltammetry studies[6] have shown that the insertion of lithium ions into LiMn$_2$O$_4$ is reversible, and hence the interest in LiMn$_2$O$_4$ as a solid-solution electrode for secondary batteries. This material has recently been utilised in a secondary lithium 'AA' cell, based on organic liquid electrolytes, from Moli Energy.[7]

X-ray diffraction at elevated temperature (~80°C) has been employed to confirm the MnO$_2$ to Li$_{1+x}$Mn$_2$O$_4$ transformation occurring within the polymer cell. Figure 11 shows the XRD pattern of a MnO$_2$ composite cathode from an undischarged polymer electrolyte cell after 2 h at 120°C. The structure of the MnO$_2$ is that of pyrolusite (β-MnO$_2$) represented by the lower, computer generated, line pattern. The nickel peaks originate from the cell current collector, and the copper lines from the hot stage in the sample chamber.

The XRD pattern of an MnO$_2$ composite cathode taken from a fully discharged cell is shown in Fig. 12. The main MnO$_2$ peak at $2\theta \sim 29°$ is no longer present and the diffraction profile is similar to that of the tetragonal Li$_2$Mn$_2$O$_4$ phase (calculated pattern shown below) resulting from the lithiation of LiMn$_2$O$_4$. The line at $2\theta \sim 18°$ is characteristic of the [Mn$_2$]O$_4$ framework of the spinel LiMn$_2$O$_4$. These XRD data confirm that the MnO$_2$ to Li$_{1+x}$Mn$_2$O$_4$ transformation is taking place during the initial discharge of the Li/PEO/MnO$_2$ cell. Lithiation of MnO$_2$ results in significant line broadening, presumably due to the reduction in particle size within the composite cathode.

David et al.[8] have previously reported that lithium insertion with n-butyl lithium into the rutile phase β-MnO$_2$ results in a product

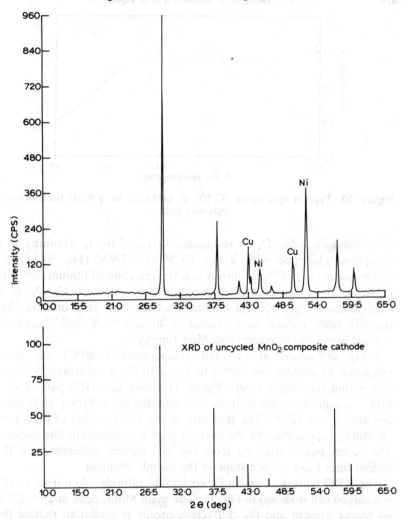

Figure 11. XRD pattern of uncycled MnO_2 composite cathode and calculated β-MnO_2 pattern.

isostructural with the tetragonal $Li_2Mn_2O_4$. They suggested a mechanism for the rutile to spinel transformation via an essentially diffusionless transition; the driving force being the electrostatic repulsion between Li^+ and $Mn^{4+/3+}$ ions in the lithiated MnO_2. A similar transformation appears to be occuring within the high temperature Li/MnO_2 polymer electrolyte cell.

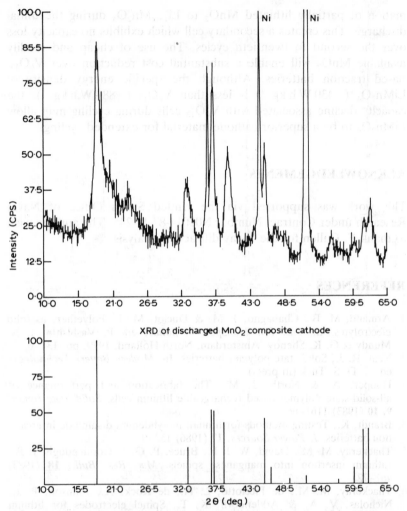

Figure 12. XRD pattern of discharged 'MnO$_2$' composite cathode and calculated Li$_2$Mn$_2$O$_4$ pattern.

CONCLUSIONS

The conversion of battery grade MnO$_2$ from a primary to a highly reversible secondary material within the polymer electrolyte cell is a significant discovery. This conversion results from an in-situ transformation of partially lithiated MnO_2 to $Li_{1+x}Mn_2O_4$ during the initial discharge. This creates a secondary cell which exhibits no capacity loss over the second to twentieth cycles. The use of cheap and readily available MnO_2 will enable a substantial cost reduction over V_6O_{13}-based traction batteries. Although the specific energy density of $LiMn_2O_4$ ($\sim$430 W h kg^{-1}) is less than V_6O_{13} ($\sim$880 W h kg^{-1}), the capacity decline associated with V_6O_{13} cells during cycling may allow $LiMn_2O_4$ to be a superior cathode material for extended cycling.

ACKNOWLEDGEMENTS

This work was supported via the United States Office of Naval Research under Contract Number N00014-87-J-1245. The authors wish to thank F. Cullen for the X-ray diffraction analysis.

REFERENCES

1. Armand, M. B., Chabagno, J. M. & Duclot, M. J., Polyethers as solid electrolytes. In: *Fast Ion Transport in Solids,* ed. P. Vashishta, J. N. Mundy & G. K. Shenoy. Amsterdam, North-Holland, 1979, pp. 131–6.
2. Neat, R. J., Solid state polymer batteries. In: *Modern Battery Technology,* ed. C. D. S. Tuck (in press).
3. Hooper, A. & North, J. M., The fabrication and performance of all-solid-state polymer-based rechargeable lithium cells. *Solid State Ionics,* **9, 10** (1983) 1161–66.
4. Brandt, K., Testing methods for lithium molybdenum disulphide intercalation batteries. *J. Power Sources,* **17** (1986) 153–9.
5. Thackeray, M. M., David, W. I. F., Bruce, P. G. & Goodenough, J. B., Lithium insertion into manganese spinels. *Mat. Res. Bull.,* **18** (1983) 461–72.
6. Thackeray, M. M., de Picciotto, L. A., de Kock, A., Johnson, P. J., Nicholas, V. A. & Addendorff, K. T., Spinel electrodes for lithium batteries—a review. *J. Power Sources,* **21** (1987) 1–8.
7. Brandt, K., Electrical characteristics and safety of rechargeable manganese oxide lithium batteries. In: *Proc. Fourth International Seminar on Lithium Battery Technology and Applications,* Deerfield Beach, Florida, March 1989.
8. David, W. I. F., Thackeray, M. M., Bruce, P. G. & Goodenough, J. B., Lithium insertion into β-MnO_2 and the rutile-spinel transformation. *Mat. Res. Bull.,* **19** (1984) 99–106.

Electrochromic Windows and Displays Using Polymer Electrolytes

Daniel Deroo

Laboratoire d'Ionique et d'Electrochimie du Solide de Grenoble, ENSEEG, BP75, 38402 St Martin d'Hères Cedex, France

ABSTRACT

The use of polymer electrolytes and more particularly anhydrous polymer electrolytes in ECC devices is examined. After a short review of the desirable characteristics of the devices, the possible processes and the electrochromic materials, the properties of anhydrous polymer electrolytes are discussed. Some devices, using these electrolytes, including displays, thermo-electrochromic displays and variable transmission windows are compared. It is shown, in view of cycling numbers, contrast, memory effect and response time that anhydrous polymer electrolytes are one of the choices for electrochromic devices.

INTRODUCTION

Much interest has developed over the last two decades in electrochromic devices (ECD) including displays (from simple level indicators to complex alphanumeric and graphic information) and adjustable transmission or reflection windows ('smart windows'). Electrochromism, in the terminology of display technology, is broadly defined as the production of a colour absorption band in a display medium by an applied field or current. A large number of solid and liquid materials exhibit an EC effect. Nevertheless, the physical mechanisms involved may be different: electronic phenomena (EC systems) or electrochemical effects (ECC systems).

New types of devices have been proposed which are based on

electrochemical effects.[1] The electrophoretic device employs a dye suspension in a contrasting colour liquid and the colour change results from electrophoretic (EP) migration of dye particles. The electrochemichromic effect (ECC) is the direct result of a colour change induced by an electrochemical redox reaction (including metal deposition). Electrochemiluminescence (ECL) is produced by recombination of two species reduced and oxidized at a cathode or an anode, respectively, in a chemical solution. Fluorescence quenching (FQ) occurs when a fluorescent species is reduced or oxidized to a non-fluorescent species.

The first two electrochemical processes (EP and ECC) and the last two (ECL and FQ) give respectively passive and active devices. The EP device has very poor characteristics which would make multiplexing difficult and requires a high voltage. The three others have a low threshold voltage and a similar electric behaviour. The most attractive one is the ECC system because it has fewer chemical constraints, simpler device operation and better performance in terms of memory capability and high contrast. Nevertheless, in the case of ECC systems, the write speed and the erase speed (around 10 ms) given by Chang *et al.*[1] is only restricted to the case of proton diffusion in liquid electrolyte. When the electrochromic effect is due to the change of oxidation degree in an insertion material, the diffusion coefficient of the intercalated proton (or alkaline ion) becomes lower and the response time could range from 1 to 100 s.

Chang *et al.* have distinguished four types of electrochemical reactions for use in ECC systems: these reactions are given in Table 1. For example, type I can use species dissolved in an electrolyte such as polytungsten anion (PTA), dimethyl viologen dichloride . . . , type II,

Table 1. Different Types of Electrochemical Reactions in ECC Systems[1]

Type I	Simple redox reaction: $A \pm ne \rightarrow B$ (coloured species or metal deposit)
Type II	$A \pm ne \rightarrow B$ (coloured)
	$B + Z \rightarrow A + Z'$
	$Z' \pm ne \rightarrow Z$ The memory effect could be controlled
Type III	$A \pm ne \rightarrow B$
	$B + C \rightarrow BC$ (precipitation)
Type IV	Host–insertion compound
	$\leftrightarrow + xM^+$ (or H^+) $+ xe \leftrightarrow \{xM\}$
	bleached coloured
	(or coloured) (or bleached)

Glass
Electronic   conductor
Electrochromic   material
Electrolyte
Counter   electrode

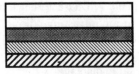

Figure 1. Schematic structure of an electrochromic display.

the reaction PTA (A) + H$_2$O$_2$(Z) . . . , type III, diheptyl viologen dibromide and type IV, tungsten trioxide, iridium oxide, molybdenum trioxide . . . .

Before examination of the different materials used in ECC displays and windows, it is interesting to remember the structure of these corresponding devices and their characteristics necessary for use. An electrochromic cell is a battery. Its capacity is not considered in terms of energy but in terms of degree of colouration or optical transmission or reflection. The most important characteristic of electrochromic devices is the use of thin film materials (100–500 nm). This fact, kinetically important, shows the difference between electrochromic devices and batteries. Figures 1 and 2 show the structure of displays and windows. Reflecting electrochromic devices could be obtained by intercalation of a reflecting layer between the electrochromic electrode (or the counter electrode) and the electrolyte.[2] Tables 2 and 3 give the ideal characteristics that these devices should possess.[3,4]

One of the problems with electrochromic windows is the choice between reflectivity or absorption processes. A major part of the solar spectrum is situated in the IR region. The visible zone for the human eye is between 0·4 and 0·7 μm. If the reflectivity for a coloured state is 100% in the visible region, the window will be opaque, but for energy management, it is only necessary to create a variation in the nearest IR region.[5] For buildings, the emission gives a spectrum in a range over 0·3 μm. The bleached state will stop the radiation and will minimize the losses of energy of the building. In the case of

Glass
Electronic   conductor
Electrochromic   material
Electrolyte
Counter   electrode
Electronic   conductor
Glass

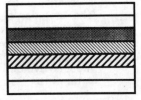

Figure 2. Schematic structure of a variable transmission window.

Table 2. Characteristics of an Electrochromic Display[3]

Contrast:	The light transmission could vary from 5% (bleached state) to 75% (coloured state).
Write speed:	Some ms for electronic applications. Some s for large displays
Lifetime:	10^7 colouration-bleaching cycles.
Temperature:	−20 to 100°C depending on applications
Memory:	Yes, refreshment possible
Grey scale:	Preferable
Energy:	Some mJ/cm^2

absorption, the windows are submitted to frequent thermal shocks, limiting the lifetime. For this reason it will be important to add a reflecting film for the farthest IR.

Electrochromism can be obtained with several organic and inorganic materials. The preparation modes are various: anodic oxidation, vacuum evaporation, cathodic pulverization or dissolution of a salt in the electrolyte. Several organic electrochromic materials are possible: rare earth phthalocyanines, conducting polymers (polypyrrole,

Table 3. Characteristics of an Electrochromic Window[4]

Contrast:	the light transmission could vary from 5% (bleached state) to 75% (coloured state) in the spectrum region: $0.35-2 \mu m$
Write speed:	It could be relatively long (around 300 s). Electrochemical reactions could not be fast. This is an essential difference with displays which need very weak response time (some ms). Uniform colouration on large areas
Lifetime:	10 years of use (around 3 cycles of colouration-bleaching a day). The problem of the degradation of the constituting materials is important
Temperature:	−20 to 100°C. However, in case of slow rates of colouration, a previous heating could be possible
Energy:	1 Wh/m^2 per cycle
Memory:	More than one week
Moisture:	The operation of the window must not depend on moisture percentage and no degradation must occur in the presence of water
Security:	A degradation of the window due to a high voltage must be avoided
Cost:	Around 10 $US\$/m^2$

polythiophene, polyaniline . . .), functionalized polymers (tetrathiaful-valen, polythiazoline, polyviologens . . .), viologens, anthraquinones, phenothiazine . . . .[4,6] Phthalocyanines are studied for polychromic applications. For materials in solution in the electrolyte (polytungsten anion for example), a separator is needed to keep the memory effect. This effect will be determined by the diffusion speed of the coloured species through the electrolyte. In the case of viologens, this problem is avoided by precipitation of the coloured species on the electrode. However, a residual colouration occurs after several cycles.

Inorganic materials are generally used as insertion materials. The insertion of lithium or proton is reversible. The transition metal oxides are the most used in electrochromic devices. Lampert,[6] in an excellent review has also presented the inorganic electrochromic materials which are or can be used:

—for alkaline ions devices: WO_3 and tungsten bronzes, V_2O_5, Nb_2O_5, TiO_2, MoO_3, prussian blue;
—for protonic devices (basic or acid): WO_3, $Mn(OH_2)$, $Ni(OH)_2$, IrO_2-nH_2O, MoO_3.

Amorphous tungsten trioxide, in which the diffusion coefficient of Li^+ and H^+ is respectively $2 \cdot 5 \times 10^{-11}$ and $\approx 10^{-10}$ cm^2/s at room temperature (given an acceptable response time) is the most employed. In the case of aqueous electrolytes, involving proton conduction, the poor stability of some electrode materials such as tungsten trioxide and the bad memory effect are an obstacle for the realization of electrochromic devices other than organic or solid electrolyte devices.[7]

ANHYDROUS POLYMER ELECTROLYTE DEVICES

Polymer Electrolytes

The choice of the electrolyte depends on the type of application. In the case of electrochromic displays, the low response time required pre-supposes high current densities. Liquid electrolytes which have high conductivities are preferable but for some specific applications (airports or train stations, publicity . . .) involving large area displays and not needing a fast writing speed (near 1 s), lower conductivity materials, like polymer electrolytes, could perform well.

On the other hand, for electrochromic windows, electrolytes with a conductivity of 10^{-5} S/cm at 20°C could be used (response time around 100 s). The ohmic drop depends on the thickness of the electrolyte film.

We must distinguish between anhydrous solid polymer electrolytes and polyelectrolytes conducting only in the presence of water. Among these last electrolytes, polyvinyl alcohol–H_2SO_4 aq., polyvinyl sulphonic acid (PVSA), polystyrene sulphonic acid (PSSA), polyperfluorosulphonic acid and poly-2-acrylamido-2-methyl propane sulphonic acid (Poly AMPS) have already been used in electrolytes for displays or windows.[4] However, the influence of moisture on the conductivity and the semi-solid state of these electrolytes are disadvantages.

Anhydrous polymer electrolytes have already been used for 10 years in thin film solid state batteries of high energy density. They have also been associated with electrochromic materials.[8–12] Elastomeric properties of polymer electrolytes give a better contact than other solid electrolytes ($LiNb_2O_5$ for example).[13] They permit the mechanical absorption of the volume change of the electrodes when insertion occurs. Another advantage is the possibility to realize easily thin films (20–100 μm) for use in thin film cells, such as electrochromic devices.

Some polymer electrolytes such as polyethers having solvating properties can make ionic conducting complexes with alkali metal salts (with low lattice energy[14]) or acids.[15] In these electrolytes, the anionic and cationic conductivities are comparable ($0.3 < t^+ < 0.6$).[16] Polyethylene oxide (PEO), the most studied, is a good candidate but other electrolytes such as polyethylene imine (PEI),[17] polyvinyl pyrrolidone (PVP),[15] modified polyethers (crosslinked POE-based polyurethanes)[18] or polysiloxanes[10] are promising materials.

Complexes with lithium salts such as PEO–$LiClO_4$ or PEO–$LiCF_3SO_3$ are now currently used. It is well known that the conductivity of these electrolytes is related to a phase diagram (Fig. 3). Between the glass transition temperature and the eutectic temperature, there are a crystalline phase and a metastable amorphous phase. For the complexes PEO–Li^+ (Ref. 19) or PEO–H^+ (Ref. 15), the conductivity is given by the amorphous phase. It is preferable to have a fully transparent electrolyte (and a high conductivity), so that the complex does not contain crystallites. The conductivity of some anhydrous polymer electrolytes is given in Fig. 4.

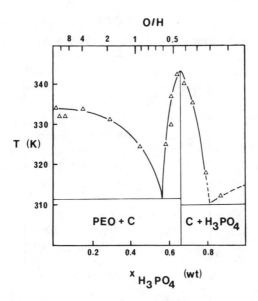

Figure 3. Phase diagram of the system PEO–H$_3$PO$_4$ (from Ref. 15).

PEO Complexes

For the complex (PEO)$_8$LiClO$_4$, crystallization occurs and the conductivity at room temperature is not sufficient for electrochromic displays. PEO–H$_3$PO$_4$ (O/H = 0·7) has a high conductivity and a low speed of crystallization (not visible with the naked eye after some months). For the PEO–LiTFSI complex,[20] the T_g is less than that of the perchlorate complex: the conductivity can be compared to those of protonic complexes and the transparency is good.

PVP–Acid Complexes

The most interesting is the PVP–H$_3$PO$_4$ complex. This electrolyte is amorphous at all temperatures and has an acid content. The change of conductivity versus temperature is due to the crossing of the glass transition temperature (reversible process). The variations of the conductivity versus temperature are given in Fig. 4 (curves 2 and 4). The conductivities of the polymer complexes vary rapidly with the temperature (depending on the ratio O/H) and follow the law of free volume. For the ratio 0·22 the conductivity at room temperature is

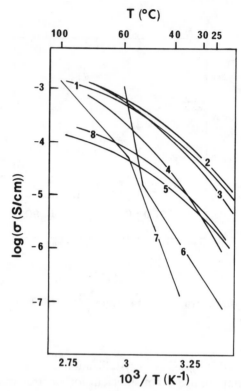

Figure 4. Variations of the conductivity of some anhydrous polymer electrolytes: 1, PEO/LiTFSI;[20] 2, PVP/H₃PO₄(O/H = 0·22);[15] 3, PEO/H₃PO₄(O/H = 0·66);[15] 4, PVP/H₃PO₄(O/H = 0·16);[15] 5, PEO triol + diisocyanate/LiClO₄;[18] 6, PEO/H₃PO₄(O/H = 0·4);[15] 7, PEO/LiClO₄(O/Li = 8);[14] 8, PMEO₇/LiClO₄.[24]

near 5×10^{-5} S/cm. The stability of amorphous tungsten trioxide in contact with PVP or PEO–acid electrolytes has been demonstrated by Armand et al.[8] They have also obtained a memory effect of more than 6 months.

Anhydrous Polymer Electrolyte Displays and Thermoelectrochromic Displays

Some authors have presented displays using PEO-based electrolytes, LiClO₄ or LiCF₃SO₃, which work at temperatures higher than room temperature. The possibility of acting at room temperature with LiTFSI has been demonstrated in Ref. 4 for electrochromic windows (10 000 cycles). A simple display using a PVP (or PEO)–acid electrolyte, WO_3 and a stainless steel counter electrode has been cycled 10^5 times without degradation and colouration-bleaching losses.[3] Another display acting at room temperature with $LiCF_3SO_3$–PEI has also been proposed, but the authors[17] did not give the number of cycles obtained; they have obtained 64 000 cycles but with a PEI–acid electrolyte which is not anhydrous.

Another variant of electrochromic displays has been proposed by three groups.[7–9,12] The principle is the following: the conductivity of the chosen polymer complex must be low at room temperature ($<10^{-7}$ S/cm) so the colouration will be very low (or does not appear). If the conductivity of the complex varies rapidly with temperature to reach 10^{-3} S/cm, a sudden increase of the temperature is able to induce the colouration (or bleaching) if the device is polarized. This type of device has been called a thermoelectrochromic display. Figures 5 and 6, which give respectively examples of the variations of the voltammograms and of the response times versus temperature are a good illustration of this phenomenon. Armand et al.[8] have shown the feasibility of the colouration of a pinpoint by local heating. An example of the results is given in Table 4 in which the local heating has been realized by a laser beam.

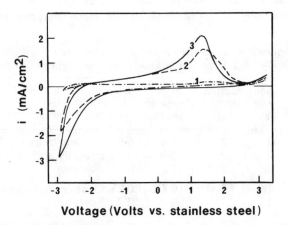

Figure 5. Cyclic voltammetry of a stainless steel/PEO–$LiClO_4$/WO_3 cell at various temperatures (potential sweep, 100 mV/s): 1, 25°C; 2, 60°C; 3, 80°C (from Ref. 7).

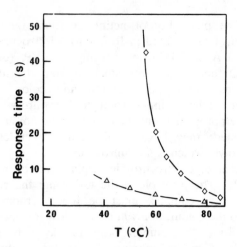

Figure 6. Response time of two thermo-electrochromic displays using PVP–H₃PO₄ (Ref. 3) and PEO–LiClO₄ (Ref. 7) electrolytes: WO₃ potential: ◇, −0·5 volt/NHE; △, 2 volts/Li.

Anhydrous Polymer Electrolyte Windows

The Problem of the Counter-Electrode

The counter-electrode must be transparent and colourless in the reduced state. The electrode reaction must be rapid and reversible with the same magnitude as those of the electrochromic electrode

Table 4. Colouration of a Pin Point by Laser Beam Heating[8] Cell: WO₃/PVP–H₃PO₄ (O/H = 0·2)/ITO laser: red (krypton 647 nm)

Applied potential	Laser power (W)	Heating time (s)	Spot diameter (mm)	OD	Observations
−500 mV	2	2	0·4	0·45	Spots with increasing
−750 mV	2	2	0·9	0·56	intensity and increasing
−1 V	2	2	1·2	0·65	diameter
−750 mV	1	2	0·4	0·25	Clear
−1 V	1	2	0·4	0·35	Spots

The three first and the two last spots have been bleached at exactly reverse potential (for example +750 mV if colouration at −750 mV) and same heating conditions.

response time $= 100\,\mathrm{s}$ and $10\,000$ cycles). Two types of electrode are possible:

—The material must be transparent and colourless in the reduced state and in the oxidized state. The variations in the optical transmission will be only due to the electrochromic electrode.
—The material must be colourless in the reduced state and coloured in the oxidized state. This configuration is interesting because of the addition of the colouration of the two electrodes. The contrast is higher than in the preceding case. The only example of this configuration using a polyether-based electrolyte is the system realized by Tada *et al.*[11] (tungsten bronze/polyurethane + KCF_3SO_3/prussian blue). However, the ratio of the thicknesses of the two electrodes must be chosen to avoid no residual colouration in the bleached state. Another problem is the possible presence of parasitic reactions which could induce an imbalance and consequently decrease the contrast.

Lithium or proton insertion in host materials has a good electrochemical reversibility. However most of them are transition metal oxides, sulphides or selenides. The orbitals which participate in the chemical link are 'd' orbitals with a strong overlap. The possible electronic transitions are situated in the visible part of the solar spectrum. These materials are coloured. In fact, an insertion material must be a mixed conductor. Properties of electronic conduction and non-absorbance in the visible are not compatible. To have a colourless material, it is necessary for it to have a gap higher than $3\,\mathrm{eV}$.

Several oxides and transition metal oxides are possible counter-electrode material candidates: In_2O_3, SnO_2, VO_2, V_2O_5, IrO_2.[4] Some authors have reported lithium insertion in ITO.[21] They have shown that the reversibility decreases if the tin content increases. It seems that the electrochemical properties of this material depend on the microstructure (and the preparation mode) and the insertion is not higher than $Li_{0.15}$ (ITO). The diffusion coefficient of Li^+ is less than $6 \times 10^{-13}\,\mathrm{cm^2/s}$ at $20°C$ (15 times less than in WO_3). Baudry[4] has shown, using impedance spectroscopy, that the diffusion coefficient of Li^+ in V_2O_5 is $2.5 \times 10^{-12}\,\mathrm{cm^2/s}$ at $20°C$ (10 times less than WO_3). The diffusion kinetics are comparable and a window using V_2O_5 and WO_3 is realizable.[22] Figure 7 shows the variations of the diffusion coefficient of lithium in different materials versus temperature.

The thin film structure allows materials with lower conductivities

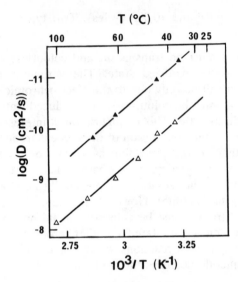

Figure 7. Variations of the diffusion coefficient of Li in WO$_3$ and V$_2$O$_5$ versus temperature with PEO–LiClO$_4$ electrolyte (Ref 4). △, V$_2$O$_5$; ▲, WO$_3$.

than in the case of composite electrodes for batteries. Baudry[4] has studied a new family of possible insertion materials, respectively for batteries and electrochromic windows: rare earth oxides which are weakly coloured (CeO$_2$, Tb$_2$O$_3$, Pr$_2$O$_3$) but poor electronic conductors. The diffusion of lithium in CeO$_2$, which has given the best results, seems not very fast but the determination of the diffusion coefficient is complicated by an interface phenomenon, giving a strong capacitance and limiting the electrochemical reaction speed. Figure 8 shows the cyclic voltammograms of CeO$_2$ and WO$_3$ (for comparison).

Some Examples of Anhydrous Polymer Electrolyte Windows
Table 5 gives some non-anhydrous and anhydrous polymer windows described in the literature and their known characteristics. All these devices use alkaline ion conduction. The contrast change is smaller than in the case of a protonic window like those proposed by Cogan *et al.*,[23] WO$_3$/poly AMPS/IrO$_2$. However, they present a relatively fast response time (1–300 s) and a memory effect longer than 2 weeks.

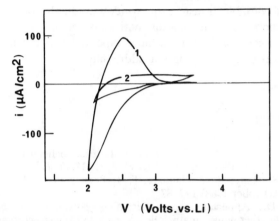

Figure 8. Comparison between voltammograms of WO_3 and CeO_2 thin films in PEO–$LiClO_4$ electrolyte (potential sweep 300 mV/mn) (from Ref. 4). 1, WO_3; 2, CeO_2.

CONCLUSIONS

This paper has provided a brief review of all solid state electrochromic devices using amorphous polymer electrolytes. Amorphous polymers and more particularly PEO–LiTFSI, PVP and PEO–acid electrolytes, and cross-linked polyurethanes are promising materials. The recent data given in the literature on systems using tungsten oxide and tungsten bronzes and these electrolytes show a better memory effect than for aqueous systems. The response time is sufficient for large

Table 5. Electrochromic Windows using Polymer Electrolytes

Inserted ion	Electrochromic material	Counter electrode	Electrolyte	Contrast (%)	Number of cycles	Ref.
Window with polyelectrolyte conducting in presence of water:						
H^+	WO_3	IrO_2	Poly-AMPS	7/70	100 000	23
Anhydrous polymer electrolytes windows:						
Li^+	WO_3	V_2O_5	POE-LiTFSI	13/41	10 000	22
Li^+	WO_3	CeO_2	POE-LiTFSI	36/50	400	4
K^+	Tungsten bronze Prussian blue		Modified polyether	14/56	Unknown	11

displays and variable transmission windows. On the other hand, the large variations of the conductivities of some polyether-based electrolytes have resulted in a new type of device which can be controlled thermally with pinpoint addressing.

REFERENCES

1. Chang, I. F., Gilbert, B. L. & Sun, T. I., Electrochemichromic systems for display applications. *J. Electrochem. Soc.*, **122**(7) (1975) 955–62.
2. Baucke, F., Range, K. & Gambke, T., Reflecting electrochromic devices. *Displays* (October 1988) 179–87.
3. Pedone, D., Application d'un électrolyte polymère protonique à la commande thermique matricielle d'afficheurs électrochromes. Thèse INPG, Grenoble, 1987.
4. Baudry, P., Etude de vitrages electrochomes à électrolyte polymère. Thèse INPG, Grenoble, 1989.
5. Truong, Vo-Van, Des fenêtres intelligentes pour la gestion de l'énergie. *Interface*, **7**(3) (1986) 15–21.
6. Lampert, C. M., Electrochromic materials and devices for energy efficient windows. *Solar Energy Materials*, **11** (1984) 1–27.
7. Bohnke, O., Bohnke, C. & Amal, S., Polymeric solid state electrochromic display. *Materials Science and Engineering* (1989) (in press).
8. Armand, M., Deroo, D. & Pedone, D., Solid state thermo electrochromic display based on PVP-H_3PO_4 electrolytes. *Proc. International Seminar on Solid State Ionic Devices*. World Scientific Publishing Co., Singapore, 1988, pp. 515–20.
9. Pantaloni, S., Passerini, S. & Scrosati, B., Solid state thermo-electrochromic display. *J. Electrochem. Soc.*, **134**(3) (1987) 753–5.
10. Stevens, J. R., Svensson, J. S., Granqvist, C. G. & Spindler, R., Electrochromism of WO_3-based films in contact with a solid Li-doped siloxane elastomer electrolyte. *J. Appl. Opt.*, **26**(8) (1987) 1554–7.
11. Tada, H., Bito, Y., Fujino, K. & Kawahara, H., Electrochromic windows using a solid polymer electrolyte. *Solar Energy Materials*, **16** (1987) 509–16.
12. Bohnke, O. & Bohnke, C., Solid state electrochromic display device. *Proc. International Congress on Optical Science and Engineering*, International Society for Optical Engineering, Bellingham, USA.
13. Goldner, R. B., Electrochromic smart windows glass. *Proc. International Seminar on Solid State Ionic Devices*, World Scientific Publishing Co., Singapore, 1988, pp. 379–88.
14. Chabagno, J. M., Etude de la conductivité des systèmes polymères organiques—sels de métaux alcalins. Thèse, Grenoble, 1980.
15. Defendini, F., Complexes polymère-acide à conduction protonique. Synthèse et caracterisation. Thèse INPG, Grenoble, 1987.

16. Armand, M., Polymer electrolytes, an overview. *Solid State Ionics,* **9–10** (1983) 745–54.

17. Akhtar, M., Paist, R. M. & Weakliem, H. A., Thin film tungsten oxide electrochromic display. *J. Electrochem. Soc.,* **135**(5) (1988) 1597–8.

18. Le Nest, J. F., Etude fondamentale des relations entre la structure de réseaux macromoléculaires et leur propriété de conduction ionique. Thèse, Grenoble 1985.

19. Robitaille, C. D. & Fauteux, D., Phase diagrams and conductivity characterization of some PEO–LiX electrolytes. *J. Electrochem. Soc.,* **133**(2) (1986) 315–25.

20. Armand, M., Gorecki, W. & Andreani, R., Perfluorosulfonimide salts as solute for polymer electrolytes. *Second International Symposium on Polymer Electrolytes,* ed. B. Scrosati. Elsevier Science Publishers, London, 1990, pp. 91–7.

21. Golden, S. & Steele, B. C. H., Indium oxide counter electrode for electrochromic applications. *Solid State Ionics,* **28–30** (1987) 1733–7.

22. Baudry, P. & Deroo, D., 176th Meeting of the Electrochemical Society, Hollywood (Florida) October 15–20, (1989), to be published.

23. Cogan, S. F., Plante, T. D., McFadden, R. S. & Rauh, R. D., Design and optical modulation of a-WO_3/a-IrO_2 electrochromic windows. *Proc. SPIE Vol. 823 Optical Materials Technology for Energy Efficiency and Solar Energy Conversion,* **6** (1987) 106–12.

24. Kobayashi, N., Uchiyama, M., Shigehara, K. & Tsuchida, E., Ionically high conductive solid electrolytes composed of graft copolymer–lithium salt hybrids. *J. Phys. Chem.,* **89**(6) (1985) 987–91.

Optical Properties of Thin Film Oxide Bronzes for Window Applications

Mino Green, H. I. Evans & Z. Hussain

Department of Electrical Engineering, Imperial College, London SW7 2BT, UK

ABSTRACT

Three different window configurations of the electrochemical cell M, ITO, $G_xOx(1)$ |G^+X^-| $G_xOx(2)$, ITO, M are considered, where G_xOx is an oxide bronze in which the guest atoms, G, are either H, Li, or possibly Na and the host, Ox, is either WO_3, MoO_3, V_2O_5 or some solid solution of these.

The paper is largely concerned with the optical properties of oxide bronzes relevant to window structures. Only fine grained (~ 5 nm) polycrystalline thin films are considered in the composition range x up to 0·2.

Two bands are identified, the main band associated with electron transitions from the valence to empty conduction band states, and the 'blue band', which has a peak in the very near infrared, and is associated with the guest atoms.

Data on H, Li and Na tungsten bronzes and H and Li molybdenum bronzes are reported. Values of the absorption edge. E_O, are obtained. E_O versus x shows a broadly 'saw-tooth' decrease for all five bronzes investigated.

The blue band has been shown to increase with x value. The position of the peak of the blue band is not x sensitive, but the peak is sensitive to the nature of the guest atom. Thus the peak for Na and Li tungsten bronzes is about 1·3 to 1·4 eV while for hydrogen tungsten bronze it is 1·2 to 1·25 eV. For hydrogen and lithium molybdenum bronzes it is 1·4 and 1·7 eV, respectively. These subtle variations from bronze to bronze are seen as variations in the shade of blue in a display or window.

The absorption spectra of lithium vanadium bronzes are reported; no separate electrochromic band is observed in the range 0·5 to 4·0 eV. It is observed that E_O is 2·18 eV for pure host and increases monotonically with x. Such an electrochromic material can act as the inverse counter electrode.

Finally, the optical properties of $1\cdot5(V_2O_5)$, MoO_3 is shown to give a bronze which at low x value is pale pink and is colourless at x value $> 0\cdot1$.
It is concluded that the prospects for windows are excellent.

INTRODUCTION

Windows of electrically controllable optical density are needed, and are slowly but surely being seen to be needed. There are a number of possible material combinations that can be used in the fabrication of electrochromic cells, which form the basis of these devices, and we will here consider some of the oxide bronzes as the desired optically colourable material; leaving systems using materials such as the viologens or nickel compounds or liquid crystals, to name but a few, for others to discuss.

There is a considerable 'check list' of materials and systems properties that a material must have in order to be a useful candidate for windows. Without going into excessive detail, we estimate that the important optical entries on such a list would be the following:

Minimum optical density	$< 0\cdot1$
Maximum optical density	$1\cdot0$
Maximum time to write or erase	50–100 s
Maximum energy consumption acceptable	$1\,kWh/m^2/day$
Colour desired	Grey, blue and pink
Temperature range of operation	$-40°C$ to $+80°C$
Uniformity of colouration over area	$\pm10\%$
Life time of system	20 years

When one considers such materials as thin films of hydrogen, lithium or sodium bronzes of tungsten, molybdenum or vanadium, or solid solutions of these, in the light of this partial listing one notes three significant shortcomings. The first is that the write/erase times below about $-10°C$ may be longer than 50 s. The second restriction is that it may be hard to make the windows colour and uncolour uniformly. The third restriction is that most, but not all, the bronzes are blue at low guest atom concentrations (x value) and we need other colours. The first problem can be overcome by heating during write or erase; the second problem is discussed briefly, but it is with the third problem (colour range) that this paper mostly deals.

In this work we are considering electrochemical cells of the general

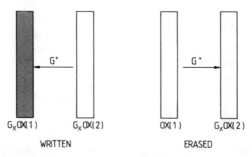

Figure 1. Window cell with transparent source/sink counter electrode.

type:

$$M, ITO, G_xOx(1) \,|G^+X^-|\, G_xOx(2), ITO, M$$

Here $G_xOx(1)$ is an oxide bronze which has an optical density which varies monotonically with x value and where the guest atom G may be H or Li (or possibly but not likely, Na) and $Ox(1)$ is a transition metal oxide in its highest oxidation state, most likely WO_3 or MoO_3. $G_xOx(2)$ has the same guest atom as $G_xOx(1)$ and the electrolyte, but may be a different host oxide. The electrolyte G^+X^-, whose properties are absolutely central to the fabrication of successful windows, will most likely be a polymer material or an inorganic solid thin film. In either case the electrolyte is much more likely to be the cause of restriction to direct operation at low temperatures.

There are three types of electrochemical cell configuration suitable for windows, which are illustrated in Figures 1–3. All the cells require an electrochromic electrode, which for convenience we call the 'front electrode'. The first class of cell (Fig. 1) requires a counter electrode

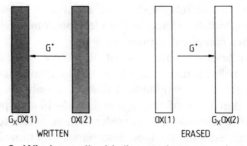

Figure 2. Window cell with 'inverse' counter electrode.

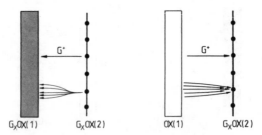

Figure 3. Window cell with dot matrix source/sink counter electrode.

which, while being a source or sink for guest atoms, is transparent at all times. The second class of cell (Fig. 2) requires a host material which is coloured when containing guest atoms (the 'normal' electrochromic oxide bronze) and a host oxide which is coloured at zero x and becomes colourless beyond some convenient value of guest atom concentration.[1] Finally the arrangement shown on Fig. 3 consists of a fine dot matrix (e.g. 50–150 μm diameter dots) of fairly high x value oxide bronze covering about 10% of the counter electrode area (giving about 90% transmission). The counter electrode is a source/sink for guest atoms.[2] Transparent island structures could also be used.

The problem of non-uniform colouration arises from the non-uniform potential distribution associated with '$I \times R$' drop. Thus the variation of potential drop across a cell of width, L, is given by

$$V(x)/V(o) = (e^{\alpha x} + e^{\alpha(L-x)})/(1 + e^{\alpha L})$$

where $\alpha = \sqrt{2R/\rho d}$, and R is the sheet resistance of the ITO, ρ is the resistivity of the polymer electrolyte of thickness d. The island structure shown in Fig. 3 increases the effective value of ρ (typically about tenfold) and will therefore ease the uniformity problem. In addition to the above relationship, we require to know how current and voltage are locally related in the electrochromic cell, and that of course is a more complicated problem.

We will now briefly report some of our recent work on the optical properties of oxide bronzes relevant to the kinds of window cells just considered. It should be emphasised that we are always discussing the properties of thin films of fine grained, c. 5–10 nm, polycrystalline, water free, oxide bronze. No consideration is given to either large single crystal material or so-called amorphous thin film. Because these are window applications, we consider almost exclusively the dilute bronzes, i.e. x up to 0·2.

TUNGSTEN AND MOLYBDENUM BRONZES

In considering the optical absorption properties of these substances there are found to be two bands of importance.[3] There is the main band, which is associated with electronic transitions from valence states to empty conduction band states. The second band is associated with transitions of the free electrons produced by the guest atoms: this band is called the blue band because of the colour it confers upon the oxide bronze.

Our recent results[3,4] for H, Li and Na tungsten bronzes and H and Li molybdenum bronzes show that the variation of the absorption coefficient with photon energy can be well represented by relations which are consistent with *indirect* transitions between parabolic bands. Extrapolation of the curves of $(\alpha h v)^{1/2}$ versus $h v$, where α is the absorption coefficient and $h v$ is the photon energy, to zero α gives the indirect bandgap energy E_O shown in Fig. 4. Figures 5 and 6 show the

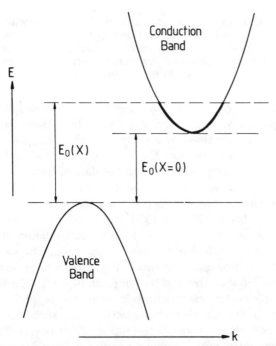

Figure 4. Schematic and simplified $E-k$ diagram for tungsten and molybdenum bronzes showing band partially filled by electrons derived from guest atoms.

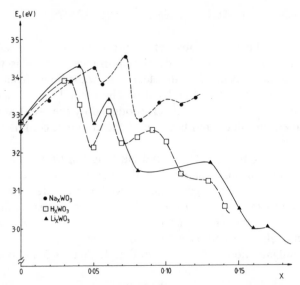

Figure 5. Variation of E_O, the indirect band gap energy, with x value for hydrogen, lithium and tungsten bronzes. (The slightly smaller value of E_O for pure WO_3 in the sodium series is ascribed to a slightly larger mean grain size, there being an inverse relation between grain size and E_O in this regime.)

values obtained of E_O versus x for tungsten bronzes and molybdenum bronzes respectively. The remarkable thing about these data, bearing in mind that E_O is obtained by a fairly extensive extrapolation, is that the seemingly wild variations of E_O versus x are reproduced in each set of data, thereby lending confidence to the data obtained.

Consider Fig. 5, the initial rise in E_O versus x for Na_xWO_3 was taken by Green and Travlos[3] to be associated with electron filling of the conduction band within a rigid band approximation. The subsequent variations are taken to be due to a combination of structural change and electron filling of the conduction band. The variations for the molybdenum bronzes, as shown in Fig. 6, appear to be even more structure sensitive than those of the tungsten bronzes. The significance of these findings for electrochromic windows is probably to be associated with the ultra-violet protection that such films provide, particularly important for the preservation of art objects rather than the small, but fluctuating, increase in overall optical density in the visible spectrum.

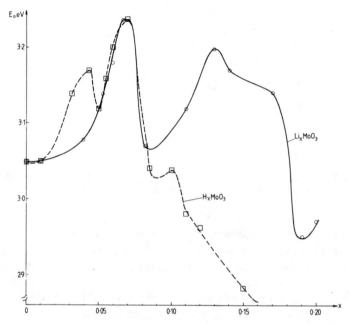

Figure 6. Variation of E_0 with x for hydrogen and lithium molybdenum bronzes.

BLUE BAND

A typical blue band is bell-like in shape with a peak in the near infra-red, see Fig. 7. The width of the bell increases with increasing x value. The bell is somewhat irregular since there are contributions to α on the high energy side from the main band and within the bell there are transitions to various bands. It has been proposed[3] that the electronic transitions of importance are between the lower and higher conduction band states and between the lower conduction band state and the guest ion state as shown on Fig. 8. Free carrier and poloron effects are absent in these fine grained polycrystalline thin films.

The question arises, does the position of the blue absorption band depend upon the nature of the guest atom? The guest atom might be expected to affect the band position if significant transitions to guest ion states are occurring. It should also be of importance if structural changes are affecting the band structure of the host. It is certainly true

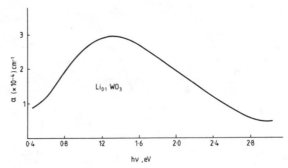

Figure 7. Typical blue band absorption spectrum in the range 0·4–3·1 eV for a $Li_{0·1}WO_3$ thin film of mean grain size 5 ± 1 nm.

that the energy corresponding to the maximum in absorption coefficient for the blue band, E_p, depends upon the nature of the guest atom. E_p for dilute sodium and lithium tungsten bronzes is consistently higher (abour 1·3–1·4 eV) than E_p for dilute hydrogen tungsten bronze (about 1·2–1·25 eV). While in the case of the dilute hydrogen and lithium molybdenum bronzes the E_p values are about 1·4 and 1·7 eV respectively. Not only does the host material affect the value of E_p, but the width of absorption band appears affected by the host, which one would expect because of the different $E-k$ relations. These subtle variations from bronze to bronze are perceived as variations in the shade of blue seen in a display device or window.

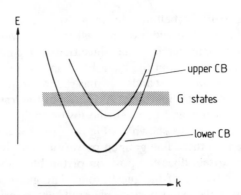

Figure 8. Proposed schematic band diagram for a dilute tungsten or molybdenum bronze. Note the guest atom states high in the conduction band.

It seems clear that variations in shade and colour of dilute oxide bronzes should be obtainable, and this subject is again addressed in Section 4. We now briefly consider a vanadium bronze.

LITHIUM VANADIUM BRONZE[5,6]

We have investigated[7] the absorption spectra of thin films of V_2O_5 and dilute $Li_xV_2O_5$. No blue type band was found. Instead we observed that the main absorption band (also indirect in nature) shifted from low energies, where it appears yellow/green, to higher values with increased lithium concentration. Table 1 is a partial listing of the data showing the values of E_O and B obtained by extrapolation using the relation

$$\alpha = B(h\nu - E_O)^2/h\nu$$

The experimental curves are shown in Fig. 9.

The shift of the main band for the tungsten and molybdenum bronzes is attributed to electron filling of the conduction band states. The absence of a guest atom band is taken to connote the possibility that it is 'lost' in the main band: more detailed spectroscopic studies are needed here.

The relevance of these data for electrochromic windows is that the lithium vanadium bronze can be used as the 'inverse' counter electrode in the scheme shown in Fig. 2. We have in fact made small windows (75 mm × 75 mm) using tungsten trioxide and lithium vanadium bronze: the electrolyte was a polymer containing a lithium salt.

Table 1. Optical Properties of Vanadium Bronze Thin Films (mean grain size 5 ± 1 nm)

x	E_O (eV)	B (cm^{-1} eV^{-1})
0	2·18	9·1 × 10^5
0·012	2·23	6·5 × 10^5
0·029	2·17	4·4 × 10^5
0·056	2·24	4·1 × 10^5
0·135	2·38	3·4 × 10^5
0·312	2·69	6·1 × 10^5
0·64	2·91	5·6 × 10^5

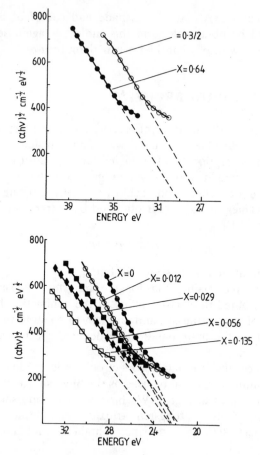

Figure 9. Absorption spectra for a number of lithium vanadium bronze thin films.

MIXED OXIDES

We have investigated the absorption spectrum of several mixed lithium vanadium/molybdenum bronze thin films. There is observed an intermediate value of the energy gap: the results are summarised in Table 2. Some of these films are pink in colour and some are colourless and can therefore be used in the scheme of Fig. 1. Once again, as for the $Li_x V_2 O_5$ bronzes the blue band is shifted out of the

Table 2. Optical Properties of Lithium Vanadium–Molybdenum Bronze Thin Films (mean grain size, 10 ± 1 nm)

x value	Molecular ratio $\frac{1}{2}V_2O_5/MoO_3$	E_0 (eV)	B (cm^{-1} eV^{-1})
0	3·53	2·32	3.2×10^5
0·11	3·98	2·54	3.4×10^5
0·17	3·40	2·50	2.1×10^5
0·27	3·27	2·61	1.9×10^5
0·35	3·81	2·95	1.5×10^5
0·51	3·27	2·86	1.0×10^5

infra-red. It would be interesting to study the MoO_3 rich end of the composition range in order to follow the shift of the blue band.

CONCLUSIONS

Study of the optical properties of thin films of polycrystalline oxide bronzes and especially their solid solutions has been amply rewarded by a deeper understanding of the nature of the transitions involved, as well as the discovery of new effects useful in electrochromic devices. Further study is highly desirable.

REFERENCES

1. Cogan, S. F., Nguyen, N. M., Perrotti, S. J. & Rauk, R. D., Electrochromism in sputtered vanadium pentoxide. *Proc. SPIE,* **1016** (1989).
2. Green, Mino, Variable transmission optical device. UK Patent GB 2164170B, 1988.
3. Green, Mino & Travlos, A., Sodium–tungsten bronze thin films, I. Optical properties of dilute bronzes. *Phil. Mag. B,* **51,** (1985), 501–20.
4. Green, Mino & Hussain, Z., unpublished work.
5. Aita, C. R., Liu, Y-L., Kao, M. L. & Hansen, S. D., Optical behaviour of sputter-deposited vanadium pentoxide. *J. Appl. Phys.,* **60** (1986) 749–53.
6. Fujita, Y., Miyazaki, K. & Tatsuyma, C., On the electrochromism of evaporated V_2O_5 films. *Jap. J. Appl. Phys.,* **24** (1985) 1082–6.
7. Evans, H. I., Optical properties of oxide bronzes. PhD Thesis, University of London, 1987.
8. Green, Mino & Kang, K., unpublished work.

An Adhesive Polymer Electrolyte for Electrochromic Smart Windows: Optical and Electrical Properties of PMMA–PPG–LiClO$_4$

W. Wixwat, J. R. Stevens

Department of Physics, University of Guelph, Guelph, Ontario, Canada N1G 2W1

&

A. M. Andersson, C. G. Granqvist

Physics Department, Chalmers University of Technology, S-412 96 Gothenburg, Sweden

ABSTRACT

Optical and electrical data for a novel PMMA–PPG–LiClO$_4$ polymer electrolyte are reported. The best ionic conductivity is ~$2 \times 10^{-6}\,\Omega^{-1}\,cm^{-1}$ at 290 K with corresponding solar transmittance ~89% for a thin layer. This electrolyte has adhesive properties which make it an ideal candidate for a lamination material for electrochromic smart window applications.

Recent advances in large-area chromogenics[1] give prospects for electrochromic smart windows[2] whose transmittance of radiant energy can be modulated by low voltage pulses. Among the many possible applications[1] we may mention energy-efficient architectural windows and anti-dazzling devices for automobiles. One of the critical material issues is an adhesive transparent solid electrolyte. This paper reports on the optical transmittance and ionic conductivity of a novel polymer electrolyte[3] with potential for smart window utilization.

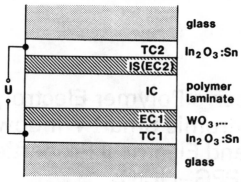

Figure 1. Design principle for an electrochromic smart window showing transparent electronic conductors (TC), electrochromic layer(s) EC1(2), ionic conductor (IC), and ion storage (IS). Transmittance is changed by voltage (U) pulses.

Transparent electrochromic devices can be constructed in several different ways. A suitable design for large area applications is depicted in Fig. 1. It shows two glass panes, each having a two-layer coating, laminated together by the polymer electrolyte. The layers denoted TC1 and TC2 are transparent electronic conductors (for example In_2O_3:Sn). EC1 is an optically active electrochromic layer (for example WO_3-based). IS is an ion storage which, ideally, could be an electrochromic layer (EC2) which colours and bleaches in synchronization with the optical modulation of EC1. The ion conductor IC is at the focus of the present study. Our material comprises a poly(methylmethacrylate) (PMMA) network incorporating poly(propyleneglycol) (PPG) units of molecular weight ~4000 complexed with lithium perchlorate ($LiClO_4$). The fabrication of this PMMA–PPG–$LiClO_4$ material will be described below.[4] When a voltage U is applied between TC1 and TC2, lithium ions are moved from IS via IC, and inserted into/extracted from EC1. The ensuing optical effect is a colouration/bleaching which can change the overall transmittance between 20 and 80%, say.

Reagents used in the synthesis of the electrolyte were prepared with great care. The PPG (Polyscience Inc., MW = 4000) was degassed and dried in a sealable flask, using a freeze–thaw technique. While a vacuum of 10^{-4} Torr was maintained, the sample was repetitively quenched using liquid nitrogen, and then allowed to thaw. Once a minimum number of cracks and no bubbles form upon thawing, the

PPG is assumed to be both dry and free of O_2, which would inhibit polymerization. The PPG still under vacuum is then transferred to a large dry argon glove box.

The methyl methacrylate (MMA) (Aldrich) was dried under vacuum and over CaH_2 for 24 h to remove all water before distilling. The MMA was then distilled and used directly. The salt, $LiClO_4$ (98% Aldrich) was used as received.

The salt was dissolved into the PPG, without the use of external solvents, under an argon blanket. To avoid any decomposition, temperatures were maintained below 100°C. Once the salt is dissolved in PPG, MMA along with 2% (w/w) Azo isobutyronile (AIBN) is added to the mixture. An argon atmosphere is maintained over the reaction mixture at all times. The polymerization was complete after 2 h.

To separate the PMMA from the PPG–PMMA–$LiClO_4$ solution for molecular weight determination, the polymer blend was dissolved in hot methanol. PPG and salt being completely soluble and PMMA insoluble, a simple filtration allowed for complete separation. Gel permeation chromatography (GPC) was then used to determine the number average molecular weight, $M_n = 106\ 773$, with a distribution of 2·2.

Table 1 shows data for four different PMMA–PPG–$LiClO_4$ samples prepared with the shown weight ratios of PPG to PMMA and ratios of oxygen atoms to lithium ions. The oxygen atoms include the ether oxygens in the PPG backbone and the oxygens in the PMMA ester groups. The optical transmittance through these polymer electrolytes is of obvious interest for smart window uses. Figure 2 shows spectral transmittance in the 0·3–2·5 μm interval for a 1 mm thick layer of electrolyte P3 placed between 1 mm thick plates of Corning 7059 glass. The solid curve shows normal transmittance measured on a Perkin-

Table 1. Data for PMMA–PPG–$LiClO_4$ Polymer Electrolytes. Integrated Solar Transmittance T_{sol} and Ionic Conductivity σ at 290 K are Reported

Sample	PPG/PMMA	O/Li$^+$	T_{sol}	$\sigma\,(\Omega^{-1}\,cm^{-1})$
P1	8·2	9·9	86·9	$1·6 \times 10^{-7}$
P2	25·1	9·3	88·7	$7·4 \times 10^{-7}$
P3	18·7	8·6	88·7	$2·9 \times 10^{-7}$
P13	10·6	12·1	88·7	$2·1 \times 10^{-6}$

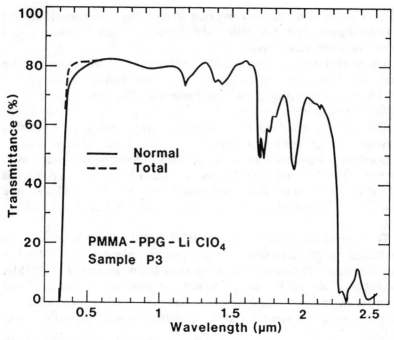

Figure 2. Spectral transmittance through a 1 mm thick PMMA–PPG–LiClO₄ sample.

Elmer Lambda 9 spectrophotometer. The transmittance is ~80% in the luminous range. Characteristic absorption peaks appear in the infra-red. The dashed curve, which overlaps the solid one over most of the spectrum, shows total transmittance measured on a Beckman ACTA MVII spectrometer with integrating sphere. We conclude that a PMMA–PPG–LiClO₄ layer much thinner than 1 mm can be regarded as non-scattering for practical purposes. The solar transmittance was obtained by integrating the spectral data over the AM2 solar irradiance spectrum; data are shown in Table 1 for 25 μm thick PMMA–PPG–LiClO₄ samples. The solar transmittance is seen to be as large as ~89%.

The ionic conductivity was determined on ~1 mm thick electrolyte layers placed between steel electrodes in a quartz cylinder. Complex impedance was recorded in the frequency domain by applying a sine-wave voltage between the electrodes and measuring the current through the electrolyte. Measurements were performed in the 30–4000 Hz range using a function generator, a current amplifier, and a computer with an analogue-to-digital interface input.[4] The conductivity of sample P13, given in Table 1, is higher than that reported[5] for PPG of molecular weight 2880.

In conclusion, we have reported optical and electrical data for a novel PMMA–PPG–LiClO$_4$ polymer electrolyte. The ionic conductivity can be at least ~$2 \times 10^6 \, \Omega^{-1} \, cm^{-1}$ at 290 K and the corresponding solar transmittance is ~89% for a thin layer. The high conductivity, high transmittance, low optical scattering and its usefulness as a lamination material, makes PMMA–PPG–LiClO$_4$ of great interest for electrochromic smart window applications.[3]

REFERENCES

1. Lampert, C. M. & Granqvist, C. G. (eds) *Large-area Chromogenics: Materials and Devices for Transmittance Control.* Optical Engineering Press, Bellingham, 1989 (in press).
2. Svensson, J. S. E. M. & Granqvist, C. G., *Solar Energy Mater.,* **12** (1985) 391.
3. Andersson, A. M., Granqvist, C. G. & Stevens, J. R., *Proc. SPIE,* **1016** (1988) (in press).
4. Brantervik, K., Signal processing for automatic complex AC-measurements in the range 10^{-4} to 10^4 Hz. Chalmers University of Technology, GIPR-275, 1986, unpublished.
5. Watanabe, M. & Ogata, N., In: *Polymer Electrolyte Reviews-1,* ed. J. R. MacCallum & C. A. Vincent. Elsevier Applied Science, London, 1987, p. 39.

Index of Contributors

Subject Index